De Gruyter Studies in Mathematical Physics 2

Gene Byrd
Arthur D. Chernin
Pekka Teerikorpi
Mauri Valtonen

Paths to Dark Energy
Theory and Observation

De Gruyter

Physics and Astronomy Classification 2010: 95.30.Sf, 95.35.+d, 95.36.+x, 98.62.Gq, 98.62.Ck, 98.80.-k, 98.80.Bp, 98.80.Cq, 98.80.Es, 98.65.Dx, 98.80.Jk.

ISBN 978-3-11-025854-7
e-ISBN 978-3-11-025878-3

Bibliographic information published by the Deutsche Nationalbibliothek

The Deutsche Nationalbibliothek lists this publication in the Deutsche Nationalbibliografie; detailed bibliographic data are available in the internet at http://dnb.d-nb.de.

Library of Congress Cataloging-in-Publication Data

A CIP catalog record for this book has been applied for at the Library of Congress.

Typesetting: RoyalStandard, Hong Kong, www.royalstandard.biz
Printing and binding: Hubert & Co. GmbH & Co. KG, Göttingen
♾ Printed on acid-free paper
Printed in Germany

www.degruyter.com

Preface

It is fascinating to see how major advances in modern cosmology have been linked to advances in both theory and observational techniques and instrumentation. The universe of galaxies was discovered observationally during the first decades of the 20th century thanks to astrophotography and large optical telescopes. The cosmological redshift (in the 1910s) and then Hubble's distance-redshift law (in 1929) were revealed when astronomical spectroscopy was sufficiently developed, leading to the picture of the expanding universe. The new field of radio astronomy made it possible to discover the ocean of microwave photons in the 1960s. This highly isotropic thermal radiation quickly made the Big Bang model the ruling paradigm. At the end of the 1990s, with the development of large charge-coupled diodes to search for supernovae effectively, distant supernova observations showed that the universe is expanding at an accelerated rate (a surprising result, contrary to the expected gravitationally retarded expansion). Widely hailed as a revolution in cosmology, this discovery has greatly influenced the course of both observational and theoretical studies of cosmology.

It is also quite interesting how, over the same time, the development of theoretical cosmology has paralleled new observations, sometimes even anticipating them. After his General Relativity was published, Einstein quickly assumed, in 1917, that the universe might be regular, finite and static. For this he had to add into his field equations the now famous cosmological constant Λ, which has now become among cosmologists as the major candidate for the enigmatic "antigravitating" dark energy observed to be speeding up the universal expansion. In the same year, 1917, de Sitter derived another solution of Einstein's equations, where the matter density was made to approach zero (and Λ was included). In this special model a distance dependent cosmological redshift was predicted. De Sitter's "pure vacuum" model has experienced a renaissance in modern cosmology. Following Einstein's Cosmological Principle of homogeneous matter distribution, Friedmann then showed that within General Relativity it is possible to have expanding (or contracting) universe models. A few years later Lemaître showed that a linear distance-redshift law would result from the expansion and gave the modern dark energy interpretation as the reason for this expansion. The linear distance-redshift law was observationally discovered by Hubble in 1929. Finally, Gamow "back extrapolated" the physical state of the universe to ascertain the creation of light elements and predict the observation today of photons emitted not long after the Big Bang.

The Big Bang concept and the discoveries of dark matter and dark energy have shown how large scale "macroscopic" and very small scale "microscopic" observations are equally important for understanding the universe. It is usually assumed that non-baryonic dark matter is made of weakly interacting massive particles

(WIMPs) predicted by modern theories of elementary particles. The enigmatic dark energy is often viewed as a cosmological manifestation of the physical vacuum of quantum fields. One line of thought, pioneered by Gliner, regards the initial state of the universe as a kind of very dense ancient vacuum, out of which matter was born into an expanding, even accelerating state, leading to an inflationary period in early cosmic history, and explaining, among other things, why Einstein's very useful cosmological principle of homogeneity and isotropy is observationally valid. Thus dark energy could be a much diluted "fossil" from the times truly primordial, though the problem is still quite open.

In the present book, we present an overview of the observations and theory of the phenomenon of dark energy. We do not presuppose advanced knowledge of astronomy, and basic mathematical concepts used in modern cosmology are presented in a simple, but rigorous way. In writing this treatise, we have benefitted from our research experience in a variety of fields in cosmology, extragalactic astronomy, our Local Group of galaxies, and down to the Solar System. Hopefully, all this will make the book useful for both astronomers and physicists, and also for university students of physical sciences.

The title indicates that there can be different paths to dark energy. This repulsive force was first discovered in the very large-scale universe where the antigravity of dark energy presently dominates the gravity of dark and baryonic matter. However, we emphasize that dark energy can have also detectable effects in smaller scale systems. For example, our research has found its effect on close-by groups and clusters of galaxies, even within our own Local Group. This local path to dark energy forms an important test of the standard cosmological model where dark energy (in the form of Einstein's cosmological constant in the form of dark energy) should have a uniform distribution and constant density everywhere, both on very large and very small scales.

It is interesting that, first, Newton applied the law of gravity to the fall of an apple on Earth, the Earth's Moon, then other Solar System objects, then it was later applied to the motions of other stars. Then Einstein's General Relativity was used to extend the explanatory power of gravitation to the immense depths of the universe of galaxies. Contrary to gravity, dark energy whose "antigravity" increases in proportion to the distance, was first proposed and discovered in its natural realm of very large scales, beyond the largest gravitationally bound structures over billions of light years. However, the influence of dark energy has now been found in many scales—at distances only several million light years away at the edge of our own Local Group and of other small systems of galaxies.

One big question of modern cosmology is the presence of different kinds of substances, each with their own equation of state, in the universe. After having gone through the discoveries of the galaxy universe with its *baryonic matter* and non-baryonic *dark matter*, the thermal background *radiation*, and finally *dark energy*, we consider these four "elements" as a possible manifestation of a deep cosmic internal symmetry (COINS) which links the four cosmic energies together despite differences in their physical properties. This symmetry suggests a new unifying approach to some old and modern cosmological problems, as we discuss in the concluding chapter of the book.

The importance of dark energy in modern fundamental physics research was splendidly recognized when the Nobel Prize 2011 in Physics was handed to Saul Perlmutter, Adam G. Riess, and Brian P. Schmidt. We wish to note that the remarkable work of these Nobel Laureates and their research groups, the study of the Hubble diagram (distance-redshift law) of very distant Type Ia supernovae, in a sense brought to completion the observational program formulated by Allan Sandage in 1961 when he considered classical cosmological tests as a way to determine the true geometry of the universe. During his extraordinary career, Sandage worked on many fields some of which we also discuss in this book, from the ages of the oldest stars and the evolution of the Milky Way to the classification of galaxies and the expansion rate of the Universe.

Sandage also pioneered in the problem of the very local Hubble law close to galaxy systems, now seen to be a way to study the local energy density of dark energy. He was also deeply aware of the mystery in the regular linear Hubble law within the very lumpy local universe and already in the 1970s recognized it as a possible manifestation of some dense unknown homogeneous medium. In a 2006 paper on the global and the local Hubble constant, Sandage and collaborators commented on their similar values to the local dark energy effect proposed by us as: "No viable alternative to vacuum energy is known at present. The quietness of the Hubble flow lends strong support for the existence of vacuum energy". All this said, we wish to dedicate this book to the memory of Allan Rex Sandage (1926–2010).

Readers can contact the authors at byrd@bama.ua.edu and see a list of additional remarks, corrections etc at:

http://bama.ua.edu/~byrd/PathsToDarkEnergyAddendum.doc

For a non-technical book on the history of physics and cosmology, see Teerikorpi et al. (2009) The Evolving Universe and the Origin of Life. We wish to mention with gratitude several people whose collaboration over the years with some of us has helped us to write this book.

Ron Adler, Yurij Baryshev, Gennadiy Bisnovatyi-Kogan, Lucette Bottinelli, Valentin Dolgachev, Ljudmila Domozhilova, Yury Efremov, Timo Ekholm, Tarsh Freeman, Erast Gliner, Lucienne Gouguenheim, Alexander Gromov, Mikko Hanski, Sethanne Howard, Toivo Jaakkola, Igor Karachentsev, Dmitry Makarov, Dmitry Nagirner, Georges Paturel, Alexander Silbergleit and Gilles Theureau.

January 2012   Gene Byrd, Arthur D. Chernin
                                                                    Pekka Teerikorpi and Mauri Valtonen

Contents

Chapter 1
The start of the paths

Modern measurements of the expansion of the universe have brought to light an odd "substance", which with its antigravitation is even stranger than ordinary matter or even that of the unknown dark matter. This substance has **negative** pressure plus its energy is currently larger than that of both ordinary and even dark matter so that it controls the dynamics of the whole universe. Because of its unusual physics it is called "dark energy" to distinguish it from dark matter and ordinary luminous matter. It had long been thought that the gravity of matter, both dark and the ordinary luminous one, is slowing down the universal expansion. So it came almost like a bolt from the blue when astronomers studying distant supernovae inferred that the expansion is actually **accelerating**. What is speeding up the universe? Most theorists ascribe the acceleration to the "antigravity" of dark energy. In the following chapters we will discuss historically the theoretical and observational background of this concept. We shall see that even though dark energy is exotic compared with ordinary laboratory physics it rather naturally arises in classical General Relativity, in particular in a cosmological context.

1.1   Newton's absolute space and time

To explain dark antigravitating energy we must start with Isaac Newton's (1643–1727) system of gravitational dynamics. Newton described universal gravitation and the three laws of motion, which dominated the scientific view of the universe for almost three centuries. His theory is today still excellent for many applications in physics and astronomy. However, fundamental difficulties with this system as applied to the whole universe as well as to particular observations require concepts of relativity and, as modern observations seem to show, ultimately dark energy.

Newton showed that the motions of objects on Earth and of celestial bodies are governed by the same set of natural laws by demonstrating the consistency between his theory of gravitation and Kepler's laws of planetary motion. His universe consisted of an absolute space and time which existed quite independent of the matter in it. Absolute space has three Cartesian dimensions; absolute time had one. Time and space were independent. These coordinates formed a natural grid to study physical events. The universe's space was infinite in extent. The "speed" of gravity was infinite. The square of the distance ds^2 between two points is given by

$$ds^2 = dx^2 + dy^2 + dz^2 \tag{1.1}$$

where x, y, z, are the usual Cartesian coordinates. This is the **metric** (the ruler that allows one to measure distances) of Newtonian space.

Correspondence between Newton and a theologian, Richard Bentley, in 1692–1693 revealed problems in Newton's conception of an infinite Euclidean universe. Bentley looked to science for arguments against atheism. Science reveals laws of nature (such as that of gravitation), but do these negate the existence (or intervention) of a supernatural being? Newton, a deeply religious man and also the greatest expert in physical science, was the natural person to consult.

At that time, the composition of the universe would be considered as a distribution of stars extending in all directions equally. Today, we might substitute galaxies for stars with the same results. Bentley asked Newton how matter distributed uniformly in space would behave. Newton answered that matter could be in equilibrium only if the gravitational pulls from different directions on every particle exactly cancelled out. However, Newton pointed out that this balance is extremely unstable. Newton compared the situation to needles standing on their points. Even a tiny imbalance would lead to a catastrophic collapse. Thus the past and future existence of a universe of stars (today we would say galaxies) together with gravity appeared a rather finely tuned mystery. Bentley accepted this as a need for divine intervention. Thus, this correspondence turned Newton's thoughts to cosmology, a topic which he had neglected in his younger years.

In 1895, Hugo von Seeliger similarly concluded that a gravitating infinite Euclidean universe with stars uniformly distributed cannot be stable. To resolve this, von Seeliger proposed a small extra weakening in the gravitational force in addition to the inverse square law. As we shall later see mathematically, this modification has a parallel in Einstein's later proposal to add the so called cosmological constant to his equations of General Relativity, in order to make his own universe stay in a state of rest. This cosmological constant is of the same mathematical form as today's dark energy but today, rather than being "fine-tuned" to produce a static universe, it actually dominates the universe's expansion. Thus the "roots" of the need for dark energy even extend back to Newton.

1.2   Light versus absolute space and time

As measurements grew more precise it became clear that physical reality did not always agree with Newtonian dynamics. In the 19th and 20th centuries new measurements showed that Newton's laws of motion and gravitation had to be flawed. For example, by the late 1800s observations of the orbit of Mercury showed an inconsistency with Newton's laws. There was a tiny discrepancy in the precession of Mercury's orbit. As we shall see, it requires Einstein's General Relativity to explain that discrepancy.

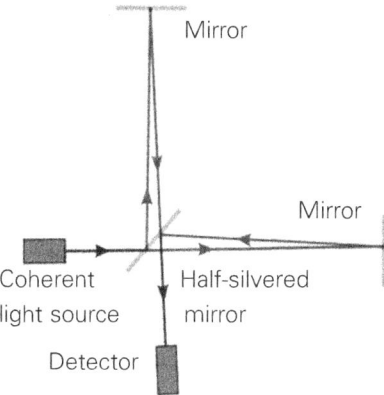

Figure 1.1. Michelson's interferometer. Light from a source is split into two beams using a semitransparent mirror. Two beams which travel in perpendicular directions are reflected back from two mirrors. The reflected beams are directed back through the same semitransparent mirror so they are combined going to a telescope. Analyzing the fringes arising from the interference of the two determines how the speed of light, c, depends on the motion of the Earth through space.

First, we will discuss the disagreement that led to the Special Theory of Relativity. Consider the Earth moving through space towards a light source, then light approaching the Earth will appear to move faster than light leaving the Earth in the Newtonian approach. Analogously to sound waves in air, it was believed that light moved through a fixed reference frame known as the aether.

In 1887 two American physicists Albert Michelson (1852–1931) and Edward Morley (1838–1923) tested this hypothesis. They tried to measure the motion of the Earth through space by studying from which direction light comes at highest speed. Michelson and Morley calculated that light travel time along a path to and from a mirror should be at its greatest when the line connecting the centres of the mirror and the observer is parallel to the orbital motion of the Earth. The time should be at its smallest when the ray of light between the two travels 90 degrees to this motion (see Figure 1.1). Even treating light as a photon (analogous to a bullet rather than a wave), ought to produce a measureable difference.

Michelson and Morley's estimate of the travel time difference in their experiment was small, due to the orders of magnitude difference between the known orbital velocity of the Earth ($\sim$30 km/s) and the speed of light ($c \approx 300,000$ km/s), but easily measurable. No travel time difference was observed (Michelson and Morley 1887). This was verified by other experimenters. Michelson and Morley's experiment demonstrated that light is not a wave which propagates in a medium called the aether. The aether is now taken to be non-existent. The conclusion is that light travels always at the same speed independent of the motion of the measuring apparatus. An observer at rest measures the same speed of light as another going towards the source of light, or another going away from the source. The speed of light is independent of the motion of the observer. It exists outside of absolute space and time.

1.3   Space-time events and intervals

In the 1905 Special Theory of Relativity, Albert Einstein (1879–1955) accepted **as a postulate** Michelson and Morley's observation that the speed of light is a constant, independent of the state of motion of the observer. Einstein did not ask why there was constancy of the speed of light independent of the motion of the observer, but instead determined what consequences derive from this fact. He realized the speed of light can be the same for everybody only if space and time are entangled in a way which nobody had anticipated. He made use of what he called the Relativity Principle: All observers moving at constant speed should observe the same physical laws. It is interesting that the Relativity Principle was already stated a little earlier (1904) by the French mathematician Henri Poincaré. In fact, originally, in 1902, Poincaré spoke about "the principle of relative movement", which shows the root of the word "Relativity"—we study phenomena measured by observers moving at different uniform speeds relative to each other.

Einstein's Special Theory of Relativity ties time and space together to make a four dimensional space-time. This space-time has four coordinates (x, y, z, t). Its geometry is flat, i.e., it resembles a three dimensional Euclidean geometry. There are a number of interesting consequences of the Special Theory of Relativity which are better discussed by introducing new concepts—**events** and **intervals** in space-time.

In the Special Theory of Relativity, one uses space and time together to describe an **event**. Let the event be, for example, the signing of a document. Then the "coordinates" of the event are written like this: in Turku, Finland on 26th of March 1987. The coordinates in space are given by the name of the city "Turku" while the time coordinate is the day, month and year. More mathematically, we label an **event** by 4 coordinates, for example t, x, y, z where t is time and x, y, z are the three Cartesian coordinates of space.

Even though events can be labeled by many different coordinate systems, the **interval** between neighboring events is independent of the choice of coordinates. The **spatial distance** between two locations is independent of the Newtonian reference system. Let us consider the coordinates of two cities and then calculate their differences. If the north-south difference is 40 km and in the east-west 15 km and both are at the same altitude, via the Pythagorean Theorem the distance between them is $\sqrt{40^2 + 15^2}$ km, which is 43 km. This distance is the same whether we use local Cartesian (x, y, z) or spherical coordinates of radius and two angles.

In classical dynamics, spatial and temporary intervals can be treated separately, because of the absolute space and time arena. In Special Relativity, it is essential that **time is included in interval calculations**. The square of the interval (the **metric**) is defined as:

$$ds^2 = c^2 dt^2 - dx^2 - dy^2 - dz^2 \tag{1.2}$$

where c is the speed of light. We take the sign convention of Equation (1.2) so that we have a positive sign in front of the squared timelike interval and a negative sign in front of the squared spacelike interval of Equation (1.1). The metric conforms to the measured Michelson and Morley experiment results, the constancy of the speed of light, independent of the motion of the observer. Strictly speaking, this is the line element, not the metric, but for the purpose of this book we shall use the words interchangeably.

Note how the interval is a combination of space and time—now called space-time. Light travel time becomes the fundamental unit of measure. The space distance unit is the light second, the distance travelled by light in one second. A similar much larger unit is the familiar light year (c yr).

The calculation of the space-time interval between two events can be shown by the following example. Consider a difference in time of 40 seconds and a difference in position of 15 light seconds, then the interval is the square root of $40^2 - 15^2$, i.e., 37 s. The interval is independent of the coordinate system used. Note that when calculating the space-time interval, one must use a **minus sign** inside the square root of the spatial coordinates.

The special nature of the interval between two events can be clarified by the example of the nearest star which is about 4 light years away. Let event number one be the occasion when a beam of light starts its journey from this star, and let event number two be when that same light arrives here on Earth. Then using Equation (1.2) the square of interval between the events is

$$ds^2 = c^2 dt^2 - dx^2 - dy^2 - dz^2 = c^2 (4 \text{ yr})^2 - (4 \ c \text{ yr})^2 = 0. \tag{1.3}$$

This surprising zero result emphasizes the difference between the definition of this interval and the Newtonian concepts in three spatial dimensions.

For a stationary observer, $x, y,$ and z are constant, and the interval is purely timelike

$$ds^2 = c^2 dt^2. \tag{1.4}$$

In this case the time coordinate is called the **proper time** τ of the observer. As proper time increases, the observer progresses along a **worldline** through space-time. The proper time is the worldline parameter.

1.4   Space-time measurements and Lorentz transformations

Aside from the postulate that everyone will measure the same speed of light, Special Relativity assumes uniform motion (no acceleration and deceleration) as felt by any observer. With these two assumptions, one can derive the **Lorentz transformation**.

The Lorentz transformation describes how, according to the theory of Special Relativity, two observers' varying measurements of space and time can be converted into each other's frames of reference. It is named after the Dutch physicist Hendrik Lorentz. It reflects the surprising fact that observers moving at different velocities may measure different distances, elapsed times, and even different orderings of events.

The Lorentz transformation was originally the result of attempts by Lorentz and others to explain how the speed of light was observed to be independent of the reference frame, and to understand the symmetries of the laws of electromagnetism. Einstein later re-derived the transformation from his postulates of Special Relativity. The Lorentz transformation supersedes the Galilean transformation of Newtonian physics, which assumes an absolute space and time. According to Special Relativity, the Galilean transformation is a good approximation only at relative speeds much smaller than the speed of light.

To see what this means, consider two systems, one primed (t', x', y', z') and one non-primed (t, x, y, z). The axes are aligned with each other. An observer in the primed system moves relative to another in the non-primed system at a rate u along the same x', x axis direction. For this case, the Lorentz transformation becomes

$$\begin{aligned} t' &= \gamma(u)\big(t - ux/c^2\big) \\ x' &= \gamma(u)(x - ut) \\ y' &= y \\ z' &= z \end{aligned} \qquad , \tag{1.5}$$

where

$$\gamma(u) = \frac{1}{\left(1 - u^2/c^2\right)^{1/2}} \tag{1.6}$$

First, for small u relative to c, the transformations match those of Newtonian physics, $t' = t$ and $x' = x - ut$. Second, Special Relativity is necessary for calculations when speeds are significant compared to the speed of light. Consider the speed of an object which may be comparable to that of light as seen by observers in both systems. If the object is moving with speed $x/t = v$ in one system, then its speed is $x'/t' = v'$ in the other. The two systems are moving with speed u with respect to each other. Using the Lorentz transformations divide x' by t' and simplifying, we get how the measured speeds of the object relative to the two systems are related:

$$v' = \frac{v - u}{1 - vu/c^2}. \tag{1.7}$$

For small u relative to c, the transformation above matches that of "everyday" Newtonian physics, $v' = v - u$. We see however, that for speeds v comparable to c, a more complicated formula applies. In particular, if the moving object is a light photon then in the x coordinate $v = c$, and the Equation (1.7) becomes

$$v' = \frac{c - u}{1 - cu/c^2} = \frac{c - u}{1 - u/c} = c. \tag{1.8}$$

This is, of course, one of the assumptions by which the Lorentz transformations were derived. Thus if the x system is a super highway in which a law officer points a laser beam at an automobile receding with speed u, then a scientist in the car will measure the same speed c for the beam as another observer standing on the roadside as the automobile passes. Should the auto be a police car and a partner in it sends a laser beam back to a standing pedestrian, the measured speed will again be c.

Taking the differentials of the first two Lorentz transformations, we obtain

$$dx' = \frac{dx}{\sqrt{1 - u^2/c^2}}$$

$$dt' = dt\sqrt{1 - u^2/c^2}$$

so that

$$dx = dx'\sqrt{1 - u^2/c^2}$$

$$dt = \frac{dt'}{\sqrt{1 - u^2/c^2}}. \tag{1.9}$$

For the first equation in Equations (1.9), dx can be the observed length of a short stick which is dx' long in the moving system. We see that the fixed observer measures this length as shortened by $1/\gamma(u)$ so that the observed length dx approaches zero as u approaches c. An object in the moving system will appear squashed to the stationary observer. This is called **length contraction**. Since everything is relative in Special Relativity, an observer in the moving system x' would observe a similar squashing of objects at rest in the x system.

Similarly, for the second equation in Equations (1.9), dt, can be an observed short time interval (tick-tick) dt' of a clock moving with the moving system. We see that the fixed observer measures this length as lengthened by $1/\gamma(u)$ so that the observed length of the tick-tick dt approaches infinity as u approaches c. The primed system clock will appear to be running very slowly. Since everything is relative in Special Relativity, an observer in the moving system x' would observe a similar slowing of clocks at rest in the x system. This is called **time dilation**. If we consider the observer and the clock as both fixed in the moving system, i.e., moving together, the clock indicates for the observer his/her proper time τ (of which dt' above is a short interval). Thus, in terms of proper time, the time dilation is $dt = \gamma(u)d\tau$. The proper time is related to the rate of processes in the rest frame of a physical system. A well-known example is the subatomic particle π meson which in its rest frame decays at its proper rate, but laboratory observers measure for high-speed mesons a slower decay rate exactly as predicted by Equations (1.9).

1.5   The Minkowski diagram

It is very useful to use a **Minkowski diagram** (Figure 1.2), a diagram of space-time events in the pseudo-Euclidean system that is described by Equation (1.2). The defining feature of a Minkowski diagram is that light rays are drawn at a 45 degree angle to the line or plane respresenting space. This means that if the time axis is measured in seconds, then the space axes are measured in light seconds (the distance light can travel in one second). It is hard to draw four dimensions in a two-dimensional picture. Thus two of three space coordinates are usually left out,

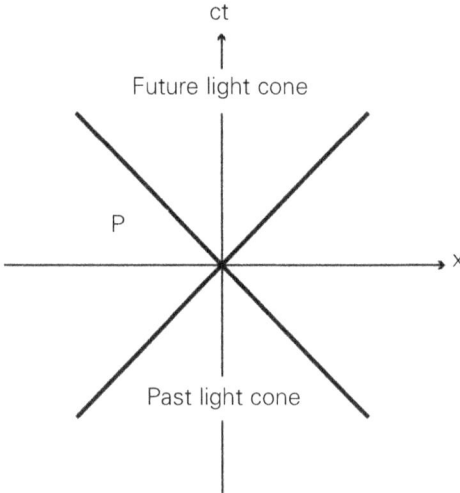

Figure 1.2. A two-dimensional Minkowski diagram. An inertial observer is at **P**. His/Her worldline is the time axis, ct. Light rays move at a 45° angle, along a line corresponding to $x/t = c$.

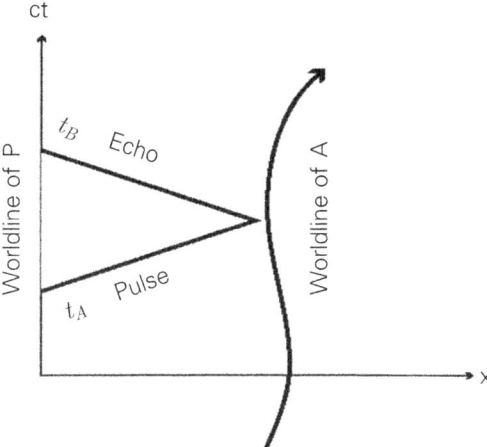

Figure 1.3. An inertial observer **P** sends a pulse to a reflector **A** at time t_A and recording an echo at time t_B. The observer determines that reflection took place at a distance $r = c(t_B - t_A)/2$ at time $t = t_A + (t_B - t_A)/2$.

but retaining time. An inertial observer **P** receives information from the past coming along the past **light cone**, and he/she may send signals to the future light cone. He/she may also send radar signals towards a body **A** that he/she receives at a later time (Figure 1.3). Thus the parts of the space-time inside the light cones are accessible to communication. No information can be sent outside the future light cone or be received from outside the past light cone to its left and right in Figure 1.2.

The interval of Equation (1.2) has a different appearance in different coordinate systems. In many problems, it is better to use spherical polar coordinates. Using the transformation

$$t = t$$
$$x = r \sin \theta \cos \phi$$
$$y = r \sin \theta \sin \phi \tag{1.10}$$
$$z = r \cos \theta$$

we see that Equation (1.2) becomes

$$ds^2 = c^2 dt^2 - dr^2 - r^2 d\theta^2 - r^2 \sin^2 \theta d\phi^2. \tag{1.11}$$

More generally, we may write

$$ds^2 = g_{\alpha\beta} dx^\alpha dx^\beta \tag{1.12}$$

or

$$ds^2 = \sum_{\alpha=0}^{3} \sum_{\beta=0}^{3} g_{\alpha\beta} dx^{\alpha} dx^{\beta}$$

where α and β are summation indices and $g_{\alpha\beta}$ gives the coefficients that define the space-time. Commonly we leave out the summation symbols in this context to interpret a repeated index meaning summation from $0 \ldots 3$ over this index. Thus x^{α} and x^{β} are coordinates that may correspond to any of the symbols $x^0 \ldots x^3$ in the different terms of the sum. In Equation (1.2), $x^1 \ldots x^3$ are the three spatial coordinates, x, y, z, and $x^0 = ct$ is the time coordinate.

The well known relation in Special Relativity, the observed mass m of a particle moving with speed v relative to an observer compared to the rest mass m_0 of a the same particle is

$$m = \frac{m_0}{\sqrt{1 - v^2/c^2}}. \qquad (1.13)$$

This relation indicates that the fixed observer measures an increased mass m for a moving object whose mass would be m_0 if it were at rest. This has been observed. The total energy of the body of mass m is

$$E = mc^2. \qquad (1.14)$$

Thus matter and energy are equivalent in Special Relativity. This equivalence has been observed and verified many times.

Box 1. Tensors.

Tensors are geometric entities introduced into mathematics and physics to extend the notion of scalars, geometric vectors, and matrices to higher orders. A scalar, for example, is a tensor of rank 0. In the three dimensional space of Newtonian physics it is convenient to use vectors. Vectors are absolute entities that are independent of the coordinate system. For example, we may rotate the coordinate axes, and obtain new values for the vector components without changing the vector itself.

In Special Relativity, we raise the order from three to four and use four-vectors. Generally, the Newtonian three-vector is complemented with a suitable fourth component. In order to maintain the "absolute" quality, the four-vector must not depend on the observer, and thus it must be invariant under Lorentz transformation. For the position four-vector,

$$\mathbf{R} = (ct, x, y, z) \qquad (B1.1)$$

the Lorentz transformation becomes

$$
\begin{aligned}
t' &= \gamma(t - vx/c^2) \\
x' &= \gamma(x - vt) \\
y' &= y \\
z' &= z
\end{aligned}
\tag{B1.2}
$$

where the Lorentz factor is

$$
\gamma = \frac{1}{\sqrt{1 - v^2/c^2}}.
\tag{B1.3}
$$

Here the (t', x', y', z') system moves with the velocity v in the direction of the positive x-axis. The $\gamma = \gamma(v)$ is given by Equation (B1.3). The velocity four-vector $\mathbf{U}$ is defined as

$$
\mathbf{U} = \frac{d\mathbf{R}}{d\tau} = \left(\frac{cdt}{d\tau}, \frac{dx}{d\tau}, \frac{dy}{d\tau}, \frac{dz}{d\tau} \right) = \gamma(u)c(1, \mathbf{u}/c).
\tag{B1.4}
$$

Here $\mathbf{u}$ is the velocity three-vector with components $u_x = dx/dt$, *etc.*, and $dt/d\tau = \gamma(u)$.

The magnitude or norm of the four-velocity is given by

$$
U^2 = c^2 U_0^2 - U_1^2 - U_2^2 - U_3^2 = \gamma(u)^2 c^2 (1 - u^2/c^2) = c^2.
\tag{B1.5}
$$

We see that the magnitude of the velocity four-vector is always c. As stated earlier, this is the reason for the choice of Equation (1.2), to conform with the constancy of the speed of light, c.

A three-dimensional tensor example is the viscous stress tensor S_{ik}, defined as the force per unit area acting in the i-direction across a surface 90° to the k-direction. This could be for, say, a rubber medium that can be compressed but also resists strain. Note that these directions are two general directions and not necessarily those of the conventional x and y. Its components are specified in a matrix

$$
S_{ik} = \begin{bmatrix} S_{xx} & S_{xy} & S_{xz} \\ S_{yx} & S_{yy} & S_{yz} \\ S_{zx} & S_{zy} & S_{zz} \end{bmatrix}.
\tag{B1.6}
$$

In the definition of pressure P the i- and k-directions are the same, i.e., S_{xx}, S_{yy}, and S_{zz} represent pressure in three orthogonal directions. In an isotropic medium such as a gas that does not support any strain, the matrix is simple:

$$S_{ik} = P \begin{bmatrix} 1 & 0 & 0 \\ 0 & 1 & 0 \\ 0 & 0 & 1 \end{bmatrix}. \tag{B1.7}$$

In four dimensional relativistic physics, tensors play the role of "absolute" entities. The tensors are defined so that they satisfy certain transformation laws during a change of coordinates. Four-vectors are tensors of rank one. Their transformation is an example which is generalized to higher rank tensors. The rank of a tensor tells the number of degrees of freedom, which shows up immediately as the number of not repeated (or free) indices in the tensor symbol.

A tensor's rank can be reduced by two in an operation called **contraction**. Consider the second rank tensor S_{ik} as an example. Contraction is performed by taking a trace of the component matrix which is defined as

$$S = \sum_{i=1}^{3} S_{ii}. \tag{B1.8}$$

The result is a single number S, a scalar, a tensor of rank 0. Usually, the summation signs are not written explicitly. However, the fact that the index i is repeated is understood to imply a sum over this index: $S_{ii} = S_{11} + S_{22} + S_{33}$. In Box 2 we will encounter a fourth rank tensor $R^{\mu}_{\nu\rho\sigma}$ which is contracted to a second rank tensor $R_{\nu\rho} = R^{\mu}_{\nu\rho\mu}$ by putting $\mu = \sigma$ with the implied summation over μ.

Now, the indices of a tensor appear sometimes in the upper corner, sometimes in the lower corner, and often there is a mixture of the two. These refer to different coordinate representations (i.e., co-variant and contra-variant). For our purposes, we need to know only that the physical quantity represented by the tensor is not altered by the varying positions of the indices. In tensor equations, expressions on both sides should have the same relative positions of the indices.

We will encounter the metric tensor $g_{\mu\beta}$ frequently. The positions of the indices indicate that its components are given in covariant form. The alternative is the contravariant form $g^{\nu\alpha}$ (vectors are contravariant). The two forms are related by

$$g^{\mu\alpha} g_{\mu\beta} = \delta^{\alpha}_{\beta} \tag{B1.9}$$

where δ^{α}_{β} is the unit tensor

$$\delta^{\alpha}_{\beta} = \begin{bmatrix} 1 & 0 & 0 & 0 \\ 0 & 1 & 0 & 0 \\ 0 & 0 & 1 & 0 \\ 0 & 0 & 0 & 1 \end{bmatrix}. \tag{B1.10}$$

The 4×4 matrix $g_{\alpha\beta}$ in Equation (1.12) is called the **metric tensor**. The metric tensor captures all the geometric and causal structure of space-time defining the way the length is measured. For the Special Relativistic interval of Equation (1.2), the metric tensor is

$$g_{\alpha\beta} \equiv \eta_{\alpha\beta} = \begin{bmatrix} 1 & 0 & 0 & 0 \\ 0 & -1 & 0 & 0 \\ 0 & 0 & -1 & 0 \\ 0 & 0 & 0 & -1 \end{bmatrix}. \tag{1.15}$$

Now for the case of spherical coordinates the interval of Equation (1.11)

$$g_{\alpha\beta} = \begin{bmatrix} 1 & 0 & 0 & 0 \\ 0 & -1 & 0 & 0 \\ 0 & 0 & -r^2 & 0 \\ 0 & 0 & 0 & -r^2\sin^2\theta \end{bmatrix}. \tag{1.16}$$

These two metric tensors above are purely diagonal but in general all elements could be nonzero. As a simple example of a non-diagonal metric tensor, the interval for a rotating (angular speed ω) cylindrical coordinate system is

$$ds^2 = \left(c^2 - \omega^2 r^2\right) dt^2 - dr^2 - r^2 d\phi^2 - 2\omega r^2 d\phi dt - dz^2. \tag{1.17}$$

Problem. Use the transformation

$$x = r\cos\bar{\phi}$$
$$y = r\sin\bar{\phi}$$
$$z = z$$
$$\phi = \bar{\phi} - \omega t$$

to transform the interval Equation (1.2) to Equation (1.17) for a coordinate system that rotates around the z-axis with angular speed ω.

We will later need the determinant g composed of the elements of the metric tensor. It is required in the calculation of the volume element of the space-time. We see easily that the value of the determinant $\det(g_{\alpha\beta})$ in Equation (1.16) is $-r^4\sin^2\theta$ and the value of the determinant in Equation (1.15), $\det(\eta_{\alpha\beta})$, is -1. In this context the determinant g is always a negative number, and thus when we later take its square root, we take it from $-g$. In the volume element, the product of the coordinate differentials is multiplied by this square root.

The space-time represented by the metric tensor in Equation (1.15), $\eta_{\alpha\beta}$, is termed **Minkowski space-time** which was introduced by the German physicist Hermann Minkowski in 1908. It is Euclidean except for the different sign of the time coordinate (i.e., pseudo-Euclidean). This space-time is described as being **flat**. When the interval between points is zero, then in Euclidean space, the points are identical. In a Minkowski space-time the points can be connected by a light ray. Consider our earlier example of a light ray from the nearest star.

Mathematically, the space-times of Equation (1.11) and Equation (1.17) are defined to be flat because transformations exist that take them back to the basic form Equation (1.2). The Special Theory of Relativity deals exclusively with flat space-times. One must go beyond the Special Theory to include the curvature of space-time.

Chapter 2

General Relativity: apparent acceleration of gravity

2.1 Gravitation as an apparent force

The starting point of Einstein's Theory of General Relativity was Galileo's (1564–1642) well-known observation that falling bodies of different masses accelerate equally under gravity (ignoring the friction of air). A contemporary, less well known, performance of this experiment was carried out by the Dutch-Belgian mathematician Simon Stevinus in 1586. Two different bodies dropped from the same height at the same time reach the ground at the same time. Newton's second law of motion (force equals mass times acceleration) and his law of gravity (gravitational force is proportional to the mass of the body) mathematically combine to explain Galileo's observation. In both these laws, the force is proportional to the mass of the body, so the acceleration of a falling body is independent of its mass. Since we are dealing with two independent laws of nature, **we have to wonder how both of them happen to have the same proportionality to mass**.

To Einstein, this common proportionality was not accidental. Galileo's observation shows that gravity is not a true force but instead is an "apparent force". A familiar apparent force is the Coriolis force explained by the French physicist Gaspard de Coriolis (1792–1843). In Earth's northern hemisphere, south flowing winds tend to turn toward the east and north flowing winds toward the west; causing a counterclockwise rotation of winds around low pressure areas into which winds flow. A clockwise rotation of winds results around high pressure areas from which winds flow outward. The Coriolis force is simply a reflection of Earth's rotation around its axis. It is not a true force in the sense that it is connected with our own motion along with the surface of the Earth. Typically, apparent forces accelerate all bodies equally, independent of properties such as mass, size, or electric charge.

Similarly, the acceleration by gravity is independent of the bodies' properties. An apparent force is easy to eliminate. If we float above the Earth's North Pole observing a southward flowing wind toward a low pressure area, it will be seen to move in a straight line while an Earth bound observer would disagree. Similarly, if we go to free fall, gravity disappears. In a freely falling capsule **on Earth or in space**, we experience weightlessness, e.g., in an orbiting space station or the Earth-bound case of an elevator whose cable snaps and whose brakes fail. It is interesting that the Russian "father of astronautics", Konstantin Tsiolkovsky (1857–1935) in his writings in the beginning of the 20th century accurately described the phenomenon of weightlessness inside a freely moving spacecraft and the apparent gravitation-like force felt in an accelerated rocket.

Far from Earth, a force like gravity at the Earth's surface may be generated artificially in a spacecraft which accelerates by 9.8 m/s^2, equal to the acceleration of gravity that we normally experience. As an example, assume both Newton and Einstein observe the falling of an apple. Both are in identical enclosed rooms, Newton at the Earth and Einstein in a distant spacecraft whose rockets accelerate it by 9.8 m/s^2. For both, the fall of the apple is observed to happen in the same way.

2.2   Principle of Equivalence

The idea that it is always possible to find a locally flat space-time, called a Lorentz frame, in a curved space-time is usually stated as the **Principle of Equivalence**. Einstein postulated that local, freely falling, nonrotating laboratories are fully equivalent in the performance of all physical experiments.

In a freely falling, nonrotating laboratory one can establish the laws of physics. Each freely falling, nonrotating laboratory constitutes a **local inertial frame**. An inertial frame is where no effects of gravity forces exist. Our falling elevator is an example. In such a frame, a particle at rest remains at rest and a particle in motion continues in motion. There is no way of detecting a uniform gravitational field in such a laboratory. On the other hand, an observer who is not in free fall, but for example, stands on the surface of the Earth, feels the pull of gravity. The reference frame transformation that puts the observer into free fall removes the pull of gravity.

In 1911 Einstein added a corollary to handle an accelerated frame:

"Whenever an observer detects the local presence of a force that acts on all objects in direct proportion to the inertial mass of each object, that observer is in an accelerated frame of reference."

The fact that gravity can be made to disappear by a transformation to a freely falling system caused Einstein to propose that gravity is not a real force but a property of space-time akin to Coriolis forces in a rotating coordinate system. According to Einstein, gravity is due to curvature of space-time which does not show up in a local Minkowski frame but which is very real on a larger global scale. In other words, the General Theory of Relativity is so constructed that locally the Special Theory of Relativity applies. In other words, matter causes curvature in the surrounding space-time. Bodies react to this curvature so that there appears to be a gravitational attraction which causes acceleration of the body.

In Minkowski spacetime the motion of a free particle takes place along a straight line. In the curved space-time of General Relativity the free particle motion is along the straightest possible worldline. Such a line is called the **geodesic**. The straightest possible worldline means that the interval ds is an extremum (usually the minimum) value along the geodesic, among all other possible routes. A two dimensional every-day example is the great circle route on the Earth. A great circle route is the shortest possible distance between two points on the spherical Earth. The plane of the great circle cuts through the center of the Earth.

The Principle of Equivalence is one tenet of the General Theory that is represented by the Einstein equation of General Relativity. We will initially concentrate on metrics and integrals of motion, presenting the Einstein equation when we discuss the scale factor of the universe and the inclusion of dark energy. The metrics we will discuss are solutions to the Einstein equation.

2.3   Lagrangians and motion of bodies

A similar principle to the great circle analogy occurs in classical mechanics. There we define the Lagrangian function L as the difference between two energies, the kinetic energy, E_{KIN}, and the potential energy of a dynamical system, W. The Lagrangian is represented $L = E_{KIN} - W$. Mathematically, the **Hamiltonian principle** (introduced by the Irish mathematician William Rowan Hamilton in 1834) stated that the dynamical system evolves in such a way that the **action integral**

$$\int L \, dt \tag{2.1}$$

is at its extremum along the path defined by the time parameter t. Phrased differently, the evolution of a system between two states at two specified times is a stationary point at which the variation is zero of the action integral. For a physical system there exists an invariant integral—the action. When the system evolves from state A to state B, it does in a way so that the action is minimized.

The Hamiltonian principle leads to Lagrangian equations of motion. As defined by the French mathematician and astronomer Joseph L. Lagrange (*Mècanique analytique*, 1788) the equation of motion is

$$\frac{\partial L}{\partial q_i} - \frac{d}{dt}\frac{\partial L}{\partial \dot{q}_i} = 0 \quad i = 1, \ldots, m \tag{2.2}$$

where q_i, $i = 1, \ldots, m$ are coordinates of the system. The dot refers to the derivative.

Moving from classical to relativistic dynamics the next step also starts with the Lagrangian; however, instead of $E_{KIN} - W$, the Lagrangian is defined by the metric tensor

$$L = g_{\mu\nu} \dot{x}^{\mu} \dot{x}^{\nu}. \tag{2.3}$$

The Lagrangian is thus proportional to the square of $\dot{x}$. Given that dx^{μ} represents the coordinates and $g_{\mu\nu}$ is the metric, then $ds^2 = dx^{\mu} g_{\mu\nu} dx^{\nu}$ is the square of the infinitesimal distance along a curve parameterized by t or explicitly

$$ds^2 = \frac{dx^{\mu}}{dt} g_{\mu\nu} \frac{dx^{\nu}}{dt} dt^2$$

The action is

$$\text{action} = \int ds$$

Then the interval s between events σ_1 and σ_2 is

$$s = \int_{\sigma_1}^{\sigma_2} \sqrt{L}\, d\sigma. \tag{2.4}$$

Here σ is the parameter of the curve; usually it can be chosen as the proper time along the path. The dot signifies a derivative with respect to the curve parameter.

Just as in the case of classical mechanics, the requirement that s obtains its extremum value leads to the equations of motion. The L in Equation (2.2) is replaced by $\sqrt{L}$ and σ replaces t:

$$\frac{d}{d\sigma}\left(\frac{1}{\sqrt{L}}\frac{\partial L}{\partial \dot{x}^\mu}\right) - \frac{1}{\sqrt{L}}\frac{\partial L}{\partial x^\mu} = 0. \tag{2.5}$$

Two different cases must be considered here:

(a) the **timelike geodesic**, $L > 0$ which represents the worldline of a particle with non-zero rest mass. Taking L to be a constant along the trajectory, e.g., $L = c^2$ makes the curve parameter $\sigma = \tau$, the proper time along the curve. Because the curve parameter can be chosen freely, this choice is always possible when $L > 0$. Substituting, Equation (2.5) then simplifies to the equations of motion:

$$\frac{d}{d\tau}\frac{\partial L}{\partial \dot{x}^\mu} - \frac{\partial L}{\partial x^\mu} = 0. \tag{2.6}$$

(b) the **null-geodesic**, $L = 0$ which is appropriate for a zero rest mass particle such as a photon moving with the speed of light. In this case the idea of proper time along the worldline is meaningless.

2.4   Integrals of motion

In classical mechanics, **integrals of motion** are very useful. They are functions of the coordinates which are constant along a trajectory in phase space. From Equation (2.6) we get the first integral of motion as follows. For material particles, L is not explicitly time dependent and is conserved, and so we have:

$$L = c^2. \tag{2.7}$$

If the metric $g_{\mu\nu}$ is independent of one coordinate, say u, then $\partial L/\partial u = 0$. Then by Equation (2.6)

$$\frac{d}{d\tau}\frac{\partial L}{\partial \dot{u}} = 0 \tag{2.8}$$

that can be integrated to obtain

$$\frac{\partial L}{\partial \dot{u}} = \text{constant} \tag{2.9}$$

along the geodesic parameterized by τ. The coordinate u is considered **ignorable**. This first integral of motion is referred to as **momentum** by analogy with the momentum of Newtonian physics.

As an example, consider the motion in the equatorial plane $\theta = \pi/2$ in the flat Minkowski metric using spherical polar coordinates (Equation (1.11)). The Lagrangian constant is

$$L = c^2\dot{t}^2 - \dot{r}^2 - r^2\dot{\phi}^2 = c^2. \tag{2.10}$$

The ignorable coordinates, t and ϕ, have two other corresponding integrals which are

$$\dot{t} = \gamma \tag{2.11}$$

and

$$r^2\dot{\phi} = J. \tag{2.12}$$

Here γ is the familiar Lorentz factor of Special Relativity

$$\gamma = \frac{1}{\sqrt{1 - v^2/c^2}}. \tag{2.13}$$

This factor relates the proper time τ of a body moving with speed v to the coordinate time t of an inertial observer. From here on, γc^2 will simply be called "energy per unit mass". The angular momentum per unit mass is represented by J.

The significance of γ is seen more clearly by eliminating $\dot{t}$ and $\dot{\phi}$ from the above equations to get

$$\dot{r}^2 + \frac{J^2}{r^2} = c^2(\gamma^2 - 1). \tag{2.14}$$

At the classical limit of low speeds, $\gamma \approx 1$ and $\gamma^2 - 1 = (\gamma + 1)(\gamma - 1) \approx 2(\gamma - 1)$. The orbital energy is defined as the total energy minus rest mass energy: $E = \gamma m_0 c^2 - m_0 c^2 = (\gamma - 1)m_0 c^2$. By multiplying Equation (2.14) by $\frac{1}{2}\, m_0$, Equation (2.14) then states that the radial kinetic energy + the potential energy of the centripetal force = total orbital energy. Thus at the limit we have Newton's theory.

Chapter 3
Tests of General Relativity

3.1 The Schwarzschild metric and the gravitational redshift

The **Schwarzschild metric** is an exact solution to the Einstein equation of General Relativity in the case of a point mass. As stated earlier, here we concentrate on metrics and integrals of motion, saving a detailed discussion of the Einstein equation when we discuss the scale factor of the universe and the inclusion of dark energy. We will use this particular metric during our discussion of dark energy effects around our Local Group of galaxies in a later chapter finding that dark energy must be included. On the distance scales discussed in this chapter, dark energy is not significant.

As our next case study, consider the metric of space-time surrounding a point of mass M. The German astronomer and physicist Karl Schwarzschild (1916) showed that the appropriate metric is:

$$ds^2 = c^2 \left(1 - \frac{2GM}{c^2 r} \right) dt^2 - \frac{dr^2}{1 - \frac{2GM}{c^2 r}} - r^2 d\theta^2 - r^2 \sin^2 \theta d\varphi^2. \tag{3.1}$$

Here r, θ, and ϕ are the usual spherical polar coordinates. Although Equation (3.1) looks rather complicated, one can see that the combination

$$r^2 \left(d\theta^2 + \sin^2 \theta d\phi^2 \right) \tag{3.2}$$

in the metric shows that r labels a sphere of surface area $4\pi r^2$. However, we should make clear that r is different from the usual Euclidean radial distance. Also, from the first term of Equation (3.1), timekeeping now depends on r. For $M = 0$ we get the Minkowski metric in the flat space spherical coordinates we discussed earlier (Equation (1.11)). For $M > 0$ the space-time is curved. The phenomenon may be illustrated by the following experiment which qualitatively matches in two dimensions the effects of mass in General Relativity. Stretch a flexible rubber sheet so that it is tight and lies horizontally, like the surface of a table. Place an iron ball in the middle of the surface; its weight causes a dip or curvature in the surface around the iron ball. The rubber space-time is Minkowski (analogous to a flat sheet) at infinitely large distances from the central body even for non-zero mass of the ball.

To study the dynamics in this metric, consider a particle which is in free fall towards the point mass. For purely radial motion we set $\phi = \theta = $ constant. Then the Lagrangian is (from Equation (2.3))

$$L = c^2 \left(1 - \frac{2GM}{c^2 r} \right) \dot{t}^2 - \frac{\dot{r}^2}{1 - \frac{2GM}{c^2 r}} = c^2 \tag{3.3}$$

where derivatives are taken with respect to the proper time τ defined in case (a) of Equation (2.6). The t coordinate is obviously ignorable; therefore we have our first integral γ:

$$\left(1 - \frac{2GM}{c^2 r} \right) \dot{t} = \gamma. \tag{3.4}$$

Substituting in Equation (3.3) we get after some rearrangement

$$c^2 \gamma^2 - \dot{r}^2 = c^2 \left(1 - \frac{2GM}{c^2 r} \right). \tag{3.5}$$

Differentiating with respect to τ:

$$\ddot{r} = -\frac{GM}{r^2}. \tag{3.6}$$

Equation (3.6) is Newton's law of gravity. Notice, however, that the time co-ordinate τ and the radial coordinate r are not their Newtonian equivalents. At large distances r, where the Schwarzschild metric of Equation (3.1) approaches the Minkowski metric of Equation (1.11), we do have a close correspondence between the universal time t and the proper time τ. Gravity is obtained from the Schwarzschild metric as a **limiting case** which is Newtonian gravity.

Complications arise from the definition of r in Schwarzschild metric. When we measure the radial distance between two simultaneous (in coordinate time t) events on the same radial line, the interval between events (t, r_1, θ, ϕ) and (t, r_2, θ, ϕ) is not $r_2 - r_1$ as in Newtonian physics. Instead it contains a logarithmic term:

$$s = \int_{r_1}^{r_2} \frac{dr}{\sqrt{1 - 2GM/c^2 r}} \approx \int_{r_1}^{r_2} \left(1 + \frac{GM}{c^2 r} \right) dr = r_2 - r_1 + \frac{GM}{c^2} \ln \frac{r_2}{r_1}. \tag{3.7}$$

At large distances, the Newtonian limit $GM/c^2 \ll r_2 - r_1$. Thus the logarithmic term may be neglected.

Now that the Schwarzschild metric has a singularity $(g_{00} = 0; g_{11} = \infty)$ at

$$r_s = 2GM/c^2 \tag{3.8}$$

which is known as the **Schwarzschild radius** of the body. This singularity can be removed mathematically by another choice of coordinates. However, the Schwarzs-child radius is extremely important physically. It is the scale within which extreme curvature effects dominate and the Newtonian approach to dynamics is not possible any more. The Newtonian methods are applicable only when $r \gg r_s$.

The Schwarzschild radius is very small for common celestial bodies. For a hypothetical point mass equal to the mass of the Earth $r_s \cong 9$ mm while for a point mass equal to that of the Sun it is $r_s \cong 3$ km. In comparison, the radius of the Earth is 6 378 km and the radius of the Sun is 696 000 km. The Newtonian theory is thus generally adequate to describe the gravitational field outside these celestial bodies. Either long time span or very precise short term measurements are required to detect the weak effects of General Relativity in the Solar System. Such short term measurements do exist and are used routinely for Solar System bodies as well as global positioning system satellites.

In contrast, **neutron stars** are not much bigger than their Schwarzschild radii and thus are prime examples of "relativistic" celestial bodies. Another prime example is the **black hole** where the whole mass of the body has collapsed to the center. The Schwarzschild radius is a true boundary for communicating information outwards. For an outgoing radial light beam, $ds^2 = 0$, $d\theta^2 = d\phi^2 = 0$ in Equation (3.1), and

$$c^2 \left(1 - \frac{2GM}{c^2 r} \right) dt^2 = \frac{dr^2}{1 - \frac{2GM}{c^2 r}}$$

i.e.,

$$\frac{dr}{dt} = \pm c \left(1 - \frac{2GM}{c^2 r} \right).$$

The speed of the beam or signal goes to zero at $r = r_s$. It takes infinite time (measured with t, the time coordinate at large distance) for the signal to escape from $r = r_s$ to infinity. Any event occurring at $r = r_s$ or less cannot propagate information via light to a large distance. As a result, the Schwarzschild radius is called the **event horizon**, the "surface" of a black hole (which we have assumed not to be rotating).

As second simple example of dynamics in the Schwarzschild metric is a particle in a circular orbit about a central body. The coordinates of the particle are:

$$\begin{aligned}
t &= \gamma \tau \\
r &= \text{constant} \\
\theta &= \pi/2 \\
\phi &= \omega \tau
\end{aligned} \tag{3.9}$$

where ω is the angular velocity. The constant γ is to be determined for this case. For an observer far away, the particle's velocity will be measured as

$$v = r \frac{d\phi}{dt} = r \frac{\omega}{\gamma}. \tag{3.10}$$

The Lagrangian gives another equation:

$$L = c^2 \left(1 - 2GM/c^2 r\right)\dot{\gamma}^2 - r^2 \omega^2 = c^2. \tag{3.11}$$

Eliminating ω from the last two equations gives

$$\gamma = \frac{1}{\sqrt{1 - \left(\frac{v^2}{c^2} + \frac{2GM}{c^2 r}\right)}}. \tag{3.12}$$

In the equation above, the reader will note that the usual time dilation due to orbital motion at speed v, given by the Lorentz factor of Equation (2.13), is augmented by time dilation in the gravitational field.

We see that the time dilation of a stationary clock ($v = 0$) in the gravitational field of the body of mass M is

$$\gamma = \frac{1}{\sqrt{1 - \frac{2GM}{c^2 r}}}. \tag{3.13}$$

Time dilation can be observed as a redshift in spectral lines from a source. Suppose n waves of frequency ν_e are emitted in a proper time interval $\Delta\tau_e$ by a particle near a massive body. Here the subscript e stands for "emitted". Then

$$n = \nu_e \Delta\tau_e.$$

The observer at a distance will receive the same n waves. However, the frequency and the time of duration of the wave train will be changed. Denote the subscript a as "arrival". For the receiver, the equation

$$n = \nu_a \Delta\tau_a$$

is valid, where $\Delta\tau_a$ is the proper time interval of the receiver, and ν_a the frequency observed by the receiver. Since the numbers are the same,

$$\nu_e \Delta\tau_e = \nu_a \Delta\tau_a$$

and the frequencies are

$$\nu_a = \nu_e \Delta\tau_e / \Delta\tau_a.$$

Taking $\tau_e = \tau$, $\tau_a = t$, this becomes

$$\nu_a = \nu_e \gamma^{-1}. \tag{3.14}$$

A distant observer will find that the frequency of light decreases, and there is a shift toward the red end of the spectrum compared to, say, the same spectral line of an element near the observer. Using γ from Equation (3.13) we call this the **gravitational redshift** which was predicted by Einstein (1907).

So according to General Relativity, time slows down in curved space-time, in a strong gravitational field of attraction. Thus time goes a little more slowly in the basement than in the attic of a house since the attic is more distant from the center of the Earth and gravity is slightly less higher up. In 1960 the American physicists R.V. Pound and G.A. Rebka (1959, 1960) measured the difference over a vertical distance of 22.5 m. They obtained a result which agreed with Einstein's theory within 10% and subsequently verified it to the accuracy of 0.01%.

3.2   Orbits in General Relativity

Einstein concluded that matter causes curvature in the surrounding space-time, and bodies react to this curvature in such a way that there appears to be a gravitational attraction which causes acceleration. From the geometry of space-time, it is possible to calculate the orbits of bodies which are influenced by gravity. In flat space-time the force free motion happens on a straight line, but in a space-time curved by mass the force free motion can create practically closed orbits. An example is the motion of a planet around the Sun. A planet moves forward as straight as possible, but because the large mass of the Sun has curved the space-time around it, the planet's orbit becomes an ellipse in space. Returning to our stretched rubber sheet, place an iron ball in the middle of the surface; its weight causes a dip in the surface around the iron ball, the curvature mentioned earlier. Now roll a small light ball (like a child's marble) along the surface. With a suitable push in the right direction, one can get the marble to roll around the iron ball, possibly in an elliptic orbit. It appears as if there is a central force that attracts the two balls. In fact the closed orbit arises from the geometry of the surface distorted by the mass of the large ball, a good analogue to General Relativity.

A planet is an example of a body that is in free fall in the gravitational field of the Sun. In the example of planetary orbits around the Sun, both Newton's theory and Einstein's theory give practically the same result. The most important difference arises for the close-in orbit of the planet Mercury. Over a long time, the major axis of Mercury's ellipse approximating the actual orbit precesses slowly due to gravitation of other planets. But Einstein's theory gives an additional precession of $43''$ per century compared to Newton's theory. This observationally solid but small bit extra had already been observed and was a problem for Newtonian theory at the time. We shall see that Einstein's theory gives a force law which is almost but not quite the inverse square law proposed by Newton. We will now investigate this mathematically.

An orbit is not a worldline but rather the projection of the worldline to a surface $t = $ constant. The shape of the orbit is obtained by eliminating time from the equations of motion. We will consider an orbit in the $\theta = \pi/2$ plane. Therefore the shape of the orbit is described by the polar coordinates r and ϕ.

In Newton's theory the equation of motion is

$$\frac{d^2 r}{dt^2} = -\frac{GM}{r^2} + r\left(\frac{d\phi}{dt}\right)^2.$$

$$(3.15)$$

The last term is called centripetal acceleration. The first integral of the motion J, the angular momentum per unit mass, is

$$J = r^2 \frac{d\phi}{dt}.$$

$$(3.16)$$

We introduce a new variable u by setting

$$u = \frac{1}{r}$$

$$(3.17)$$

whereby

$$\frac{dr}{d\phi} = -\frac{1}{u^2} \frac{du}{d\phi}$$

$$(3.18)$$

and

$$\frac{dr}{dt} = \frac{dr}{d\phi} \frac{d\phi}{dt} = -J \frac{du}{d\phi}.$$

$$(3.19)$$

The left side of Equation (3.15) is now

$$\frac{d^2 r}{dt^2} = -J \frac{d}{dt} \frac{du}{d\phi} = -J \frac{d\varphi}{dt} \frac{d^2 u}{d\phi^2} = -J^2 u^2 \frac{d^2 u}{d\phi^2}$$

$$(3.20)$$

while the right side of Equation (3.15) becomes

$$-\frac{GM}{r^2} + r\left(\frac{d\phi}{dt}\right)^2 = -GMu^2 + J^2 u^3.$$

$$(3.21)$$

Equating the two, we get

$$u'' + u = \frac{GM}{J^2}$$

$$(3.22)$$

where primes $(')$ designate derivatives relative to ϕ.

The well-known solution to this equation is an ellipse with one focus at $r = 0$. If the eccentricity of the ellipse is e and the inverse radial coordinate is u_1, the orbit is

$$u_1 = \frac{GM}{J^2}(1 + e \cos \phi).$$

$$(3.23)$$

That is the familiar Newtonian approach. Moving to relativistic dynamics, we can write the Lagrangian using Equation (3.1) as

$$L = c^2\left(1 - \frac{2GM}{c^2 r}\right)\dot{t}^2 - \frac{\dot{r}^2}{1 - \frac{2GM}{c^2 r}} - r^2\dot{\phi}^2 = c^2. \tag{3.24}$$

Here t and ϕ are ignorable coordinates, the corresponding conserved quantities are as before,

$$\left(1 - \frac{2GM}{c^2 r}\right)\dot{t} = \gamma \tag{3.25}$$

and (as in Equation (2.12))

$$r^2\dot{\phi} = J, \tag{3.26}$$

where J is the angular momentum integral. The quantity γc^2 is called "energy at infinity". Eliminating $\dot{t}$, $\dot{r}$ and $\dot{\phi}$ from the above equations, and also using

$$\dot{r} = \frac{dr}{d\phi}\dot{\phi} \tag{3.27}$$

we get after rearranging

$$\frac{J^2}{r^4}\left(\frac{dr}{d\phi}\right)^2 = c^2\left(\gamma^2 - 1\right) + \frac{2GM}{r}\frac{J^2}{r^2} + \frac{2GM}{c^2}\frac{J^2}{r^3}. \tag{3.28}$$

Again changing the variable r to u:

$$\left(\frac{du}{d\phi}\right)^2 = \frac{c^2\left(\gamma^2 - 1\right)}{J^2} + \frac{2GM}{J^2}u - u^2 + \frac{2GM}{c^2}u^3. \tag{3.29}$$

Differentiate Equation (3.29) relative to ϕ, and again denote the differentiation by primes. After dividing out u' from both sides we are left with

$$u'' + u = \frac{GM}{J^2} + \frac{3GM}{c^2}u^2. \tag{3.30}$$

Equation (3.30) is identical to the Newtonian case except for the last term. It may be written in the form $3/2(r_s/r)u$ which is small as long as $r \gg r_s$. This is true, for example, in the Solar System. The ratio of the two terms on the right hand side of Equation (3.30) is 3×10^{-8} for the Earth's orbit.

The relativistic solution must be close to the elliptical orbit. We already have the elliptical solution, which is u_1 of Equation (3.23). This solution may be substituted for u in the last term of Equation (3.30) leading to an approximate orbit equation

$$\frac{d^2 u}{d\phi^2} + u = \frac{GM}{J^2} + \frac{3(GM)^3}{J^4 c^2}\left(1 + 2e\cos\phi + e^2\cos^2\phi\right). \tag{3.31}$$

Construct the solution by solving the following particular integrals $u_2 \ldots u_4$:

$$\frac{d^2 u_2}{d\phi^2} + u_2 = \frac{3(GM)^3}{J^4 c^2} \tag{3.32}$$

$$\frac{d^2 u_3}{d\phi^2} + u_3 = \frac{3(GM)^3}{J^4 c^2} 2e\cos\varphi \tag{3.33}$$

$$\frac{d^2 u_4}{d\phi^2} + u_4 = \frac{3(GM)^3}{J^4 c^2} e^2\cos^2\phi. \tag{3.34}$$

When these three equations are added together with the Equation (3.23) for u_1,

$$\frac{d^2 u_1}{d\phi^2} + u_1 = \frac{GM}{J^2} \tag{3.35}$$

It can be verified that we get Equation (3.31) for

$$u = u_1 + u_2 + u_3 + u_4. \tag{3.36}$$

We already know u_1. What are $u_2 \ldots u_4$? The solutions of Equations (3.32)–(3.34) are

$$u_2 = \frac{3(GM)^3}{J^4 c^2} \tag{3.37}$$

$$u_3 = \frac{3(GM)^3}{J^4 c^2} e\phi\sin\phi \tag{3.38}$$

$$u_4 = \frac{1}{2}\frac{3(GM)^3}{J^4 c^2} e^2 - \frac{1}{6}\frac{3(GM)^3}{J^4 c^2} e^2\cos 2\phi. \tag{3.39}$$

These are added to u_1. We see that u_2 adds a minute constant to u_1, the first term in u_4 has the same effect. The second term of u_4 produces a very small periodic wiggle on u_1. However, u_3 is proportional to ϕ and can become large in comparison with other corrections to u_1. Leaving out insignificant terms, our next approximation is

$$u = u_1 + u_3 = \frac{GM}{J^2}\left(1 + e\cos\phi + \frac{3(GM)^2}{J^2 c^2} e\phi\sin\phi\right). \tag{3.40}$$

A factor in last term in the parentheses

$$b = \frac{3(GM)^2}{J^2 c^2} \phi \tag{3.41}$$

is generally small due to the coefficient in front of ϕ. Thus $\sin b \approx b$ and $\cos b \approx 1$. Using the identity

$$\cos(\phi - b) = \cos \phi \cos b + \sin \phi \sin b \approx \cos \phi + b \sin \phi \tag{3.42}$$

we transform Equation (3.40) to

$$u \approx \frac{GM}{J^2} \left[1 + e \cos \left(\phi - \frac{3(GM)^2}{J^2 c^2} \phi \right) \right]. \tag{3.43}$$

The radial coordinates start to repeat themselves after one revolution (2π) plus a small additional bit, i.e., when

$$\phi = \frac{2\pi}{1 - \frac{3(GM)^2}{J^2 c^2}} \approx 2\pi \left[1 + \frac{3(GM)^2}{J^2 c^2} \right]. \tag{3.44}$$

The "bit" is then

$$\Delta\phi = \frac{6\pi(GM)^2}{J^2 c^2}. \tag{3.45}$$

Since the orbit is very close to a Keplerian orbit, we may express J^2 in terms of the semi-major axis a_B and eccentricity e of the ellipse:

$$J^2 = GMa_B(1 - e^2) \tag{3.46}$$

which leads to

$$\Delta\phi = \frac{6\pi GM}{c^2 a_B(1 - e^2)} = 3\pi \frac{r_s}{a_B(1 - e^2)}. \tag{3.47}$$

We will now use this equation to calculate a theoretical value for the relativistic **perihelion shift** $\Delta\phi$ for Mercury. Substitute the Schwarzschild radius for the Sun, $r_s = 2.96$ km, the semi-major axis of Mercury's orbit $a_B = 57.91 \times 10^6$ km, and its $e = 0.2056$ into Equation (3.47). Multiply the result by 206 265 to go from radians to arcsec and by 415, the number of revolutions per 100 years to obtain

$$\Delta\phi = 43''/\text{century}. \tag{3.48}$$

The mathematician Urbain Le Verrier of Paris Observatory concluded in 1855, after other causes of perihelion shift (i.e., the influence of other planets) have been deducted, that there was an observed shift of 35″/century in the perihelion of Mercury. Since General Relativity was not known, he ascribed this to an unknown planet inside Mercury's orbit. This presumed planet was never found despite extensive searches (Le Verrier 1859). In 1882, the United States Naval Observatory's Simon Newcomb (1895) recalculated the observed perihelion shift and obtained 43″/century (also see Hall 1894, Shapiro et al. 1976). This early test of General Relativity gave great confidence in the theory (Einstein 1915).

Another more extreme and more recent example is the radio pulsar PSR 1913 + 16, a binary system of two neutron stars. We can consider the pulsar to be a "particle" orbiting another neutron star of similar mass. Although Equation (3.47) has been derived for a particle in the gravitational field of a massive body, it also applies to a binary of more equal masses. If the orbital motion of the bodies in a binary system is not relativistic, Newtonian mechanics is valid. In this case, the problem of the binary is reduced to the problem of one body in the central gravity field. Using the semi-major axis of the pulsar relative to the center of mass of the binary $a_B \approx 9.7 \times 10^5$ km, the Schwarzschild radius of the other neutron star $r_s \approx 4.17$ km, eccentricity $e \approx 0.617$, we find that in one year (1131 orbital revolutions) the periastron advances

$$\Delta\phi \approx 4.2°/\text{year}. \tag{3.49}$$

We call it periastron instead of perihelion since we have a star which is not the Sun. The orbital parameters in General Relativity result in this value of the precession which reproduce the observations of the pulses from the system (Hulse and Taylor 1975, Taylor and Weisberg 1989, Hulse 1994).

The far more extreme object, the quasar OJ287, which is thought to harbor a binary black hole system at its center provides an even stronger test of General Relativity. The parameters in this case are $r_s \approx 4.9 \times 10^{10}$ km, $a_B \approx 1.6 \times 10^{12}$ km and $e \approx 0.7$ which give

$$\Delta\phi \approx 32°/\text{revolution}. \tag{3.50}$$

The value deduced from observations is 39°/revolution, significantly different from our theoretically calculated 32°/revolution. The difference between these two shows that the first order theory which we have derived above is not sufficiently accurate when it comes to such large precession rates. Using the higher order General Relativistic orbit theory for OJ287 has made it possible to verify Einstein's General Relativity to the second Post Newtonian order (Lehto and Valtonen 1996, Valtonen et al. 2006, 2008, 2011). The other classical tests discussed in this chapter verify the theory only to the first Post Newtonian order (Will 1993, Freire and Wex 2010).

3.3   Deflection of light

Another test of the General Theory of Relativity is the bending of light rays which pass close to the Sun. Thus, stars appear to shift away from their usual places in the sky when the Sun is in the foreground close to them on the sky (but not in distance). Normally we don't see the Sun and the stars together in the sky at the same time, but during a solar eclipse it is possible. Photographs during the eclipse with the Sun among the stars and another at night without the Sun six months later were compared. The detection of the shift of stars by the expected amount during the solar eclipse of 1919 was hailed as victory for Einstein's theory. Nowadays better measurements are possible using radio sources. The most accurate measurements agree with Einstein's theory with 1% accuracy. The mathematics of this bending of light leads to some interesting results. We will discuss the theory of the bending in part because of its role in acceptance of the General Theory of Relativity but also because of lensing by larger systems such as galaxies and (later in this book) clusters of galaxies. The latter will be useful in substantiating dark matter in these systems with a view of estimating the relative amounts of dark matter relative to dark energy in the universe.

A photon follows a null geodesic for which $ds = 0$. From Equation (3.26)

$$J = r^2 \frac{d\phi}{ds} = \infty \tag{3.51}$$

Thus Equation (3.30) becomes

$$u'' + u = \frac{3GM}{c^2} u^2 \tag{3.52}$$

for a photon. If the right hand side's small term is set to zero, the solution is

$$u_1 = \frac{1}{b} \sin \phi \tag{3.53}$$

where b is the minimum distance between the mass M and the photon path which occurs at $\phi = \pi/2$. Substituting u_1 into the right hand side of Equation (3.52) gives the next approximation for the path:

$$\frac{d^2 u}{d\phi^2} + u = \frac{3GM}{c^2 b^2} \sin^2 \phi = \frac{3GM}{c^2 b^2} \left(1 - \cos^2 \phi\right). \tag{3.54}$$

Its particular integral is

$$u_2 = \frac{3GM}{2c^2 b^2} \left(1 + \frac{1}{3} \cos 2\phi\right). \tag{3.55}$$

We add

$$\frac{d^2 u_1}{d\phi^2} + u_1 = 0 \tag{3.56}$$

and

$$\frac{d^2 u_2}{d\phi^2} + u_2 = \frac{3GM}{c^2 b^2}\left(1 - \cos^2 \phi\right) \tag{3.57}$$

and see that

$$u = u_1 + u_2 = \frac{1}{b}\sin\phi + \frac{3GM}{2c^2 b^2}\left(1 + \frac{1}{3}\cos 2\phi\right) \tag{3.58}$$

is the solution of the second approximation, Equation (3.54).

At large r, which means small u, $\sin\phi \approx \phi$, $\cos 2\phi \approx 1$ and

$$0 \approx \frac{\phi}{b} + \frac{2GM}{c^2 b^2} \tag{3.59}$$

or

$$\phi = \phi_\infty \approx -\frac{2GM}{c^2 b}. \tag{3.60}$$

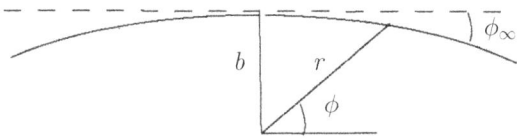

Figure 3.1. The gravitational deflection angle ϕ_∞, Equation (3.60).

The above is half of the total deflection angle (Figure 3.1). The total deflection from a straight line path is therefore (Einstein 1911)

$$\Delta\phi = \frac{4GM}{c^2 b} = 2\frac{r_s}{b} \text{ radians.} \tag{3.61}$$

For a light ray passing the surface of the Sun, $r_s = 2.96$ km, $b = 6.96 \times 10^5$ km

$$\Delta\phi = 1.75''. \tag{3.62}$$

This is in good agreement with observation. It provided the second early test of the General Relativity theory with a positive result (Eddington 1919, Dyson et al.

1920). Via the deflection of light the stars close to the edge of the Sun in the sky appear to have shifted radially outward from the center of the Sun. At that time the Sun has to be eclipsed by the Moon in order to see the nearby stars. Nowadays, the eclipse can be arranged artificially by blocking out the solar disk in space observations.

Deflection of light from background sources by massive objects is called **lensing** of the sources. Lensing can cause an apparent shift of the direction of the source, the source may appear as multiple images or strongly distorted, and the apparent source flux may change due to the lens. Multiple images of the quasar Q0957 + 561 were detected in 1979 by a team led by the English astronomer Dennis Walsh of the University of Manchester (Walsh et al. 1979) using the 2.1 meter telescope of Kitt Peak National Observatory in Tucson, Arizona. Other team members were Robert F. Carswell of Cambridge University and Ray J. Weymann of Steward Observatory in Arizona. Nowadays gravitational lenses are detected frequently, and are used in astrophysical studies in many ways some of which we will discuss later in this book.

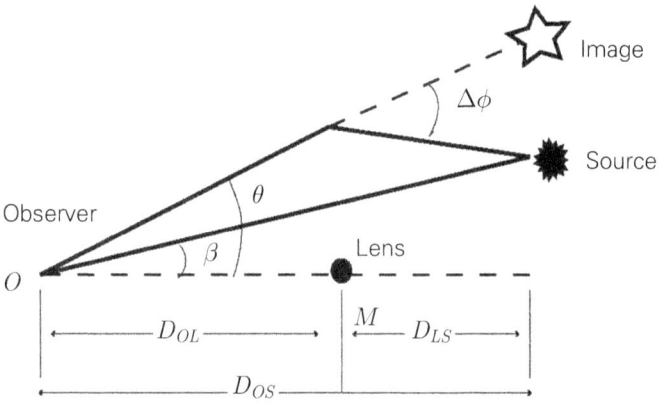

Figure 3.2. Diagram for the derivation of the lens equation, Equation (3.65).

The simplest situation is a lens formed by a point-like mass M, which is the case discussed above, and a point-like source somewhere behind it as illustrated with Figure 3.2. A source is an angle β away from the lens in the sky. Due to deflection of light, the source appears at an angle θ from the lens. The total deflection angle is $\Delta\phi$. The distances from the observer to the lens and to the source are D_{OL} and D_{OS}, respectively. The source is the distance D_{LS} away from the lens.

Since the angles θ, β, and $\Delta\phi$ are all small, it is obvious from Figure 3.2 that

$$\theta D_{OS} = \beta D_{OS} + \Delta\phi D_{LS}. \tag{3.63}$$

We define the reduced deflection angle α as

$$\alpha = \Delta\phi D_{LS}/D_{OS}. \tag{3.64}$$

Equation (3.63) may then be written

$$\beta = \theta - \alpha. \tag{3.65}$$

which is called the **lens equation**.

In our current notation

$$b = \theta D_{OL}. \tag{3.66}$$

in Equation (3.61). Therefore

$$\alpha = \frac{D_{LS}}{D_{OS}D_{OL}}\frac{4GM}{c^2\theta}. \tag{3.67}$$

Substituting this into the lens equation and using the notation

$$\theta_E = \sqrt{\frac{D_{LS}}{D_{OS}D_{OL}}\frac{4GM}{c^2}} \tag{3.68}$$

we obtain

$$\theta^2 - \beta\theta - \theta_E^2 = 0. \tag{3.69}$$

The quantity θ_E is designated as the **Einstein angle**.

If $\beta = 0$, i.e., the lens and the source are exactly aligned

$$\theta = \theta_E. \tag{3.70}$$

Because of symmetry, the source appears as a ring of angular radius θ_E around the lens. This kind of image is called an **Einstein ring** (Chwolson 1924, Einstein 1936, Refsdal 1964). Figure 3.3 is an image of an Einstein Ring. This rare almost exact alignment was discovered by the Very Large Array radio telescope in the late 1980s (Hewitt et al. 1988).

In a typical lensing situation $D_{LS} \approx D_{OL}$. Equation (3.68) may then be rewritten using the definition of the Schwarzschild radius r_s of the lensing mass M

$$\theta_E \approx \sqrt{\frac{2r_s}{D_{OS}}}. \tag{3.71}$$

The angle is exceedingly small for most cases. For stars lensing other stars in our Galaxy $\theta_E \approx$ milliarcsecond. For stars in foreground galaxies lensing much more distant quasars $\theta_E \approx$ microarcsecond. Finally, for a foreground galaxy lensing a background quasar gives a much larger $\theta_E \approx$ arcsecond. In this last example,

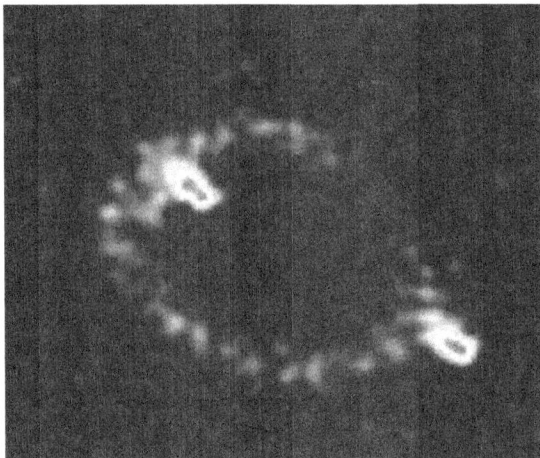

Figure 3.3. VLA image of an Einstein Ring. The field of view is 0.0014×0.0014 degrees as imaged by the core of a galaxy. Image courtesy of NRAO/AUI Twentieth anniversary press release of achievements (http://www.nrao.edu/pr/2000/vla20/background/ering/). Also see Hewitt et al. 1988.

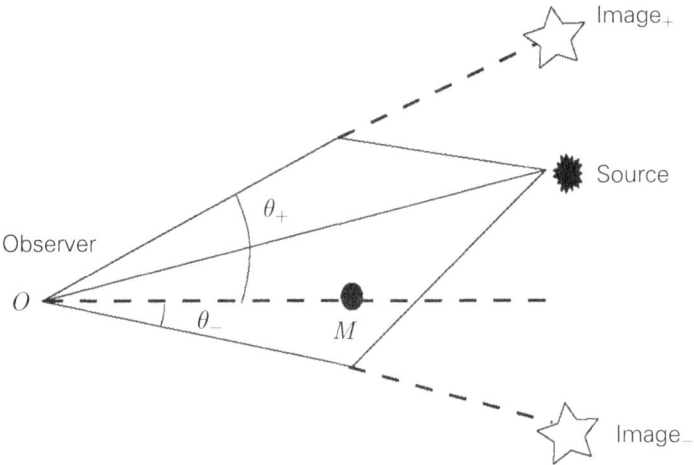

Figure 3.4. Formation of two images by the gravitational lens M.

a Galaxy is not really a point-like lens. However, the one arcsecond resolution is easily observable in the optical range as was first noted in 1937 by Zwicky at the California Institute of Technology.

In the case of Q0957 + 561, the lens and the source are not perfectly lined up causing Equation (3.69) to have two solutions θ_+ and θ_-.

$$\theta_{\pm} = \beta/2 \pm \theta_E \sqrt{1 + \frac{\beta^2}{4\theta_E^2}}. \tag{3.72}$$

These two correspond to two images of the source on opposite sides of the lens (Figure 3.4) whose angular separation is

$$\Delta\theta = \theta_+ - \theta_- = 2\theta_E \sqrt{1 + \frac{\beta^2}{4\theta_E^2}} \tag{3.73}$$

i.e., for small β approximately twice the Einstein angle.

The success of accounting for the additional precession of Mercury's orbit and predicting deflection of light are two great successes of the General Theory of Relativity placing it on a solid physical footing. Clearly the curvature of space-time due to mass can occur and can be calculated. As we shall see in the next chapter, General Relativistic non-Euclidean geometries can provide the mathematical underpinnings needed for cosmology with the mathematical formulation of dark energy making its first appearance.

Chapter 4

Curved space in cosmology

The Euclidean geometry used by Newton is a mathematical system attributed to the Alexandrian Greek mathematician Euclid, whose *Elements* is the earliest known systematic discussion of geometry of what we now call flat space.

Of the five postulates of Euclidian geometry, the most skeptically discussed (even before General Relativity) is the fifth which states that one and only one line can be drawn through a point parallel to a given line. Two lines are considered parallel if they lie in a plane and never cross each other even if infinitely extended. Euclid's followers had misgivings about the fifth postulate since there was no way of confirming it experimentally. Assume a straight line through a point P which is parallel to another line S. If we now rotate our line ever so slightly, how do we know that it does cross the line S after this rotation? In practice we are always dealing with limited segments of straight lines, and cannot observe the whole of an infinite straight line.

Modern cosmology, on the other hand, involves "non-Euclidean" space-time which may not satisfy the parallel line axiom. Einstein's Theory of General Relativity shows that the true geometry of space-time is non-Euclidean geometry—in particular, it is a four dimensional Riemannian space-time. We will first discuss curved space qualitatively and then describe it mathematically.

4.1 Non-Euclidean geometries

In the 19th century it was assumed that the fifth axiom can be replaced, resulting in other systems where geometric relations are different from the familiar Euclidian. There are two interesting cases: **hyperbolic geometry** discovered independently by Carl Friedrich Gauss (1777–1855), Nikolai Lobachevski (1793–1856), and Janos Bolyai (1802–1860), and **spherical geometry**, invented by Georg Riemann (1826–1866). For our purposes, it is important to note that besides the Euclidean **flat geometry**, these two are the only possible descriptions of a Universe which is homogeneous and isotropic where all places and directions are equivalent. Observations indicate that these requirements are satisfied by the universe at the very largest scales much bigger than galaxies. A universe which is homogeneous and isotropic satisfies the **Cosmological Principle**, an assumption necessary for current cosmology.

Riemann devised mathematical methods for calculations in non-Euclidean geometry. These methods were used by Einstein and are now known as tensor calculus. As we will discuss later in this chapter, the Riemann curvature tensor describes the way space curves or how it differs from flat Euclidean space. For four dimensions, the curvature tensor is described by 20 components.

Riemann enrolled at the University of Göttingen in 1846 to study theology, as his father wished. However, Riemann left theology and took courses in mathematics from the professor of mathematics Gauss. His Ph.D. thesis, supervised by Gauss, was submitted in 1851. Riemann was then employed by Göttingen University. To complete a higher degree, Riemann gave a lecture whose English title is *On the hypotheses that underlie geometry*. This lecture, delivered on 10 June 1854, has become a classic of mathematics. There were two parts: the first part centered on the definition of the curvature tensor, $R^{\alpha}_{\beta\mu\nu}$. The second part asked what the dimension of the real space was and what geometry described the real space. Riemann proposed that space could have measurable properties.

However, it is difficult to imagine a four dimensional space-time, not to mention its curvature. It is conceptually useful to look at two dimensional surfaces. Locally, our three dimensional universe is flat for all practical purposes, and we can visualize the curvature of two dimensional surfaces in it.

The two dimensional counterpart of a 3 dimensional spherical space is the surface of a sphere. It is not necessary to be aware of the third dimension in order to measure the curvature of the spherical surface. An imaginary creature living on the spherical surface which does not possess any third dimension off the surface, and who does not even understand what the third dimension would mean, can carry out geometrical measurements on the surface to find out the overall geometry. For example, the creature may draw a triangle, and measure the sum of the internal angles. If the result is >180 degrees, the creature thus knows that it lives on a spherical surface. See Figure 4.1 top image for a two-dimensional example (also see Figure 4.4). A triangle is drawn on the surface. Alternatively, the creature may draw a circle. If measuring along the surface of the sphere, the ratio of the circumference to the diameter is less than π, the creature would know that it lives in a spherical geometry. See one of the lines of equal latitude near the pole in the top frame of Figure 4.1. Or if it sees that all parallel lines cross each other sooner or later, it would be even more convinced that its space is spherical. This can be seen in Figure 4.1 and Figure 4.2 where constant "longitude" lines which are parallel to one another at the equator cross at the poles.

In the opposite case, when the sum of the internal angles of a triangle is <180 degrees, the ratio of the circumference of a circle to its diameter is greater than π, and one can draw any number of lines through a given point which are parallel to another line, the creature knows that it lives on a hyperbolic space. Hyperbolic space theoretically continues to an infinite distance as does the flat Euclidian sheet. A horse's riding saddle, or more precisely its central part, curves more or less like a limited area of the hyperbolic surface. In the middle, the saddle surface curves upwards when one moves along the length of the saddle, and downwards when one moves in crosswise direction; it is this opposite curvature in two perpendicular directions that is characteristic of a hyperbolic surface. See the middle image of Figure 4.1 (and also Figure 4.5).

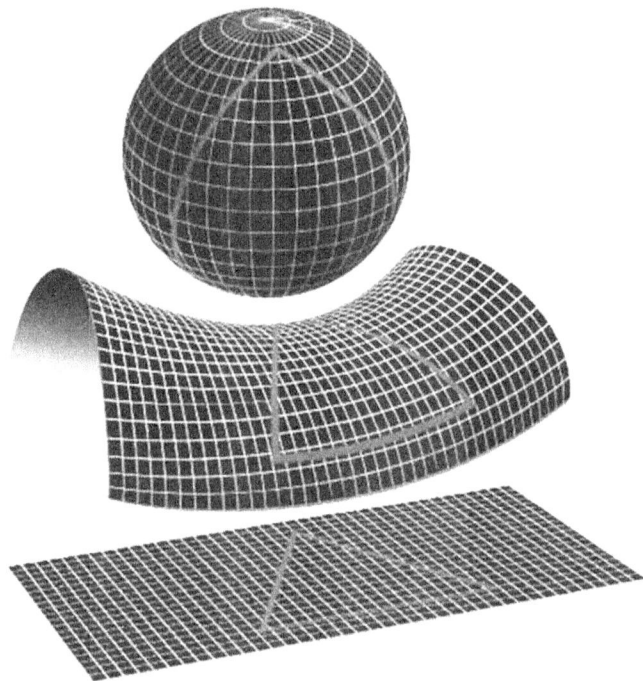

Figure 4.1. Two dimensional sphere (top), "horse saddle" infinite hyperbolic surface (middle) and flat infinite Euclidian (bottom) surfaces. Reference lines are in white and triangles in faint gray. See Figure 4.4 for corrresponding triangles drawn separately and clearly. As defined later in this chapter, the three spaces correspond to, $k = +1$ (top), $k = +1$ (middle), $k = 0$ (bottom).

Between a spherical surface and a hyperbolic surface is a flat surface, a two dimensional Euclidean space. The familiar laws of Euclidian geometry are valid in this and only in this geometry: the sum of the internal angles of a triangle is exactly 180 degrees. See Figure 4.1 (the bottom image) and also see Figure 4.4. For the flat sheet, the ratio of the circumference of a circle to its diameter is π, and one can draw one and only one straight line through a point, parallel to another straight line.

Current cosmological models assume—partly for simplicity and partly based on observations—that the universe is isotropic and homogeneous. This assumption, the Cosmological Principle, implies that at the same cosmic time or epoch the universe appears on average (in large scale) the same to every observer, wherever he/she may be or in whatever direction one looks. A consequence of this principle is that the 3-space (at a given time) **must have constant curvature**. This curvature can take one of the three different forms already described qualitatively that we will now discuss mathematically.

4.2   Curvature of 3-space

As mentioned, it is conceptually useful to consider 2-spaces and then find various ways to generalize the result to 3-space. We refer the reader to Wolfgang Rindler's *Essential Relativity* (1977, whose presentation we follow) or similar books for more details.

Figures 4.1 and 4.2 show a 2-space of constant curvature, the surface of a sphere. Anywhere on the surface you may draw two great circles mentioned earlier at right angles to each other, and find that the center of curvature of both circles is in the same direction, and both circles have the same radius from the center of the sphere. If the radius is a_K, the Gaussian curvature (named after the German mathematician Karl F. Gauss who pioneered the study of differential geometry, as well as many other fields of study) is defined as

$$K = \frac{1}{a_K^2}. \tag{4.1}$$

Figures 4.1 and 4.5 show a saddle surface, a 2-space of constant curvature. As shown in Figure 4.5, at any point in this surface you may draw two hyperbolas at right angles to each other such that one hyperbola curves one way, the other in the **opposite** direction. As local approximations of these hyperbolas you may take two circles of radius a_K. The constant curvature K is defined as having a **negative** sign in this case. For a Euclidian space shown in Figure 4.1, two lines at right angles have centers of curvature an infinite distance away and K is zero.

As stated earlier, an observer living in two dimensions can detect the curvature of his/her space by carrying out surface geometrical measurements. Understanding the 3-dimensional space is not necessary to detect the curvature K. Let us see how this might work out in case of a sphere.

Draw two nearby great circles starting from the pole an angle θ apart (Figure 4.2). At any "latitude", let the distance between these two geodesics be called η. If the geodetic deviation η is measured at distance r from the pole along the surface, and the radius of the sphere is a_K,

$$\eta = \theta a_K \sin \frac{r}{a_K}. \tag{4.2}$$

By integrating, the circumference of the circle at radius r is obtained by

$$C = \int_0^{2\pi} \eta \, d\theta = 2\pi a_K \sin \frac{r}{a_K}. \tag{4.3}$$

The surface area of the polar cap comes from integrating over r:

$$A = \int_0^r C \, dr = 2\pi a_K^2 \left(1 - \cos \frac{r}{a_K} \right). \tag{4.4}$$

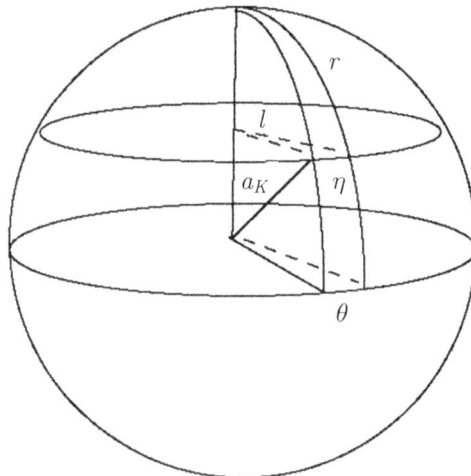

Figure 4.2. Calculation of the geodetic deviation η between two great circles separated by an angle θ. The arc η on the small upper horizontal circle is of length $l\theta$ where l is the radius of the small circle. The small circle is a distance r away from the "north" pole (measured along the surface). The corresponding angular distance in radians is r/a_K. Therefore $l = a_K \sin(r/a_K)$ from which Equation (4.2) follows.

By taking the first terms in the Taylor series with respect to r/a_K and using Equation (4.1) we get

$$\eta = \theta\left(r - \frac{1}{6}Kr^3\right) \tag{4.5}$$

$$C = 2\pi\left(r - \frac{1}{6}Kr^3\right) \tag{4.6}$$

$$A = \pi\left(r^2 - \frac{1}{12}Kr^4\right). \tag{4.7}$$

The above two equations show how the circumference is a little shorter and the area a little smaller than in a flat space with $K = 0$. Measurements like these lead to the establishment of the K value. If the surface is hyperbolic, the circumference and the area are greater than for the flat space (Figure 4.3). Thus K is a negative number.

What happens to C and A as a function of r/a_K on a sphere? Initially, they increase with r/a_K. At $r/a_K = \pi/2$ the circumference is at its maximum while the area is exactly half of the total. At $r/a_K = \pi$, C shrinks to a point while A has its maximum.

Figure 4.3. Flattening of a small circular area cut from a surface is shown. A flat surface on the left, a spherical in the middle and a hyperbolic on the right. On the left a simple flat circle results. In the middle case, the flattening requires tearing (shown schematically) since A is too small for the flat surface. In the right the flattening results in folding.

We may determine the curvature by looking at the geodetic deviation η. By differentiating the η in Equation (4.2) twice relative to r we get to the first order:

$$\ddot{\eta} = -K\eta. \tag{4.8}$$

Box 2. Riemann curvature tensor.

Equation (4.8) may be generalized to any curvature surfaces. We have to take into account that the geodesic deviation η depends on the direction towards which we measure it. Let us call its two components $\eta^\mu, \mu = 1, 2$. The direction of the two neighboring geodesics may be described by the tangent vector **V** for one of them which has components $V^\nu = dx^\nu/ds, \nu = 1, 2$. The curve parameter is s, and the two coordinates x^1 and x^2. In place of the differentiation in Equation (4.8) we must use the absolute derivative D/ds (closely related to the covariant derivative, which will be mentioned later). This derivative takes into account the usual change in vector components d/ds as well as the effects of space curvature when the vector is transported from place to place. For more details, see e.g., Rindler (1977). After these considerations, a more general version of Equation (4.8) is

$$\frac{D^2\eta^\mu}{ds^2} = (R^\mu_{\nu\rho\sigma} V^\nu V^\rho)\eta^\sigma, \quad \mu = 1, 2. \tag{B2.1}$$

As previously defined, the repeated index is a summation index; thus there is a triple summation on the right.

The coefficients $R^\mu_{\nu\rho\sigma}$ contain the information on the curvature of the surface. They can be considered as components of a tensor called the **Riemann curvature tensor**. This is named after the German mathematician Georg F. B. Riemann who in 1854 introduced the concept of a hypersphere as a model of the 3D curved space, and laid the foundation of a non-Euclidean system of geometry known as Riemann geometry. In the two-dimensional case it has $2 \times 2 \times 2 \times 2 = 16$ components. The curvature tensor measures the non-commutativity of the covariant derivative.

For a moment consider the 4-dimensional space of General Relativity. The Riemann tensor has now $4 \times 4 \times 4 \times 4 = 256$ components. Fortunately, all the individual 256 components are not needed. We may change two of the free indices of $R^{\mu}_{\nu\rho\sigma}$ to summation indices and form a second rank tensor $R_{\nu\rho} = R^{\mu}_{\nu\rho\mu}$ called the **Ricci tensor**, after Gregorio Ricci-Curbastro, an Italian mathematician instrumental in the development of the Tensor Calculus. The process has the technical name, the contraction of a tensor and $R_{\nu\rho}$ is also called the contracted Riemann tensor.

The Ricci tensor is related to Newtonian physics. Say that an observer follows a timelike geodesic $(ds = cdt)$ and measures the distance η to a nearby object. Thus the V-vector has only one non-zero component $V^0 = dx^0/ds = 1$ and Equation (B2.1) becomes

$$\frac{D^2\eta^{\mu}}{ds^2} = R^{\mu}_{00\sigma}\,\eta^{\sigma}. \tag{B2.2}$$

However, if there is a gravitational potential field Φ between the object and the observer, in the Newtonian theory, it causes a force per unit mass

$$F^{\mu} = -\frac{\partial \Phi}{\partial x^{\mu}}. \tag{B2.3}$$

The relative acceleration between the observer and the object depends on the force differential between them:

$$\frac{d^2\eta^{\mu}}{dt^2} = \frac{\partial F^{\mu}}{\partial x^{\sigma}}\eta^{\sigma} = -\frac{\partial^2 \Phi}{\partial x^{\mu}\partial x^{\sigma}}\eta^{\sigma}. \tag{B2.4}$$

A comparison of Equations (B2.2) and (B2.4) suggests that at the Newtonian limit where

$$\frac{D^2\eta^{\mu}}{ds^2} = \frac{1}{c^2}\frac{d^2\eta^{\mu}}{dt^2} \tag{B2.5}$$

we have approximately

$$R^{\mu}_{00\sigma}c^2 = -\frac{\partial^2 \Phi}{\partial x^{\mu}\partial x^{\sigma}}. \tag{B2.6}$$

Let us sum the diagonal components on both sides. Then we get (with minus sign)

$$\nabla^2 \Phi = \frac{\partial^2 \Phi}{\partial x^\mu \partial x^\mu} \qquad \text{(B2.7)}$$

on the right hand side of Equation (B2.6) which connects us with the Poisson equation of the Newtonian theory (Box 3 at the end of this chapter). On the left hand side we must have

$$R_{00} = R^\mu_{00\mu}. \qquad \text{(B2.8)}$$

We are thus led to the approximation

$$R_{00}c^2 = -\nabla^2 \Phi. \qquad \text{(B2.9)}$$

This lowest order agreement between the Newtonian potential and a component of the Ricci tensor gives motivation for the use of the Ricci tensor in General Relativity.

The rate of spreading of two geodesics from each other tells the curvature (Figure 4.4). This concept will later be crucial in examining cosmic background radiation fluctuations on the sky to deduce relative effects of dark energy and gravitating matter along with other properties of the universe to unprecedented accuracy.

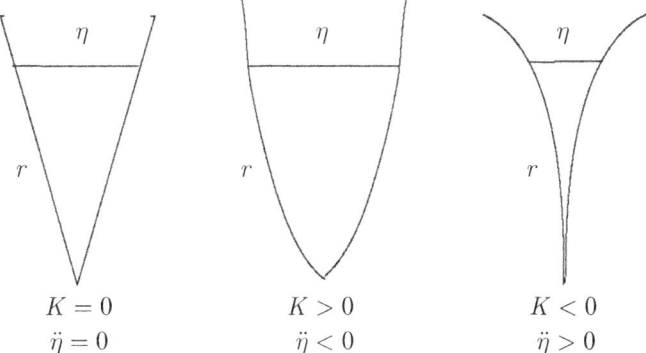

$$\begin{array}{ccc} K = 0 & K > 0 & K < 0 \\ \ddot\eta = 0 & \ddot\eta < 0 & \ddot\eta > 0 \end{array}$$

Figure 4.4. The geodetic deviation η between two geodesics increases more rapidly when $K < 0$ than when $K > 0$, while the increase is intermediate (linear) in a flat space of $K = 0$. Left to right are triangles on a flat Euclidian surface, a sphere, and a hyperboloid which were shown faintly in Figure 4.1.

Now move one dimension upwards and consider a 3-space. Just as in our two dimensional example, we can make all our measurements in the 3-space. Corresponding to the measurements of C and A in 2-space we measure the surface area S of a sphere and the volume V of a sphere as a function of its radius. Deviations from the Euclidian $S = 4\pi r^2$ and $V = 4/3\pi r^3$ tell us about the curvature of the 3-space.

Generalization to the higher dimension is obtained by replacing η/θ by $(\eta/\theta)^2$ in the equations above while 2π is replaced by 4π so that Equation (4.3) becomes

$$S = 4\pi a_K^2 \sin^2\left(\frac{r}{a_K}\right). \tag{4.9}$$

To the second order, it is

$$S = 4\pi\left(r^2 - \frac{1}{3}Kr^4\right). \tag{4.10}$$

The volume is to the second order

$$V = \int_0^r S\,dr = \frac{4}{3}\pi\left(r^3 - \frac{1}{5}Kr^5\right). \tag{4.11}$$

With accurate values of S and V, the curvature can be calculated from these equations. The total volume of the 3-sphere by integration of Equation (4.9) over r is

$$V = 2\pi a_K^2\left(r - \frac{a_K}{2}\sin\frac{2r}{a_K}\right). \tag{4.12}$$

When $r/a_K = \pi$, we get the total volume of the 3-sphere, $V = 2\pi^2 a_K^3$. It has two identical halves divided by $r = (\pi/2)a_K$. However, the 3-sphere has no boundary and no center.

Now the metric of the most general 3-space that satisfies the requirements of isotropy and homogeneity is

$$dl^2 = \frac{dr^2}{1 - kr^2} + r^2\left(d\theta^2 + \sin^2\theta\,d\phi^2\right) \tag{4.13}$$

where k may have any of the three values $k = 0, k = \pm 1$.

Justifying the above, for $k = 0$ the metric is the metric of a flat space (see the space part of Equation (1.11)). For $k = \pm 1$ we study the behavior of the geodetic deviation η at radial distance r when the deviation angle is θ at the origin.

For $k = -1$ the radial distance

$$\Psi = \int_0^r \frac{dr}{(1 + r^2)^{1/2}} = \sinh^{-1}r \tag{4.14}$$

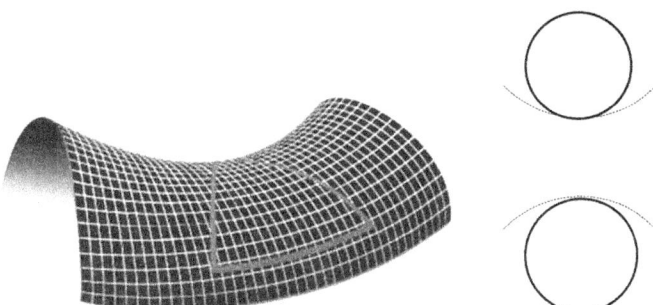

Figure 4.5. A horse saddle portion of an infinite hyperbolic surface. The surface curves upwards when moving in one direction along the saddle, and downward in the perpendicular direction. As shown on the right, the curvature of the surface is described by the radius R of the circle which locally curves at the same rate as the surface. The radius of curvature R has the same value in two perpendicular directions but the radius vectors are on opposite sides relative to the surface.

or

$$r = \sinh \Psi. \tag{4.15}$$

Thus, $\eta = r\theta = \sinh \Psi \cdot \theta$, $\ddot{\eta} = \sinh \Psi \cdot \theta = \eta$; $K = -1$ (Equation (4.8)) for a "hyperboloid." Differentiation is relative to the radial distance coordinate Ψ.

When $k = +1$, the radial distance

$$\Psi = \int_0^r \frac{dr}{(1 - r^2)^{1/2}} = \sin^{-1} r \tag{4.16}$$

or

$$r = \sin \Psi \tag{4.17}$$

and $\eta = \sin \Psi \cdot \theta$, $\ddot{\eta} = -\sin \Psi \cdot \theta = -\eta$; i.e., $K = +1$ in Equation (4.8), a "sphere".

Thus the metric of Equation (4.13) covers all the possibilities of constant curvature 3-spaces. Transformations in Equations (4.15) and (4.17) lead to another presentation of the metric:

$$dl^2 = d\Psi^2 + f(\Psi)^2 \left(d\theta^2 + \sin^2 \theta \, d\phi^2 \right) \tag{4.18}$$

where

$$\begin{aligned}
f(\Psi) &= \sin \Psi & K &= +1 \\
f(\Psi) &= \Psi & K &= 0 \\
f(\Psi) &= \sinh \Psi & K &= -1
\end{aligned} \tag{4.19}$$

When $K = -1$ we may use the coordinate transformation

$$
\begin{aligned}
x_1 &= \sinh \Psi \sin \theta \sin \phi \\
x_2 &= \sinh \Psi \sin \theta \cos \phi \\
x_3 &= \sinh \Psi \cos \theta \\
x_4 &= \cosh \Psi
\end{aligned}
\tag{4.20}
$$

to find that the metric (Equation (4.18)) represents a 3-dimensional hyperboloid

$$
x_4^2 - x_1^2 - x_2^2 - x_3^2 = 1
\tag{4.21}
$$

embedded in a 4-dimensional space of metric

$$
dl^2 = dx_1^2 + dx_2^2 + dx_3^2 - dx_4^2.
\tag{4.22}
$$

Hyperbolic geometry was first mathematically developed by the Russian Nikolai I. Lobachevsky in 1826 and independently in 1832 by the Hungarian Janos Bolyai.

Problem. Metric with $K = -1$ hyperboloid: Show that Equation (4.18)) with $K = -1$ represents a 3-dimensional hyperboloid in the 4-dimensional space of Equation (4.22).

Similarly for $K = +1$ the metric in Equation (4.18) represents a 3-dimensional sphere

$$
x_1^2 + x_2^2 + x_3^2 + x_4^2 = 1
\tag{4.23}
$$

within a 4-dimensional Euclidean space of metric

$$
dl^2 = dx_1^2 + dx_2^2 + dx_3^2 + dx_4^2
\tag{4.24}
$$

where a transformation

$$
\begin{aligned}
x_1 &= \cos \Psi \sin \theta \sin \phi \\
x_2 &= \cos \Psi \sin \theta \cos \phi \\
x_3 &= \cos \Psi \cos \theta \\
x_4 &= \sin \Psi
\end{aligned}
\tag{4.25}
$$

is used.

We have completed our qualitative and mathematical discussion of three different forms of curved space which satisfy the Cosmological Principle—that the universe is isotropic and homogeneous. For these 3-spaces, at the same cosmic time or epoch the universe appears on average the same to every observer in whatever direction one looks. These (excluding time) have constant curvature.

Box 3. Poisson equation.

For completeness, we include here a derivation of the Poisson equation:

For a body of mass m at position $\mathbf{r}$ Newton's equation of motion may be written

$$m\ddot{\mathbf{r}} = -m\nabla\Phi(\mathbf{r}) \tag{B3.1}$$

where the potential Φ is

$$\Phi(\mathbf{r}) = -\sum_i \frac{Gm_i}{|\mathbf{r} - \mathbf{r}_i|}, \quad \mathbf{r} \neq \mathbf{r}_i. \tag{B3.2}$$

contributed by masses m_i at positions $\mathbf{r}_i$. For a continuous mass distribution, use the mass density $\rho(\mathbf{r}')$ as a function of position $\mathbf{r}'$:

$$\Phi(\mathbf{r}) = -\int \frac{G\rho(\mathbf{r}')}{|\mathbf{r} - \mathbf{r}'|} d\mathbf{r}'. \tag{B3.3}$$

Then

$$\ddot{\mathbf{r}} = -\nabla\Phi(\mathbf{r}) = \int \frac{G\rho(\mathbf{r}')(\mathbf{r} - \mathbf{r}')}{|\mathbf{r} - \mathbf{r}'|^3} d\mathbf{r}'. \tag{B3.4}$$

Problem B3.1. Show that $\nabla\frac{1}{|\mathbf{r}-\mathbf{r}'|} = \frac{\mathbf{r}-\mathbf{r}'}{|\mathbf{r}-\mathbf{r}'|^3}$.

Take the gradient ∇ of both sides of Equation (B3.4):

$$\nabla^2\Phi(\mathbf{r}) = -\int G\rho(\mathbf{r}')\nabla^2\frac{1}{|\mathbf{r} - \mathbf{r}'|} d\mathbf{r}' \tag{B3.5}$$

One can verify that the gradient relative to $\mathbf{r}(\nabla)$ is connected to the gradient relative to $\mathbf{r}'(\nabla_{\mathbf{r}'})$ by

$$\nabla_{\mathbf{r}'}f = -\nabla f \tag{B3.6}$$

where f is a function of $\mathbf{r} - \mathbf{r}'$, and

$$\nabla_{\mathbf{r}'}^2 f = \nabla^2 f. \tag{B3.7}$$

Problem B3.2. Show that $\nabla_{\mathbf{r}'}^2 f = \nabla^2 f$ where $f = f > (\mathbf{r} - \mathbf{r}')$.

Therefore,

$$\nabla^2 \Phi(\mathbf{r}) = -G \int \rho(\mathbf{r}') \nabla_{\mathbf{r}'}^2 \frac{1}{|\mathbf{r} - \mathbf{r}'|} \, d\mathbf{r}'. \tag{B3.8}$$

Consider a small spherical volume of radius R around point $\mathbf{r}$. The density inside this sphere can be taken to be constant,

$$\rho(\mathbf{r}') \cong \rho(\mathbf{r})$$

so ρ can be taken outside the integral. What remains is

$$\int_{volume} \nabla_{\mathbf{r}'} \cdot \left(\nabla_{\mathbf{r}'} \frac{1}{|\mathbf{r} - \mathbf{r}'|} \right) d\mathbf{r}' = \int_{surface} \left(\nabla_{\mathbf{r}'} \frac{1}{|\mathbf{r} - \mathbf{r}'|} \right) \cdot d\mathbf{S}'. \tag{B3.9}$$

We have applied the Gauss' Theorem to this volume, and made the integral over the surface of the sphere (surface element $d\mathbf{S}'$ pointing radially outward). The gradient is

$$\nabla_{\mathbf{r}'} \frac{1}{|\mathbf{r} - \mathbf{r}'|} = -\frac{1}{R^2} \tag{B3.10}$$

which is a constant on the surface of the sphere. The remaining integral is

$$\int dS' = 4\pi R^2. \tag{B3.11}$$

Substituting these in Equation (B3.9), Equation (B3.8) becomes the Poisson equation.

$$\nabla^2 \Phi(\mathbf{r}) = 4\pi G \rho(\mathbf{r}). \tag{B3.12}$$

Chapter 5
Finite versus infinite universe in space and time

5.1   Observation of an isotropic universe

Either a finite or an infinite universe presents conceptual difficulties. Our thinking is based on Euclidean geometry which is taken for "common sense." The extension of Euclidean space to infinity is hard to visualize even in everyday geometry. The "non-common sense" non-Euclidean geometries in the 19th century permitted a spherical spatially finite universe avoiding the embarrassing question of an infinite universe. On the other hand, it is difficult to visualize in three dimensions a universe with a spherical geometry, even though its finiteness is easy to understand mathematically. Other non-Euclidian hyperbolic universes are both infinite and difficult to visualize in three dimensions.

Our real universe may be either finite or infinite. A special case is a homogeneous and isotropic universe where observers everywhere see similar things. Because we observe isotropy outside our Milky Way (equal galaxy numbers in opposite directions), it is likely that the universe is also homogeneous, unless it is centred on us (discounted in a generalization of Copernicus' ideas). Modern observations have confirmed the isotropy to very high degree. This symmetry is especially clear in the cosmic background radiation which is isotropic with an accuracy of about 0.001%. (Bennett et al. 2003). As we shall see, the isotropy of the expansion of the universe (the Hubble law) is measured at the level of 10–30% accuracy.

In the extension of space to infinity, the observation of homogeneity would imply an infinite number of galaxies or stars in a flat Euclidean universe. We should point out that a finite spherical universe of 3 space dimensions can be isotropic and homogeneous without any embarrassing edge and infinite number of galaxies. The two dimensional sphere of the Earth is a good analogy. Locally, it appears (on average) flat. Only at large distances or with high precision does the spherical nature of its surface become apparent.

5.2   A finite universe in time

In regard to the finite age of the universe, there is one cosmological observation made with the naked eye which can be easily understood; it is obviously dark at night. In 1826, the German physician and astronomer Heinrich Olbers pointed out a problem with a perpetual unchanging universe. It would incinerate the Earth! Why? The number of stars inside progressively larger imaginary spherical shells all of the same thickness increases with the radius of the shell squared. The intensity of the light from each star on Earth decreases as the radius of the shell squared. The product of these two is constant, the same for each shell, and therefore the total intensity of light increases as the number of shells going outward. Light from a large volume would be enough to burn up the Earth.

Stating this differently, if the universe were infinite in extent and full of stars, then every line of sight would sooner or later intersect the surface of a star. The surface brightness of a star per unit angle squared is constant with respect to distance. And if we are looking at the surface of stars in all directions, then the sky would be as bright as the surface of the Sun, day and night. The fact that this is not so is called Olbers' paradox. So, exactly what does the dark sky at night tell us?

In 1848, the poet Edgar Allen Poe (of all people) suggested a solution in his cosmological essay, *Eureka*, that the dark night sky implied the universe had to have a finite age. The limited speed of light would prevent the light of stars more distant than the age of the universe (in light years) from reaching us.

When we look far away, we look into the past. In order that every line of sight meets a star, the stars must have existed a sufficient time in the past. In a universe of finite age, Olbers' Paradox does not exist. Thus the universe may well be infinitely large as long as it is of finite age. The night sky is illuminated by only a finite number of stars so that actually very few lines of sight meet the surface of a star. The current view is that indeed the universe is "only" about 14 billion years old. This would be the "time edge", the time of origin of the universe. A time-like edge in the past is a concept some scientists are unwilling to accept. They would prefer the universe to extend infinitely into the past and future.

5.3   The age of the universe via its "oldest objects"

Aside from Olbers' paradox, another method of determining a finite age for the universe is to search for (or estimate the ages of) the oldest objects in it. One thus obtains a lower limit for the age. If the universe were infinitely old one would simply keep finding older and older objects.

Here we must briefly review the processes of stellar life cycles. In the process, we also introduce certain observationally distinctive variable stars which are useful in estimating cosmological distances.

Initially, after its formation, a star is predominantly composed of hydrogen and helium. Research by the Russian-American physicist George Gamow and his students Ralph Alpher and Robert Herman (1948–1953) and later work showed that hydrogen, helium, and small amounts of heavy hydrogen nuclei, deuterium, and tritium could be produced in significant amounts in the origin of the universe. More massive nuclei are made inside stars which subsequently explode. These massive nuclei still form only a small fraction of the interstellar raw material for future stars.

A solar mass star like the Sun has a central temperature of about 2×10^7 K at which rapidly moving high temperature protons are occasionally able overcome their mutual repulsive electric force, and fuse to form deuterium nuclei. After deuterium has formed, further nuclear reactions take place in rapid succession with the end result of helium nuclei. Since a helium nucleus is lighter than four protons that start the process, the energy related to the mass difference ΔM, ΔMc^2 per helium nucleus, is liberated. This energy is ultimately re-radiated providing the luminosity of the star. The actual chain of nuclear reactions is even more complicated than this. The German-American Hans A. Bethe of Cornell University, New York, was awarded the 1967 Nobel Prize in Physics for his key role in discovering the nuclear reactions which take place in the Sun and in other stars (in 1939).

A star like the Sun remains a stable star in hydrostatic equilibrium as long as the hydrogen fuel for the nuclear transformations last. A star in this phase of evolution is called a **main-sequence star**. For a solar mass star, the main-sequence lifetime is $\sim 10^{10}$ yr, a very long time. The lifetime is a rapidly decreasing function of the stellar mass: more massive stars evolve much faster than less massive ones.

When the hydrogen nuclear fuel in the center is exhausted, the star adjusts its structure radically and becomes a **red giant star**. While it is in the red giant stage, a star may pulsate and thus periodically vary in light at a rate related to its average intrinsic brightness. Stars in this distinctive stage are very useful in estimating distances of star clusters or the galaxies in which they reside. Eventually, for stars of mass comparable or smaller than our Sun, the outer layers are rather gently blown off and the central part collapses to a **white dwarf star** whose radius is typically only 1% of the main sequence radius. A main sequence star is well described as a ball of ordinary ideal gas, but the white dwarf is made of high density degenerate matter which does not contract significantly as it cools. With no nuclear energy generation, the white dwarf cools slowly and finally becomes very faint decreasing in luminosity by a factor of ten in 10^{10} yr.

In the case of stars initially much more massive than the Sun, the internal evolution proceeds through a multitude of stages of fusing of heavier nuclei. At the end there are no energy releasing nuclear fuels left, and the center of the star collapses. It can form either a neutron star, with a radius about 1 000 times smaller and with a mean density 10^9 times greater than for a white dwarf. Alternatively, the center could collapse to become a black hole. The main body of the star is violently blown outward as a supernova explosion. In the outward blast wave, a rapid build-up of heavy elements up to thorium and uranium takes place, by capture of neutrons into the already heavy nuclei more massive than iron. By "enriching" the gas clouds from which future stars will form, these heavy element nuclei find their way into other stars. Thus new generations of stars have more heavy elements on average than their predecessors.

English astrophysicist Fred Hoyle from Cambridge University with his collaborators at the California Institute of Technology showed that the origin of elements heavier than boron can be accounted for by this process of stellar evolution (in 1956). Other group members were the American nuclear physicist William A. Fowler (Nobel Prize in Physics in 1983) and the English astronomers E. Margaret Burbidge and Geoffrey R. Burbidge.

Now we will see how stellar evolution helps us in estimating ages of objects in our own Milky Way. The Milky Way consists partly of mysterious dark matter which we will discuss later and partly of visible matter. As for the visible stars, our Milky Way consists of a **stellar halo** and a **stellar disk**. The disk stars are typically like the Sun, fairly young in the galactic time scale. They are called **population I** stars. The old stars are found in the halo, either as individual halo stars or as members of **globular star clusters**. The old halo stars are called **population II** stars. To find the age of the Galaxy, we need to study the halo stars. Observationally, these populations were discovered by the German-American astronomer Walter Baade, about whom we shall say more later.

We will mainly discuss three methods of estimating ages of population II stars in our Galaxy: radioactive dating, main-sequence lifetime, and white dwarf cooling. Age determination of the oldest stars in the universe is called **cosmochronology**.

In **radioactive dating**, one compares the abundances of radioactive isotopes with the abundances of stable nuclear species. Commonly used are ^{232}Th, which has a half-life of 14×10^9 yrs, and ^{238}U, with a 4.5×10^9 yr half-life. By 1929 the Cambridge University physicist Ernest Rutherford (originally from New Zealand, Chemistry Nobel Prize winner in 1908) concluded by this method that the Galaxy must have existed billions of years before our Solar System. Burbidge et al. and the Canadian astrophysicist Alistair G. W. Cameron (in 1957) realized that the original abundance ratios of radioactive and stable nuclei could be derived from the theory of the rapid neutron capture in a supernova explosion.

As mentioned earlier, these elements get mixed into interstellar gas and are incorporated into the next generation of stars. For lower mass stars, these surface abundances are not changed by nuclear reactions in the cores of the stars which do not even use or produce these heavier elements. One can compare the formation abundances to the relative abundances on the surfaces of stars today to estimate the age. In 1987 the American-Dutch astronomer Harvey R. Butcher of the University of Groningen started spectroscopic studies of cosmochronology in individual stars via measurement of the thorium abundance. More recently, the method has been extended to uranium by Cayrel et al. (2001). In case of an old halo star CS31082-001 the age of the star was found to be $12.5 \pm 3 \times 10^9$ yr (by Hill et al. in 2002). Unfortunately, uncertainties in nuclear physics theory do not allow better than a 2×10^9 yr accuracy to date.

The **main-sequence lifetime** method uses the Hertzsprung–Russell diagram, a plot of individual stellar magnitudes versus their colors, to find out which stars in a globular star cluster are on the main sequence. In particular, the plot is used to find out the heaviest main-sequence star in the cluster. Since the main-sequence lifetime is a known function of the stellar mass, the knowledge of the mass of the heaviest star gives the age of the cluster, assuming that all stars in the cluster were formed at about the same time. It turns out that globular clusters are the oldest stellar systems in the Galaxy with the most massive main sequence stars being somewhat less massive than our Sun. For example, the main-sequence lifetime of a $0.8\ M_\odot$ star found in a globular cluster would be 14×10^9 yr while it is only 5.1×10^9 yr for a $1.1\ M_\odot$ star. The most massive main sequence star remaining in a cluster thus provides an estimate of its age as was realized by Allan Sandage in 1953.

Uncertainties in this method are related to both the distances of the globular clusters and to the theoretical models of stellar evolution. In 1970 Sandage determined an age of $11.5 \pm 0.3 \times 10^9$ yr for four globular clusters while Hesser et al. (1987) determined the age $13.5 \pm 0.5 \times 10^9$ yr for the globular cluster 47 Tucanae. Error estimates in these works are unrealistically small. Subsequent distance independent methods have been used to find the age of the oldest clusters of $13.5 \pm 2 \times 10^9$ yr (Jimenez et al. 1996), and $12.5 \pm 1.0 \times 10^9$ yr for the cluster M55 (Jimenez and Padoan 1998). For a number of clusters, an average age of $13.2 \pm 1.0 \times 10^9$ yr has been derived (Krauss and Chaboyer 2003).

The **white dwarf cooling** method makes use of the fact that the white dwarfs cool slowly, and within the age limit of the Galaxy must still be relatively hot on their surface. They have not been able to cool much below 4 000 K. The number of white dwarfs per unit volume of space in a given range of luminosities increases towards low luminosities, until there is a break and extremely sharp decline at very low luminosity. The oldest white dwarf stars have not yet had time to cool beyond the break (Liebert 1980). The luminosity at the break gives the cooling age when combined with theoretical models.

White dwarf cooling gives an age of $12.7 \pm 0.7 \times 10^9$ yr for the globular cluster M4 (Hansen et al. 2002). One of the uncertainties in this work is the fraction f_H for white dwarfs, which have hydrogen atmospheres rather than more rapidly cooling helium atmospheres. Fortunately, this uncertainty does not affect the end result very much as long as the fraction is within a reasonable range, $f_H = 0.5$–0.7 as can be seen in Figure 5.1.

Figure 5.1 outlines probability contours for the globular cluster ages. Also shown are dashed contours for Galaxy disk white dwarfs, showing that the disk is considerably younger than the halo, only $7.3 \pm 1.5 \times 10^9$ yr old. The study of open star clusters in the Galactic disk by main-sequence lifetime method has given very similar results (Janes and Phelps, 1994).

A fourth independent method is the study of bright elliptical galaxies in clusters and the modeling of the evolution of their stellar populations. Individual stars cannot be observed for these so integrated colors are used. Using a sample of 17 clusters of galaxies, Ferreras et al. (2001) determine the age of the universe to be 13^{+3}_{-2} Gyr. The uncertainty can be reduced to ± 1 Gyr by using additional information.

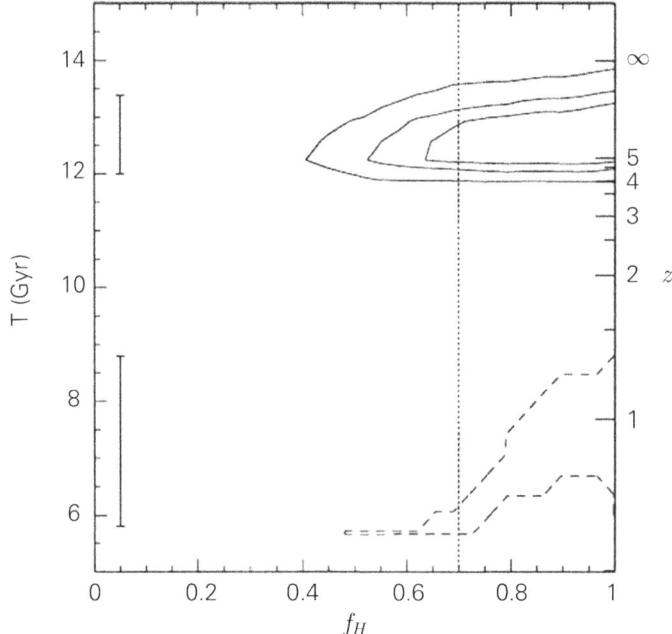

Figure 5.1. Age determination of the Galactic halo by using white dwarf cooling as a function of f_H, the fraction of white dwarfs with hydrogen atmospheres. The most values likely range from 0.5 to 0.7. Probability contours are at 1, 2, and 3 standard deviation levels. The dashed contour is for the age of the Galactic disk (Hansen et al. 2002). The right hand axis shows the fractional wavelength redshift z at which the formation of halos of similar galaxies of those ages can be observed today. The connection between redshift z and age will be derived later in this book.

Problem. Taking the white dwarf atmosphere to be made of non-degenerate gas, obtain a relation between the white dwarf's luminosity and surface temperature. Assume the star's surface to be a black body. Compute the luminosity of an old (4 000 K) white dwarf divided by an otherwise identical young one (100 000 K).

5.4   Observational discovery of the expanding universe

Sandage et al. (1998) characterized Hubble's founding work in observational cosmology as "In only 12 years from 1924 to 1936, Hubble brought to an almost modern maturity the four foundations of observational cosmology, even as its principles are practiced today." The four foundations were:

1. The discovery of the galaxy universe (nebulae are extragalactic stellar systems much like our Milky Way), which required measurements of distances to nearby galaxies.
2. The morphological classification of galaxies into a simple, but physically relevant system.
3. The initial discovery of the linear redshift-distance relation for nearby galaxies, using redshifts mainly measured by Slipher in the 1910s, and the extension of the relation into deeper space together with Humason.
4. The first large observational program for counting galaxies as a function of their apparent magnitude. This effort tested the homogeneity of the spatial distribution of galaxies and the results were also used by Hubble in an attempt to measure the curvature of space.[1]

We review briefly Slipher's and Hubble's work, with references to contributions by others. We also introduce some concepts useful for understanding some later parts of the text.

In the first two decades of the 20th century at the Lowell Observatory, Flagstaff, Arizona, the American astronomer Vesto M. Slipher (1875–1969) studied the spectra of what we now know to be galaxies. In 1914, he reported the first observation of the spectrum of what was then known as the Andromeda Nebula. The wavelengths of the spectrum were shifted to the blue (shorter) wavelengths which meant that the nebula as a whole was approaching us. He used the Doppler formula which makes use of the shift of wavelength $d\lambda$ of a spectral line in the spectrum of the nebula from its normal position λ_1 i.e., the redshift $z = d\lambda/\lambda_1$. It tells the speed V by which the nebula moves away from us: $z = V/c$. The velocity of the motion was found to be minus 300 km/s (in astronomy, the velocity is considered positive if the motion is away from us and the wavelengths are lengthened or redshifted, and negative in the opposite case).

In 1917, Slipher (1917a, b) published spectroscopic observations of other nebulae. He had managed to measure 25 spectra for faint nebulae each needing exposures of tens of hours (in comparison with ten minutes or even less with modern telescopes). Contrary to the Andromeda Galaxy, these galaxies mostly revealed redshifts in their spectra which meant that they moved away from us. By 1925, Slipher had data on 40 receding galaxies with velocities of hundreds of km/s; the largest velocity was +1 125 km/s, the fastest known speed in astronomy at that time! Slipher's data gave the first observational indication that our universe of galaxies is not stationary but expanding. But the physical nature of the nebulae (small and near our own

---

1 We can add one more technical item: 5) The introduction of first crucial cosmological tests for checking the paradigm of expanding universe (in particular the surface brightness dimming test, together with Richard Tolman).

galaxy or distant and galaxies in their own right) was not known to Slipher so his results were not conclusive. Already, in 1917, De Sitter had derived from Einstein's brand new General Relativity an interesting cosmological solution where the matter density is near to zero (and which included the cosmological constant). This model predicted the "de Sitter effect" where light from remote objects is redshifted.

To understand the real meaning of Slipher's observations, it is necessary to measure, besides the velocities, the distances to the sources of light. Edwin Hubble was able to do this, along with a few other pioneers. Almost a century earlier, around 1830, three scientists (Bessel, Henderson and Struve) had first measured distances to three different stars using the same geometric method, the annual parallax. Around 1920 three scientists (Öpik, Lundmark and Hubble) measured the (almost a million times larger) distance to one and the same galaxy, the Andromeda nebula (M31), using three different methods.

Hubble (1889–1953) was born in Missouri, and was raised in Chicago. There he obtained a degree in Mathematics and Astronomy from the University of Chicago in 1910. Hubble studied law at Oxford, returning to the United States in 1913. After practicing law half-heartedly for a year in Kentucky, he returned to the University of Chicago for graduate study. As the United States entered World War I, Hubble got an offer to join the staff of Mount Wilson Observatory. After staying up all night to finish his Ph.D. thesis, and taking the oral examination the next morning, Hubble enlisted in the infantry. After the war, he went to Mount Wilson.

First Hubble studied reflection nebulae, dust clouds which reflect light from a nearby star. Then he started using the 100 inch Mount Wilson telescope to study spiral nebulae. By 1917 Hubble had concluded that these are indeed "island universes" of stars, based on the high speeds measured by Slipher, so he starting looking for individual stars ("standard candles") which could be used to measure their distances.

With a large telescope it is possible to see distant individual bright stars. But, at a large distance even a small nebula may be mistaken for a star. Therefore it is good to look for variable stars which vary over scales of days, since nebulae which may be a light year in size or larger do not vary. The "standard candle" must also be very luminous, emitting so much energy/sec that it will be easily seen in a galaxy well outside our own. Cepheid variables are many times more luminous than our Sun. Today, a certain kind of incredibly luminous exploding star (Type Ia supernova) is useful as a standard candle.

Distances l_0 obtained with standard candles are usually expressed using the distance modulus $m - M$. Here m is the apparent magnitude of the standard candle object (star, galaxy, *etc.*) at its actual distance and M is defined as the magnitude that the object would have if it were instead at a standard distance of 10 pc from us. The absolute magnitude M of the object is related to its true brightness. One parsec is 3.26 light years, over 200 000 times the Earth's distance from the Sun.

The magnitude system is defined such that if the apparent brightness (flux) of two objects, (energy/sec/unit area), are b_1 and b_2, their magnitude difference is

$$m_1 - m_2 = -2.5 \log(b_1/b_2). \tag{5.1}$$

The apparent brightness b in transparent Euclidean space is a function of the distance, l, by $b \propto l^{-2}$ (assuming a very local universe so that we do not need to worry about redshifts and other cosmological effects). Thus for two objects which have equal intrinsic brightness or the same absolute magnitude, we have an apparent magnitude difference,

$$m_1 - m_2 = 5 \log l_1/l_2. \tag{5.2}$$

Let us measure distances in parsecs, and say that one object is brought from position 1 to position 2, the standard distance $l_2 = 10$ pc. Then by definition $m_2 = M$ and writing $m_1 = m$ and $l_1 = l_0$ (since it is the same object), we see that the distance modulus and the distance are connected as

$$m - M = 5 \log l_0 - 5. \tag{5.3}$$

One can solve Equation (5.3) for the distance in parsecs, l_0. Astronomers often simply give $m - M$, the distance modulus, for the standard candle in question. For galaxies it is usually convenient to use the larger distance unit 1 megaparsec (1 Mpc $= 10^6$ pc). Then Equation (5.3) becomes $m - M = 5 \log_{10} l_0 + 25$, and we have a helpful result: $l_0 = 10$ Mpc corresponds to a distance modulus of 30.

Hubble was searching for novae now known to be rather poor standard candle stars. These had been used already in 1919 by the Swedish astronomer Knut Lundmark (1889–1958) to derive a distance of about 175 000 pc to the Andromeda Galaxy. In the process Hubble discovered a Cepheid in the Andromeda Galaxy in 1923. Cepheids are stars of supergiant luminosity and which show periodic variations of brightness. Cepheids were already well known from observations at more modest distances within the Milky Way. What is most important about the Cepheids is that they reveal a regular dependence of the absolute magnitude on the period; as first discovered in 1912 by the American astronomer Henrietta S. Leavitt at Harvard College. Because of this, these stars can be used as standard candles. The absolute magnitude of the star can be found if the period of its light variations is measured which is usually straightforward with a sufficient number of observations over a sufficient time. Then the distance to the star follows from the comparison of the absolute magnitude with star's measured apparent magnitude using the distance modulus.

Hubble determined the period of the Cepheid to be 31 days; then using the dependence between the period and absolute magnitude for Cepheids, he calculated a 275 kpc distance for the Andromeda Galaxy. Hubble thus confirmed that the Andromeda Galaxy is clearly outside our Milky Way Galaxy disk which has a radius about 15 kpc. Later in that year Hubble found more Cepheids in the Andromeda Galaxy, which agreed with the same distance value. Hubble also found Cepheids in the nearby Pinwheel Galaxy, M33. He calculated that it is at about the same distance as the Andromeda Galaxy. Hubble went on to measure distances of a number of other galaxies. Obtaining the distances of these spiral nebulae was crucial to showing that they were large objects full of stars and thus comparable to our own Milky Way. The announcement of Hubble's results was made in the January 1925 meeting of the American Astronomical Society in Washington, where it received much attention.

Before going on with the work of Hubble, we should mention that the Estonian astronomer Ernst Öpik (1893–1985) made a remarkable determination of the distance to the Andromeda nebula using its rotation (published in the Astrophysical Journal in 1922 but reported already in 1918 in Moscow). From the data available, Öpik first calculated the distance of M31 to be 785 kpc (surprisingly close to the modern value of 770 kpc from several methods!), which value he revised to 450 kpc in the 1922 paper. Öpik's method differs fundamentally from the standard candle method. It is a prototype of dynamical cosmic distance indicators. Let us consider it in some detail, following Teerikorpi (2011).

If the object's mass is distributed spherically up to R where we measure the rotation velocity V (at the angle θ from the center) we can write for the mass:

$$M = aRV^2 = a\theta dV^2.$$

Here d is the distance of the nebula from us. For a circular orbit around the mass M, the factor a is simply G^{-1}. The observations of θ and V then give the quantity $M/d = a\theta V^2$.

This alone does not allow us to derive the distance d. because the mass M of the object is not known independently. However, if we can express M/d in another form containing observable quantities and known constants plus the unknown d, we may bypass the M/d degeneracy and derive the distance d. For example, if the mass M can be given as proportional to d^n where $n = 0, 2$ or 3, we can solve for d. In particular, in Öpik's method, the assumed mass-to-luminosity ratio M/L gives the needed auxiliary relation

$$M = (M/L)L = (M/L)4\pi d^2 f,$$

where f is the observed flux of light from the object emitted within the radius R. By assuming a value for the mass-to-luminosity ratio M/L one can now derive the distance to the rotating object. For galaxies, this method is not widely used because one does not generally know the mass-to-luminosity ratio. It evolves with cosmic time and, as we now know, a large part of the mass is in the form of dark matter. Öpik could derive a reasonable distance to the Andromeda Galaxy, because his data referred to the innermost part of the Galaxy where dark matter is not important and he could use the M/L-ratio from the solar neighborhood in our Milky Way.

In fact, all three kinds of distance indicators mentioned above (the Cepheids, the novae, the method of rotating galaxies) have their counterparts in modern extragalactic astronomy and cosmology. The period-luminosity relation of classical Cepheid stars continues to be a fundamental distance indicator in the local galaxy universe. The Tully–Fisher relation method (luminosity vs. maximum rotation speed; Tully and Fisher 1977) for spiral galaxies bypasses the need for an M/L ratio using nearby calibrator galaxies. It can be applied up to distances of hundreds of Mpc. Novae are still used as distance indicators for nearby galaxies.

When he studied the distance of the Andromeda Galaxy, Lundmark realized that in addition to novae there are much more luminous supernovae, which Fritz Zwicky and Walter Baade in 1934 linked with the death of massive stars. In 1925 Lundmark termed these giants and dwarfs among the novae as "upper and lower class", with an upper class nova reaching a luminosity comparable to that of the whole host nebula. Lundmark thus discovered supernovae which are 10 000 times more luminous than ordinary novae.

In modern cosmology, supernovae of type Ia have been especially important distance indicators. As we will discuss later, their observations have shown that the expansion of the universe is accelerating, interpreted as caused by the antigravity of dark energy.

However, these early distance measurements by Öpik, Lundmark, and Hubble were just the first parameter of the two needed to reveal the expansion of the universe. Spectral measurements were also needed, as pioneered by Slipher (see above).

After a few rather inconclusive attempts by others, in 1928 Edwin Hubble began to study whether the radial speeds depend on the distances. He used the large 100 inch Mount Wilson telescope to photograph the spectra of galaxies. Hubble had an able assistant in Milton Humason (1891–1972). A school dropout, Humason became a mule driver for the pack-trains between the Sierra Madre and Mount Wilson during construction of the Observatory. In 1911 he married the daughter of the Observatory's engineer and became a foreman on a relative's ranch. However, in 1917 he joined the staff of the observatory as a janitor and was soon promoted to night assistant. In 1919 George Hale, the observatory's director, recognized Humason's unusual ability as an observer and appointed him to the scientific staff. In total, Humason measured the redshifts of 620 galaxies.

Hubble published his relation between the redshift and the distance of a galaxy in 1929 (Figure 5.2). A rough correlation between these two quantities had been previously noticed by Knut Lundmark who used the angular size of a galaxy as a (poor) distance indicator. However, Hubble's work revealed for the first time that these two quantities are directly proportional to each other (as indicated by the lines in Figure 5.2). This has since been completely verified by using more distant galaxies. Mathematically, the speed of recession (based on the measured redshift) equals the Hubble constant × distance or

$$V = H_0 \times R. \tag{5.4}$$

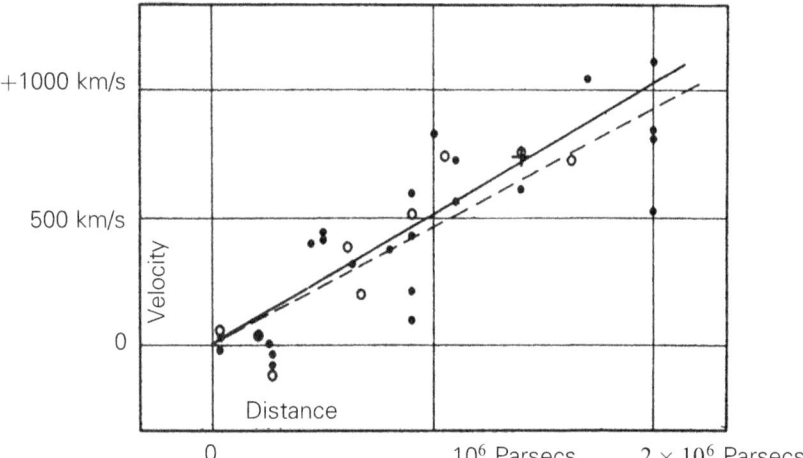

Figure 5.2. The original published diagram where Hubble plotted his distances in parsecs against Slipher's radial velocities in km/s (Hubble 1929). The distances are systematically underestimated by nearly an order of magnitude. The straight line represents the Hubble law.

This linear relation called the Hubble Law tells us that in the world of galaxies the distances are increasing, or, as stated more commonly and precisely, that the universe is expanding. The slope of the apparently straight line is designated as the **Hubble constant**, H_0, which Hubble found to be equal to about 500 (km/s)/Mpc.

The maximal fractional escape speed in Figure 5.2 is about 1 000 km/s which means that $z = V/c$ is only 0.003. The data are directly obtained in terms of the spectral redshift, and the velocity determination from these assumes that the spectral shift uses the Doppler Effect formula, but more exactly it is due to the Lemaître effect in the expanding space of the universe. These are good approximations for such a modest redshift. As we shall see, a more complex formula is needed for large redshifts which are of the order unity or even larger. They are important in the discovery of dark energy.

Distances to the nearest galaxies can be estimated using observations of Cepheid stars. However, at larger distances, Cepheids are too faint to be observed with ordinary telescopes. The Hubble Space Telescope has extended the range of the Cepheid method to greater distances up to about 20 Mpc or 30 times the distance of the neighboring Andromeda Galaxy and is obtaining more precise distances for nearby galaxies.

In Figure 5.3 we compare Hubble's velocity—distance diagram with modern measurements of some of the same galaxies as well as other more distant ones. Note that the new relation has considerably less scatter, because of more accurate measurements (smaller random errors, in addition to the correction of a large systematic error). It starts at a distance of about 1.5 Mpc, just two times the distance

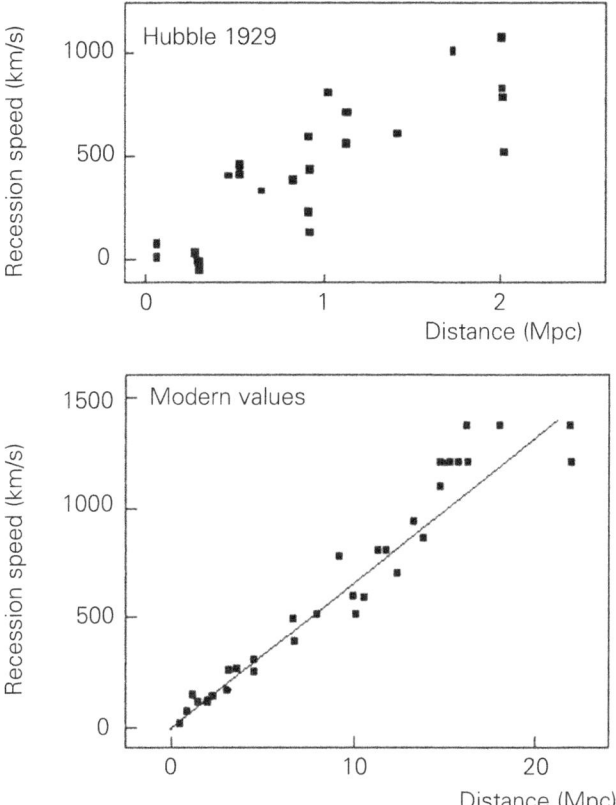

Figure 5.3. Hubble law for local galaxies. The upper panel is based on Hubble's (1929) work. The lower panel shows an updated version of the Hubble diagram for about the same speed interval. However, the new distances are almost ten times longer than Hubble's; his measurements had a large systematic error. The new Hubble relation is also less scattered than the original one because of smaller random errors. The straight line represents the Hubble law for the Hubble constant $H_0 = 65$. Actually one determines the value of H_0 from a much longer distance interval. (The data in the lower panel are from Paturel and Teerikorpi 2005).

of the Andromeda Galaxy. One may say that cosmology begins in our immediate extragalactic neighborhood. Actually, it has been a riddle how Hubble could discover such a good local relation (with small scatter and the slope about the same as globally) within our very lumpy neighborhood. Indeed, the nearest part of the expansion flow may not represent the genuine regular primordial expansion, but instead, as we shall see in Chapter 15, is observably affected by the repulsion of dark energy.

5.5   Problems with the Hubble constant and the age of the universe

The Hubble constant is a very important cosmological quantity, because it is related to the distance scale and the age of the expanding universe. The distance R to a galaxy with a measured redshift z is $V(z)/H_0$ (from the Hubble law) where $V(z)$ is the recession speed corresponding to the redshift z ($V \approx cz$ for small redshifts). The smaller the Hubble constant, the larger the distances calculated from the speed (redshift). The link to the age is also easy to see.

Assume that the expansion speed V between two galaxies has been constant throughout the history of the universe. Then the material of these galaxies was very close together a finite time ago. This age of the universe is obtained by dividing the present distance R between the galaxies by their speed of separation V. This ratio R/V is just equal to $1/H_0$. Hence the inferred age of the universe is inversely proportional to the assumed value of the Hubble constant. Recall from the earlier discussion that from Hubble's data the constant is about 500 km/s/Mpc. The galaxies escape from each other with the average speed of about 500 km/s if they are one million megaparsecs (or 3.26 million light years) from each other. The speed of one km/s corresponds to one parsec per million years; thus with the speed 500 km/s the same distance takes 2 000 years, and the million parsec distance takes 2 billion years. Thus the age of the universe, since the time when the galaxies were together, is about 2 billion years for this value of H_0. As we shall see later, the exact age depends on whether the expansion has accelerated or slowed down since the Big Bang. The modern accepted value for the Hubble constant is not nearly so large as 500 km/s/Mpc. It is about 72 km/s/Mpc with the difference due to more accurate distance determinations today. As we have already seen, the universe is much older than 2 billion years (almost 14 billion years).

However, by the late 1940s, there had developed a huge contradiction between the universe age from the Hubble constant and radioactivity estimates of the age of the Earth and the Solar System of about 4.5 billion years. The method for obtaining these radioactivity estimates is similar to those in our earlier discussion of ages of stars. A German-American astronomer, Walter Baade, "saved the Earth" by discovering that there were two kinds of Cepheid variable standard candles which differed in brightness. Baade found that this is because the two kinds of stars belonged to different star populations in our Galaxy. Astronomers were mistaking one for the other and thus greatly underestimated galaxy distances. Baade's new estimate for Hubble's constant was around 200 km/s/Mpc which did make the universe older than the Earth! This misunderstanding of a type of standard candle is a cautionary tale for the most recent discoveries about Dark Energy.

After the 1950s, difficulties with measuring the Hubble constant continued, mostly with the distance estimates. The measurement of redshift z is straightforward. However, the distance modulus has problems with both m and M. The apparent magnitude, m, is measured for the object **outside our Galaxy**, that is, the measured value of m has to be corrected for what is known about Galactic extinction in that particular direction. Experts differ in their application of this correction.

The problem with M is that we need to know the true energy output per second or luminosity of the class of objects that are used. As we have said, it is best to have a **standard candle**, a class of objects whose absolute magnitude has very narrow limits. It can be a star of a special kind or a galaxy of particular class. But even with good standard candles there still is some scatter in intrinsic absolute magnitude around an average value. The scatter introduces not only a random error but also a systematic effect or bias called the **Malmquist bias** (Malmquist 1920). This kind of problem arises commonly in astronomy. For a more detailed discussion, see Teerikorpi (1997) and chapters 3 and 4 in Baryshev and Teerikorpi (2012).

Gunnar Malmquist (1893–1982) was professor of astronomy at Uppsala, Sweden. He attempted to determine the average luminosity in space of stars belonging to a particular spectral class. This is not the same as the average luminosity of similar stars on the sky. At large distances, from where only the most luminous stars are visible, the less luminous stars will be missed. Hence the average luminosity of stars on the sky is larger than the true average luminosity in space. In 1920 Malmquist derived a formula which gives this difference in luminosities in a general case. Even earlier Eddington had derived the result for the special case of uniform spatial distribution of stars: then the average absolute magnitudes "in space" and "on the sky" differ by an amount $1.382\sigma^2$, where σ is the dispersion of the absolute magnitude distribution of the stellar class, assumed to be normal (Gaussian). Malmquist considered a general case of how the stars are distributed in space. Objects in the sky which are apparently similar (the same distance modulus!), may actually have a complicated distribution of true distances.

Malmquist's theory may be applied to galaxies as well. When an astronomer measures distances to galaxies, he can select his galaxies only from the sky, and not from space, because his ability to travel between the galaxies is badly limited. Also, he cannot see very faint galaxies. Thus the standard candles which the astronomer can reach at large distances are more luminous, on the average, than the nearby calibrator galaxies. This mismatch is the Malmquist bias which tends to creep silently into any astronomical data. It is the mean value of M that is used in the distance modulus, so that this bias leads directly to an error in distance. If the astronomer forgets this problem, he will derive too small distances to remote galaxies and he will determine too high a value of the Hubble constant and, consequently, too young an age for the expanding universe.

Even as late as 1980, the astronomical community was still greatly divided about the value of the Hubble constant. The short distance scale group, represented by the French-American astronomer (University of Texas) Gérard de Vaucouleurs, obtained

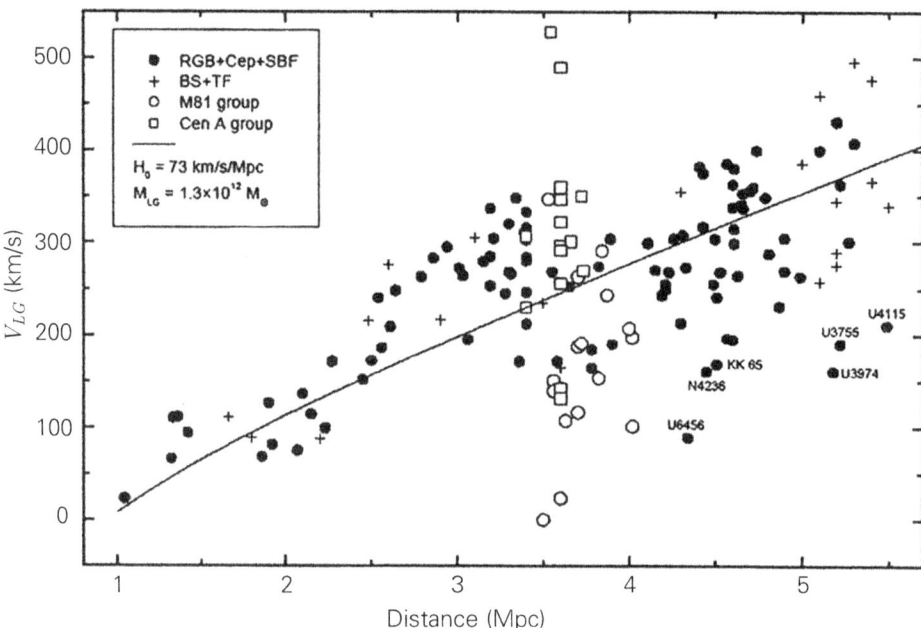

Figure 5.4. The Hubble diagram for the 156 nearby galaxies (Karachentsev et al. 2003). Velocity (km/s) and distance (Mpc) are given relative to the center of mass of the Local Group (LG) of which the Milky Way and the Andromeda Galaxy are the major members. The regression line corresponds to the Hubble constant of 73 km/s/Mpc assuming the mass of the Local Group of galaxies to be 1.3×10^{12} solar masses.

$H_0 = 100 \pm 10$ km/s/Mpc (de Vaucouleurs and Peters 1981) while the long distance scale group led by Allan Sandage obtained consistently $H_0 = 50 \pm 5$ km/s/Mpc (Tammann et al. 1980). A partial resolution of this issue came when it was shown (Teerikorpi 1975) that when allowance is made for the Malmquist bias, the higher value drops to about $H_0 = 75$ km/s/Mpc. In collaboration with Lucette Bottinelli and Lucienne Gouguenheim from Paris–Meudon Observatory as well as Georges Paturel from Lyon Observatory, (Bottinelli et al. 1986), and using the Tully–Fisher method of rotating galaxies as distance indicator, Teerikorpi found the constant to be

$$H_0 = 72 \pm 3 \, \text{km/s/Mpc}.$$

This value is also close to what is currently generally regarded as the standard value, as measured by the Hubble Space Telescope (HST) Key Project (Freedman and Madore et al. 2001; Freedman and Madore 2010). However, there are also indications that the Hubble constant might be lower (Theureau et al. 1997); e.g., the group of Sandage, also working with the HST and using supernovae of type Ia as distance indicators, derived $H_0 = 62 \pm 5$ km/s/Mpc (Sandage et al. 2006). There may also be some remaining (selection) problems in the use of the Cepheids as distance indicators in the local galaxy universe where the long-range distance indicators (such as the Tully–Fisher relation and the Ia supernovae) are calibrated (Paturel and Teerikorpi 2005). Hence, it is wise to keep in mind the possibility that the Hubble constant is not yet determined at better than 10 percent accuracy level.

As we shall see, with more precise calculations, the age of the universe for such Hubble constant values (60–70 km/s/Mpc) is close to or larger than that of the oldest objects in our Galaxy, which is reassuring.

A recent version of the Hubble diagram for the local flow of galaxies within 8 Mpc is given in Figure 5.4. This version uses high accuracy measurements of galaxy velocities and (especially) distances performed by observers led by Igor D. Karachentsev of the Special Astrophysical Observatory (SAO) in Russia using the Hubble Space Telescope, the 6 m Telescope of SAO and other large instruments. From this diagram, the Hubble constant for the local flow of expansion is $H_0 = 73 \pm 4$ km/s/Mpc.

In this diagram, we can see the effect of additional complicating factors, such as the internal motions within groups of galaxies. There is a large vertical scatter at 3.5 Mpc, the distances of the both the Cen A and M81 groups. Later in this book, we will clarify some of these factors for these two groups and our own Local Group of galaxies in terms of this internal dispersion and the effect of dark energy on outer group members.

Chapter 6
Cosmology and the "first appearance" of dark energy

The cosmological redshift and its linear dependence on distance (at not too large distances, where linearity breaks down) were the first, unanticipated discoveries in the galaxy universe. The expansion of the universe is the generally accepted explanation of both the redshift and the linear Hubble law. Redshift and its dependence on distance are also important tools for investigating the dynamics of the universe and its material constituents.

6.1 A first formulation of dark energy: Einstein's finite static universe

Now that we have an expanding universe with its Hubble constant characterizing the rate of expansion, we briefly digress to explain Einstein's finite static universe. One reason we do this is because Einstein introduced in this model what is now the mathematical formulation of dark energy. Another is that the approach Einstein used for this model is the pattern for other widely accepted models.

In 1917, Einstein, who was then at Berlin in the Prussian Academy of Sciences and the University of Berlin, published a paper on cosmology based on General Relativity. He extended the concept of curvature of space-time from the previously described application of the space-time around our Sun to the whole universe. Based on General Relativity Einstein (1917a, b) proposed a curved finite universe in space-time which was unchanging in time with no origin. Einstein's model is of a static, finite size but still boundless universe. How did Einstein arrive at this?

In regard to the large-scale structure of the universe, Einstein stated:

"...if we are concerned with the structure only on large scale, we may imagine matter as being uniformly distributed over enormous spaces...".

Recall that according to General Relativity, the geometry of space-time is determined by matter's distribution and motion. Thus a space-time in which matter is distributed uniformly must be uniformly curved everywhere. At that time, nothing in real astronomical data about the spatial distribution of stars (other galaxies beside our own had not yet been established) indicated the uniformity. This was actually a simplifying hypothesis rather than something based on any observations. Remember that already Newton in his correspondence with Bentley considered a world model where the stars are uniformly distributed in an infinite space. The hypothesis turned out to be especially promising when it was realized that the universe is populated by galaxies. The spatial distribution of galaxies and systems of them indeed approaches uniformity on the largest scales although we do not yet know accurately the scale of crossover to homogeneity. Einstein used the example of a geophysicist who describes the average shape of the Earth by a sphere, neglecting details of hills and valleys.

Einstein assumed also that the cosmic space is isotropic i.e., all directions in space are equivalent. Isotropy reveals itself through the observation that the cosmological expansion does not depend on the direction in space i.e., the Hubble constant is the same in all directions. Unfortunately, the isotropy of the Hubble expansion is measured only at the level of 10–30% accuracy. As we will see later this symmetry is especially clear in the cosmic background radiation which is isotropic to an accuracy of about 0.001%.

General Relativity treats space and time as space-time. Therefore Einstein needed an assumption about the temporal properties of the universe as a whole. Going from uniform space to time, he relied upon an old tradition of cosmology, according to which not only all spatial points but also all points in time in the history of the universe are equivalent. Thus, overall, nothing changes in the universe; it is forever the same **with no origin in time**. In this eternal and unchanging universe, time is uniform like space. The isotropic and static universe is the simplest possible construction in space-time.

Such a theoretical universe is mathematically described by an exact solution of the Einstein's General Relativity equations. This solution represents a non-Euclidian geometry of space-time and is described using the metric tensor. The analog of the spatial geometry of this isotropic and static model is very similar to the geometry of a two-dimensional surface of a sphere, which also is non-Euclidean. Of course, the space of Einstein's model is actually a three-dimensional sphere, or hypersphere. The geometry does not depend on time in Einstein's model. However, the General Relativity equations in their original (1915, 1916) form do not allow a static, time-independent cosmological solution. Einstein realized that to achieve this he needed an additional term in the equations and added the **cosmological constant**, Λ. After he did that, the equations permitted the desired static model of the maximal possible space-time symmetry. It is instructive to write down what he obtained.

As we mentioned, Einstein showed that in the case of a static hypersphere with a time-independent radius of curvature $a_K = R$ his equations when applied to ordinary matter of finite density $\rho > 0$ without pressure do not have a solution. The resulting two independent equations (which may also be obtained from Equations (8.2) and (8.3) from Chapter 8, describing expanding or contracting isotropic and homogeneous universes, for the special case of $R(t) = R = $ constant, $k = 1$, $P = 0$ and $\rho_V = 0$) were

$$\frac{3}{R^2} = \frac{8\pi G \rho}{c^2} \quad \text{and} \quad \tag{6.1}$$

$$\frac{1}{R^2} = 0. \tag{6.2}$$

In these two equations $R(t)$ is replaced R; this is how Einstein wrote them, he considered only the static situation with R as constant.

As discussed in more detail in Box 4 in Chapter 7, Einstein added to the left "geometry" side of his field equations the constant term $g_{ij}\Lambda$, leading to a new pair of equations:

$$-\frac{\Lambda + 3}{R^2} = \frac{8\pi G\rho}{c^2} \quad \text{and} \tag{6.3}$$

$$-\frac{\Lambda + 1}{R^2} = 0. \tag{6.4}$$

Now the solution exists linking the cosmological constant Λ, the constant density ρ and the radius of curvature R:

$$\Lambda = \frac{4\pi G\rho}{c^2} = \frac{1}{R^2}. \tag{6.5}$$

This remarkable equation tells us that by measuring the matter density ρ in a local, representative region, one might fulfill the old dream: to infer the size of the universe, if it indeed is a static hypersphere. The volume of a hypersphere having the radius of curvature $R(t)$ is $V = 2\pi^2 R(t)^3$.

However, observationally, the universe of galaxies turned out **not to be static** but instead it is in a state of universal expansion, as discovered by Slipher and Hubble. Thus Einstein's hypothesis of a static universe was in error (or his "biggest blunder", as Einstein himself may have once said). However, Einstein's (1917a, b) papers have great value. They show how the problem of the space-time geometry of the universe, should be formulated, stated, and solved with General Relativity. They also paved the way for deriving the values of cosmological parameters from observations, as heralded by the above formula connecting the density and the radius. Moreover, Einstein's papers enriched physics with a new grand concept—the concept of universal dark energy (antigravity) mathematically represented by the cosmological constant Einstein introduced.

6.2   Cosmological redshift and Friedmann's evolving universes

The evolving models of the universe which are the standard tools today were derived by the Russian Alexander Friedmann (1888–1925), a professor of mathematics at St. Petersburg University. In the leading journal *Zeitschrift für Physik* in 1922, he published an article, *On the curvature of space*. His second article, *On the possibility of a universe with a constant negative curvature,* appeared two years later. In the same issue where Friedmann published his paper, Einstein inserted a five sentence critique. He claimed that Friedmann had actually proved that the static model was the only possible one. In spring 1923 Einstein added four more sentences in the same journal admitting that he had been wrong and he now regarded

"Mr Friedmann's results correct and clear".

However, these papers, which formed a turning point in the study of cosmology, remained almost unnoticed. In 1927 Belgian astronomer Georges Lemaître (1894– 1966) rediscovered the same models, now known as Friedmann universes.

Friedmann (1922) followed Einstein's procedure in the statement of the cosmological problem. In his 1922 paper, he adopts Einstein's assumption that the space is isotropic. However, he made no assumptions about a static universe. He found the appropriate exact solution of the General Relativity equations which is more general and rich in content than Einstein's solution. The Friedmann solution includes the Einstein static solution as a special case in which the cosmological constant is non-zero and related to the matter density (this relation expresses the gravity-antigravity balance). However, generally, the solution puts no restrictions on the cosmological constant. It might be zero or nonzero, positive or negative, related or unrelated to the matter density. The solution depends explicitly on time, and it means that the universe is not static: it expands or contracts as a whole. In such an evolving universe, there is no uniformity with time; the states of the universe in the past differ from its states in the present and into the future. In particular, if the universe expands, matter that fills it dilutes with time. Friedmann preferred expansion over contraction and found observational evidence supporting his choice in Slipher's data on nebulae (galaxies) that are moving away from us. The data were reviewed in the *Mirovedenie* magazine in 1923. The next year they were discussed at Friedmann's weekly seminar in St. Petersburg. He died in 1925 at the age of 37 before the Hubble discovery of the redshift distance relation (Tropp, Frenkel and Chernin 1988, 1993).

Friedmann's model has features uncomfortable to contemplate. In the Friedmann model of the cosmological expansion, there is an initial moment in time when expansion starts. That was, in his words, the moment of creation of the world from nothing. Moreover, mathematically, the entire universe was a "point", at that moment! Note, that this is **not** a point in a flat Euclidean space. Playing cosmic history backward, all the galaxies come together. A physical quantity such as density becomes infinite and the space-time scale length zero; this is the cosmological singularity discovered by Friedmann. Of course, actual physical quantities cannot be infinite in reality. The singularity is a mathematical extrapolation and one expects that the physical model (General Relativity which does not include quantum theory) breaks down at some very high density.

As we now know, the early physical conditions in the universe must have been unusual. Since the galaxies are now escaping from one another, they must have been closer to each other in the past and long ago right next to each other. There must have been an initial explosion (or an extreme process), a **Big Bang** which has put the matter in the universe in the state of expansion. Needless to say, the galaxies today were not as they are today at the time of the big bang i.e., their gaseous material had not yet formed into stars, or still earlier, the gas not yet into atoms, etc.

As we have seen, the universe is apparently of finite age. How far back in time was that initial stage? Using available uncertain astronomical data—with some additional degree of uncertainty, assigned to the universe as a whole—Friedmann managed to estimate the present age of the universe as the time elapsed since the start of the expansion. With all the necessary reservations, he said in his semi-popular book, *The world as space and time*, (1923) that this time interval is measured by tens of billions of our ordinary Earth years. This was a remarkable estimate of an amazingly old universe which is comparable to today's value.

On the basis of the distance-redshift relation for galaxies, the expansion or contraction of our constant curvature 3-space can be explored using General Relativity under the assumption of universal curvature which evolves with time. We will obtain an expression for the redshift-distance relation in terms of the deceleration or acceleration of the expansion. The expansion may be represented by a **scale factor** $R(t)$ as a multiplier of the 3-space metric equation of Equation (4.13). In other words, the 3-space metric expands (scales) by $R(t)$. When we add the term for the time coordinate, the final metric is

$$ds^2 = c^2 dt^2 - R(t)^2 dl^2. \tag{6.6}$$

Examining the dates of the above mentioned publications, it is interesting to note the coincidence that theoretical ideas about an expanding universe were slightly in advance of Hubble's observations.

The coordinate system is designated as **co-moving** because observers at fixed coordinates (r, θ, ϕ) take part in the universal expansion. One may view the observers as being at rest while space itself is expanding. The observers with exactly constant positions are called **fundamental observers** (isolated galaxies nearly fulfill this requirement). Celestial bodies, galaxies, etc. are where dl^2 is from Equation (4.13) or Equation (4.18). The spatial part of the metric represented by dl^2 ranges over a spherical space, a flat space, or a hyperbolic space. This form is the **Friedmann metric** or **Robertson–Walker metric** (Friedmann 1922 and 1924, Lanczos 1922 and 1923, Robertson 1935 and 1936, Walker 1936). All of the time dependence is in the scale factor, $R(t)$. Astronomers have a special term for the motion of a galaxy relative to the fundamental observers in its neighborhood: the **peculiar velocity**. Galaxies are generally not at rest in the expanding space, but they have some peculiar velocity (a typical value could be 100 km/s), generally in motion relative to fundamental observers who represent the average motion of galaxies over large volumes of space.

6.3   The Hubble constant in the Friedmann standard model

The above metric gives an analytic solution to Einstein's equations which is sometimes called the **standard model of cosmology**. Rather than discussing it now, we will first derive the Hubble constant. In accordance with Equation (4.1) we may define a curvature parameter

$$K = \frac{k}{R(t)^2}. \tag{6.7}$$

Now consider a fundamental observer at the center of the coordinate system, $\Psi = 0$, who sends out a pulse of light. This pulse is observed by a second fundamental observer at a later time at radial distance $\Psi = \Psi_1$. The spherical coordinates θ and ϕ are constant, and since the photons follow a null-worldline

$$0 = c^2 dt^2 - R(t)^2 d\Psi^2. \tag{6.8}$$

This worldline is integrated from the initial time $t = t_1$ to the final time t_0

$$c \int_{t_1}^{t_0} \frac{dt}{R(t)} = \Psi_1. \tag{6.9}$$

We designate the current time when the light signal arrives as t_0. The departure time is designated t_1. Assume the first fundamental observer sends another pulse a little later, at $t = t_1 + \Delta t_1$, and assume it is received at time $t_0 + \Delta t_0$. Since Ψ_1 does not increase noticeably since the last pulse, we have for the second pulse

$$c \int_{t_1+\Delta t_1}^{t_0+\Delta t_0} \frac{dt}{R(t)} = \Psi_1. \tag{6.10}$$

Both integrals above can be divided into two parts and put equal to each other:

$$\int_{t_1}^{t_1+\Delta t_1} + \int_{t_1+\Delta t_1}^{t_0} = \int_{t_1+\Delta t_1}^{t_0} + \int_{t_0}^{t_0+\Delta t_0} \tag{6.11}$$

where we have not written the common integrand $dt/R(t)$. The integral from $t_1 + \Delta t_1$ to t_0 appears on both side and cancels out. The remaining integrals cover such a small time interval that $R(t)$ in the integrand can be considered a constant. Thus we can write

$$\frac{\Delta t_0}{R(t_0)} = \frac{\Delta t_1}{R(t_1)}. \tag{6.12}$$

The time intervals have a ratio

$$\frac{\Delta t_1}{\Delta t_0} = \frac{R(t_1)}{R(t_0)}.\tag{6.13}$$

Following a discussion similar to that leading to Equation (3.14) we get a change in the frequency between emission (ν_1) and reception (ν_0) of

$$\frac{\nu_1}{\nu_0} = \frac{\Delta t_0}{\Delta t_1} = \frac{R(t_0)}{R(t_1)}.\tag{6.14}$$

Substituting the corresponding wavelengths λ_0 and λ_1 and their difference $d\lambda$

$$\frac{\nu_1}{\nu_0} = \frac{\lambda_0}{\lambda_1} = \frac{\lambda_1 + d\lambda}{\lambda_1} = 1 + \frac{d\lambda}{\lambda_1} = \frac{R(t_0)}{R(t_1)}.\tag{6.15}$$

In terms of the redshift $z = d\lambda/\lambda_1$ the equation becomes

$$1 + z = \frac{R(t_0)}{R(t_1)}.\tag{6.16}$$

The received wavelength is longer than the emitted one. This is because $R(t_1) < R(t_0)$ in an expanding universe. Therefore the wavelength of the photon increases due to the expansion during the transit. It should be emphasized that this is **not** a Doppler redshift due to an individual motion of, say, a galaxy in space. Instead, it is a lengthening of the wavelength **during transit** from the galaxy to the observer as $R(t)$ increases, a **cosmological redshift**. It is not the classical Doppler Effect observable here on Earth, in our Solar System or in our own Galaxy. As such it does not represent a velocity relative to its local space. Equation (6.16) was derived by Lemaître in his classical 1927 paper and this cosmological redshift mechanism is sometimes called the Lemaître effect in order to distinguish it from the ordinary Doppler Effect. Figure 6.1 illustrates the origin of the cosmological redshift.

Space is stretched by expansion $z = (l_0 - l_1)/l_1$. Time between photons at observation is increased (and energy/sec reduced) by another factor $1/(1 + z)$.

For a particle in radial motion $(\dot{\theta} = \dot{\phi} = 0)$ the Lagrangian is

$$L = c^2\dot{t}^2 - R(t)^2\dot{\Psi}^2 = c^2.\tag{6.17}$$

Because the 3-space is isotropic and homogeneous, radial motion is representative of the motion in any other direction. The case of radial motion makes the calculation simpler.

Since Ψ is an ignorable coordinate (see Section 2.4 explanation), we have a first integral

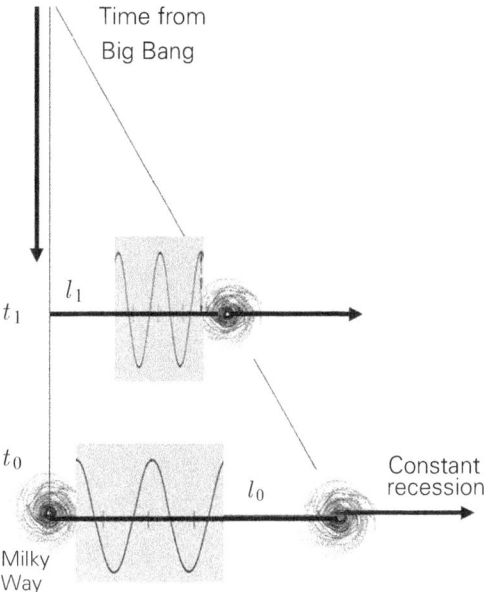

Figure 6.1. Cosmological red shift in an expanding universe. Emitted photon λ is stretched by and during the expansion. Above, $z = (\lambda_{observed} - \lambda_{emitted})/\lambda_{emitted}$. Now the observed energy of each photon is reduced by a factor $1/(1 + z)$. Successive distances, l's, are proportional to $R(t)$. Following the time convention, the subscript "o" represents "observed today". The "l_1" is at the emission time.

$$-\frac{1}{2} \frac{\partial L}{\partial \dot{\Psi}} = R(t)^2 \, \dot{\Psi} = \text{constant} = R_\alpha \, c \tag{6.18}$$

where the arbitrary constant has been denoted $R_\alpha c$. Combining Equations (6.17) and (6.18), we have

$$\dot{t}^2 = 1 + \frac{R_\alpha^2}{R(t)^2}.$$

The velocity of a particle with respect to the fundamental observers is

$$v = R(t) \frac{d\Psi}{dt} = R(t) \frac{\dot{\Psi}}{\dot{t}} = \frac{R_\alpha \, c}{\sqrt{R(t)^2 + R_\alpha^2}}. \tag{6.19}$$

Now, velocity decreases while $R(t)$ increases. Thus we see that free fall motions become steadily slower as the epoch increases.

Much observational material on the cosmological expansion comes from relatively nearby galaxies. Therefore the scale factor $R(t)$ for these galaxies is not very different from what it is today, $R(t_0)$. Thus we may use a Taylor series expansion, taking terms only up to the second order of the time difference $t - t_0$:

$$R(t) = R(t_0) + \dot{R}(t_0)(t - t_0) + \frac{1}{2}\ddot{R}(t_0)(t - t_0)^2$$

$$= R(t_0)\left[1 + \frac{\dot{R}(t_0)}{R(t_0)}(t - t_0) + \frac{1}{2}\frac{\ddot{R}(t_0)}{R(t_0)}(t - t_0)^2\right]. \tag{6.20}$$

We define the **Hubble parameter** as $H \equiv \dot{R}(t)/R(t)$. Its current observed value is

$$H_0 = \frac{\dot{R}(t_0)}{R(t_0)}. \tag{6.21}$$

This is called the **Hubble constant**. We also define $q \equiv -R(t)\ddot{R}(t)/\dot{R}(t)^2$, which has the present day value

$$q_0 = -\frac{R(t_0)\ddot{R}(t_0)}{\dot{R}(t_0)^2} = -\frac{1}{H_0^2}\frac{\ddot{R}(t_0)}{R(t_0)}. \tag{6.22}$$

This is called the **deceleration parameter**. With these definitions Equation (6.20) becomes

$$R(t) = R(t_0)\left[1 + H_0(t - t_0) - \frac{1}{2}q_0 H_0^2(t - t_0)^2\right]. \tag{6.23}$$

Using Equation (6.23) in Equation (6.9):

$$\Psi_1 \cong \frac{1}{R(t_0)}\int_{t_1}^{t_0}\frac{c\,dt}{[1 + H_0(t - t_0)]} \cong \frac{1}{R(t_0)}\int_{t_1}^{t_0}[1 - H_0(t - t_0)]\,c\,dt$$

$$\Psi_1 = \frac{1}{R(t_0)}\left[c(t_0 - t_1) + \frac{1}{2}cH_0(t_0 - t_1)^2\right]. \tag{6.24}$$

Thus, the current distance to the emitter is

$$l_0 = R(t_0)\Psi_1 = c(t_0 - t_1) + \frac{1}{2}H_0 c(t_0 - t_1)^2. \tag{6.25}$$

Recall the definition of the redshift from Equation (6.16) is

$$z = \frac{R(t_0) - R(t_1)}{R(t_1)} \tag{6.26}$$

which is obtained using Equation (6.23):

$$R(t_0) - R(t_1) \cong R(t_0)\left[H_0(t_0 - t_1) + \frac{1}{2}q_0 H_0^2(t_0 - t_1)^2\right]$$

$$R(t_1)^{-1} \cong R(t_0)^{-1}[1 + H_0(t_0 - t_1)] \tag{6.27}$$

This leads to

$$z = H_0(t_0 - t_1) + \left(1 + \frac{1}{2}q_0\right)H_0^2(t_0 - t_1)^2. \tag{6.28}$$

We get $t_0 - t_1$ by squaring l_0 in the series Equation (6.25). Since $l_0^2 \cong c^2(t_0 - t_1)^2$, it gives approximately

$$c(t_0 - t_1) \cong l_0 - \frac{1}{2}\frac{H_0}{c}l_0^2. \tag{6.29}$$

When inserted in Equation (6.28) we get finally

$$cz \cong H_0 l_0 + \frac{1}{2}(1 + q_0)\frac{H_0^2}{c}l_0^2. \tag{6.30}$$

This **redshift-distance relation** is a theoretical prediction of the Friedmann models. We see that it is linear at smaller cosmological distances corresponding to the observational Hubble law supporting the Friedmann models.

The quantity q_0 which specifies departure from a linear Hubble law may in principle be determined from observations. However, it has proven rather difficult since q_0 enters only in the second order. Much observational work in cosmology during the 20th century was devoted to determining which value applied to the real universe. For the simplest Friedmann models (where the cosmological constant is zero), the deceleration parameter q_0 and the Hubble constant H_0 together suffice to tell which kind of a (Friedmann) world we live in. That is why Sandage in the 1960s characterized observational cosmology as "the search for two numbers". Nowadays there are also other parameters in the list of "most wanted".

The measurement of redshifts from the spectra of galaxies is rather straightforward. However, as mentioned above, measuring distances to distant galaxies independently of the redshift is much more difficult, because the distance must be inferred indirectly, using various distance indicators affected by random and systematic errors which may be quite large (remember the "50–70–100 struggle" in the determination of the Hubble constant).

Continuing the work started by Hubble and Humason who used giant elliptical galaxies in galaxy clusters as standard candles, Sandage and others attempted to infer the value of q_0 from the apparent magnitude—redshift relation (where the magnitude gives the luminosity distance as will be explained in Section 9.2). The relation is rather narrow for the first-ranked cluster galaxies, suggesting that they could be good standard candles. In this way and using the 5-meter Mt. Palomar telescope, one could probe the expansion up to redshifts approaching 1. However, it was gradually realized that severe problems complicate the use of galaxies as standard candles. Selection effects and the cosmic evolution of the luminosity of galaxies easily hide from view the true value of the deceleration parameter q_0 (Sandage 1995).

Only on the verge of the 21st century, when the Hubble law could be studied at redshifts approaching and even exceeding unity, using the very luminous Ia class supernovae as good standard candles, could one detect a deviation from the linear redshift—distance relation, leading to a measurement of q_0.

As we shall see in more detail later, the current best value is $q_0 = -0.60$ which actually implies **acceleration** not deceleration of the expansion. This measurement has led to the adoption of Einstein's cosmological constant as a part of the standard world model. Note that historically q_0 is defined in Equation (6.14) so that its positive value means deceleration as earlier expected for simple Friedmann models filled by gravitating matter and with $\Lambda = 0$, i.e., with no antigravitating dark energy.

Chapter 7
Einstein's equations, critical density and dark energy

7.1 Introduction

In this chapter we will go deeper into the Einstein equations and show how they lead, together with the Cosmological Principle of homogeneous matter distribution, to a wide family of cosmological models, both without and with the Lambda term (dark energy). Friedmann assumed, as Einstein had, that matter is distributed uniformly (on average), but he did not require that the density of matter should remain constant. Thus, even though the curvature of space-time is thus everywhere the same at a given proper time, it changes with time as the universe either contracts or expands. The scale factor in the Friedmann metric defined in the previous chapter is still undetermined. We will explore its scale factor changes mathematically via the Einstein equations. So we start with them.

7.2 The path to the Einstein equations with the cosmological constant

Some of the steps leading to the final formulation of General Relativity are described in Box 4. One of the requirements was that the new theory of gravitation should approach Newton's theory at the limit of weak gravitational fields and at speeds much less than the speed of light. We have already encountered examples of this e.g., in the discussion of the Schwarzschild solution in Chapter 3. Another requirement is that the theory should also approach Special Theory of Relativity at a suitable limit; as in Special Relativity, physical quantities were written in covariant tensor form which allows them to be regarded as entities independent of the coordinate system. The last step in the derivation was carried out by David Hilbert and Albert Einstein independently in November/December 1915.

David Hilbert (1862–1943) was a German mathematician who became interested in Einstein's work on the new theory of gravitation. Einstein had developed the fundamentals of his theory already in 1907, but he then had a long struggle lasting eight years to find its final mathematical form. By the middle of 1915 also Hilbert had become interested in Einstein's struggle, and he invited Einstein to Göttingen for a week of lectures on the subject. Subsequently Einstein learned that also Hilbert was trying to complete the same scientific project, and this must have hurried Einstein to greater speed to finish his new theory. In late 1915 Einstein published his final equations in *The Field Equations of Gravitation*, but only a few months later David Hilbert's work on *The Foundations of Physics* appeared, an axiomatic derivation of the field equations. There have been many discussions of the issue of priority, who discovered the field equations first. The issue is complicated since the two scientists were in correspondence during the crucial months at the end of the year 1915. However, Hilbert never tried to discredit Einstein as the originator of the General Relativity theory, even though it is also obvious that the correspondence with Hilbert was a great help to Einstein.

An elegant way to derive the Einstein field equations is to start from the Einstein–Hilbert action:

$$S = \frac{c^4}{16\pi G} \int d^4x \sqrt{-g} (R - 2\Lambda) + S_m. \tag{7.1}$$

It is based on the linear action in terms of the Ricci scalar $R = g^{\mu v} R_{\mu v}$ where $R_{\mu v}$ is the Ricci tensor (see Box 2), and the matter action S_m. We take the variation of the action with respect to the metric $g^{\mu v}$. After some algebra we get

$$\delta S = \frac{c^4}{16\pi G} \int d^4x \sqrt{-g} \left(R_{\mu v} - \frac{1}{2} R g_{\mu v} + \Lambda g_{\mu v} \right) \delta g^{\mu v} + \delta S_m. \tag{7.2}$$

The energy-momentum tensor $T_{\mu v}$ (**Box 5**) is defined from the variation of δS_m in terms of $g^{\mu v}$:

$$\delta S_m = \frac{1}{2} \int d^4x \sqrt{-g} \, T_{\mu v} \delta g^{\mu v}. \tag{7.3}$$

Since $\delta g^{\mu v}$ is arbitrary, the condition

$$\delta S = 0 \tag{7.4}$$

leads to

$$R_{\mu v} - \frac{1}{2} R g_{\mu v} + \Lambda g_{\mu v} = -\frac{8\pi G}{c^4} T_{\mu v}. \tag{7.5}$$

With the indices μ, v ranging from 0 to 3, and considering the symmetry properties, these constitute 10 independent equations, the **Einstein equations**.

Finally we may introduce the Einstein tensor

$$G_{\mu v} = R_{\mu v} - \frac{1}{2} R g_{\mu v} \tag{7.6}$$

which allows us to write the Einstein equations more compactly

$$G_{\mu v} + \Lambda g_{\mu v} = -\frac{8\pi G}{c^4} T_{\mu v}. \tag{7.7}$$

In the mixed tensor form (one upper index, one lower index) they read

$$G^\mu_\nu + \Lambda \delta^\mu_\nu = -\frac{8\pi G}{c^4} T^\mu_\nu \quad \mu, \nu = 0, \ldots, 3 \tag{7.8}$$

In the Einstein equations we connect the metric $g_{\mu\nu}$ (Equation (1.12)), represented by the tensor G^μ_ν and Λ on the left hand side, to the **energy momentum tensor** T^μ_ν on the right hand side. Here δ^μ_ν is the unit tensor,

$$\delta^\mu_\nu = \text{diag}[1, 1, 1, 1]. \tag{7.9}$$

The tensor G^μ_ν is a complicated function of $g^{\mu\nu}$. We discuss some details of Einstein's equations in Box 4, in particular the appearance of the **cosmological constant** Λ, and the energy momentum tensor in them. Even though our derivation has been rather skimpy for the reasons of avoiding the details of tensor calculus, the various steps still illustrate the basic principles. As we have learned before in Chapter 2, we can learn a whole lot by concentrating on the Lagrangian, the integrand in Equation (7.1). For the case of an empty space containing no matter, the Lagrangian is

$$L = \frac{c^4}{16\pi G}(R - 2\Lambda). \tag{7.10}$$

Apart from the nonessential constant factor in front, it has two parts, the Ricci scalar R which is proportional to the Gaussian curvature, and the cosmological constant Λ. The action principle in classical mechanics advises us to look for the route of the least action, and from there to derive the equations of motion. Here we look for the universe with the least curvature, and finding it leads to Einstein equations. Furthermore, we have to normalize the curvature represented by R to a constant value 2Λ. Thus we see that Λ brings in an "intrinsic" curvature of the universe. Our standard for calculating the curvature is not from zero curvature but from some non-zero value represented by Λ. In our rubber sheet example, we liken this to the situation where our stretched out rubber sheet is not a level surface but slanted toward some corner. Then in the absence of heavy objects on the sheet our test body would tend to roll towards this corner.

This is the simplest explanation for the accelerating universe, the Einstein Λ explanation. In this view the value of Λ may be regarded a natural constant just like the G and c in Einstein equations, and it does not require any particular explanation more than other constants of nature do.

Box 4. How do the Einstein equations arise and how does Λ enter?

General Relativity is a metric theory of gravitation (which means that gravitational effects are described via the metric $g^{\mu\nu}$ of the spacetime so that test bodies follow geodesics of the metric $g^{\mu\nu}$). At its core are the Einstein equations, which describe the relation between the geometry of a four-dimensional, pseudo-Riemannian manifold representing spacetime, and the energy-momentum contained in that spacetime. There are two main requirements in developing the General Theory of Relativity: the theory must agree locally with the Special Theory of Relativity, and it must approach the Newtonian theory of gravity at the suitable limit. The first requirement is best achieved by writing the local physics in covariant tensor form. Then it can be readily extended to curved spaces. However, the techniques are rather cumbersome and beyond this book. For a good introduction, see e.g., Adler et al. 1975.

For a *vector* (such as a velocity vector) to be coordinate system invariant, the components of the vector must contra-vary with a change of basis to compensate. That is, the components must vary in the opposite way (the inverse transformation) as the change of basis. These are contra-variant. For a **dual vector**, (such as a gradient) to be coordinate system invariant, the components of the vector must co-vary with a change of basis to maintain the same meaning. That is, the components must vary by the same transformation as the change of basis. These are covariant. The distinction between covariance and contra-variance is particularly important for computations with tensors, which often have mixed variance. This means that they have both covariant and contra-variant components. In Einstein's notation, covariant components have lower indices, while contra-variant components have upper indices.

Here we use this statement to justify why tensors necessarily appear in the Einstein equations. The Einstein field equations were initially formulated in the context of a four-dimensional theory which requires tensors. In Special Relativity, conservation of energy-momentum corresponds to the statement that the energy-momentum tensor is divergence-free. This formula, too, is readily generalized to curved spacetime by replacing partial derivatives with their curved-manifold counterparts, covariant derivatives. The second requirement is that at the classical limit the Poisson equation is satisfied.

$$\nabla^2 \Phi = 4\pi G \rho. \tag{B4.1}$$

This connects the matter, through its density ρ, to the gravitational potential field Φ. In Newtonian gravity, the source of the gravity is mass. In Special Relativity mass is part of a more general quantity called the energy-momentum tensor, which includes both energy and momentum densities as well as stress (that is, pressure and shear). General Relativity requires a tensor on the left hand side which describes space curvature, and a tensor on the right hand side which describes the matter-energy contents of curved space. The space curvature may be taken to be the **Ricci tensor** (or the contracted **Riemann tensor**) $R^{\mu\nu}$ and the matter energy is called $T^{\mu\nu}$. See Box 2 and Equation B5.1. It can be shown that at the Newtonian limit the dominant components of these tensors are the $(0, 0)$ components:

$$T^{00} = \rho c^2 \tag{B4.2}$$

and

$$R^{00} c^2 = -\nabla^2 \Phi. \tag{B4.3}$$

Thus more generally

$$R^{\mu\nu} = -\frac{4\pi G}{c^4} T^{\mu\nu}. \tag{B4.4}$$

The above equation looks quite a bit like Equation (7.8) (for $\Lambda = 0$) except for the factor 2 difference on the right hand side and the Einstein tensor $G^{\mu\nu}$ on the left hand side in place of the Ricci tensor.

The important difference between the Ricci tensor and the Einstein tensor is that the divergence of the Einstein tensor is zero while for the Ricci tensor it is not. The Einstein tensor can be obtained from the Ricci tensor by

$$G^{\mu\nu} = R^{\mu\nu} - \frac{1}{2} R g^{\mu\nu} \tag{B4.5}$$

where R is defined as the Ricci scalar. The Ricci scalar can be calculated from the Ricci tensor by contraction $R = R^\nu_\nu$. To understand the meaning of the Ricci tensor, note that for a two-dimensional sphere $R = 2K$ where K is the Gaussian curvature. See Equation (4.1) for a definition of the curvature parameter K. For a three dimensional hypersphere of constant radius $R = 6K$. For an expanding hypersphere as defined in Chapter 6, the same result holds if the deceleration parameter $q = 1$. R is also called the **curvature scalar**. The reason why it is so important that the divergence of the left hand side of Equation (B4.4) should be zero is that the divergence of the right hand side $T^{\mu\nu}$ is zero (Equation (B5.7)). At the Newtonian limit the dominant component is $G^{00} = 2R^{00}$, which validates the argument leading to Equation (B4.4) when $R^{\mu\nu}$ is replaced by $G^{\mu\nu}$. This includes the missing factor of 2.

The steps leading to the replacement of $R^{\mu\nu}$ by $(1/2)G^{\mu\nu}$ in Equation (B4.4) were not at all obvious even to Einstein, and it took Einstein over two years to reach the final result. Other alternatives were also studied. An account of the evolution of Einstein's thoughts on this subject is given by Stachel (2002).

We do not write down the complicated expression for the Einstein tensor, but we may state some of its properties. The Einstein tensor is a function of the metric tensor $g_{\mu\nu}$, and of the first and second derivatives of the metric tensor. It has 4×4 components of which only 10 are independent, due to symmetry properties. It is linear in the second derivatives of the metric tensor. It is the most general second-rank tensor of its kind. The only way it can be further generalized is by adding a constant term (under the covariant divergence operation that is more complicated than the usual divergence in flat space), i.e., by adding an arbitrary constant Λ.

$$G^{\mu\nu} + \Lambda g^{\mu\nu}.$$

The arbitrary constant, however, does have observational consequences. It appears on the right hand side of the Poisson equation in the Newtonian limit, and thus can be interpreted as a vacuum density. We interpret it in that way in this book. From accurate observations such as those described earlier for the planet Mercury, this vacuum density does not show up in the scale of the Solar System, so its value must be very small. Therefore Λ first appears in General Relativity due to the generality requirements on the Einstein tensor, i.e., on the left hand or the geometry side of the equation. Subsequently it has become customary to shift the Λ term to the right hand or the matter-energy side of the equation. Either way, there is no **theoretical** justification in omitting it. It is a matter of experimental determination to find out its physical value. It would be nice to otherwise justify the experimental value, but attempts to this end have not been successful so far.

Box 5. The energy-momentum tensor in the Einstein equations.

In the Special Theory of Relativity the various forms of energy, momentum, and stress can be combined in a single tensor called the **energy-momentum tensor**. Start by writing that for the simplest system, termed a universe of "dust", in that we ignore pressure. It could be considered a reasonable approximation of widely spaced stars or galaxies. This dust has density ρ_0 at rest and velocity $\mathbf{v}$. In a relativistic case, the volume is reduced by the Lorentz factor γ and the mass inside the volume increased by the same factor. The relativistic energy density is

$$T^{00} = \rho c^2 = \gamma^2 \rho_0 c^2. \tag{B5.1}$$

which is one of the components of the matter tensor. Identifying other components, consider the flux of moving mass in the x-coordinate direction, ρv_x, where

$v_x = dx/dt$. The corresponding component of the matter tensor (x-component of the four-momentum density) is defined

$$T^{0x} = \rho v_x c. \tag{B5.2}$$

The momentum density flux ρv_i, $i = x, y, z$, to the x-direction, is given similarly by $\rho v_i v_x$. The component of the matter tensor T^{ix} is defined as

$$T^{ix} = \rho v_i v_x, \quad i = x, y, z. \tag{B5.3}$$

When similar components are written for other coordinates, and the symmetry $T^{ik} = T^{ki}$ is assumed, the matter tensor becomes

$$T^{ik} = \rho c^2 \begin{bmatrix} 1 & v_x/c & v_y/c & v_z/c \\ v_x/c & v_x^2/c^2 & v_x v_y/c^2 & v_x v_z/c^2 \\ v_y/c & v_y v_x/c^2 & v_y^2/c^2 & v_y v_z/c^2 \\ v_z/c & v_z v_x/c^2 & v_z v_y/c^2 & v_z^2/c^2 \end{bmatrix}. \tag{B5.4}$$

Writing more compactly in terms of the components of the four-velocity

$$V^\mu = \gamma c(1, \mathbf{v}/c) \tag{B5.5}$$

we get

$$T^{ik} = \rho_0 V^i V^k. \tag{B5.6}$$

We shall demonstrate next that the divergence of the above tensor is zero. Calculate the divergence for each of the four-vector components T^{ik}, $k = 0, \ldots, 3$, separately. Denote the first four-vector (four-momentum) $\mathbf{T}^0$ with the components $\rho c^2 (1, v_x/c, v_y/c, v_z/c)$. The divergence is

$$\nabla \cdot \mathbf{T}^0 = c \frac{\partial \rho}{\partial t} + c \left[\frac{\partial(\rho v_x)}{\partial x} + \frac{\partial(\rho v_y)}{\partial y} + \frac{\partial(\rho v_z)}{\partial z} \right] = c \left[\frac{\partial \rho}{\partial t} + \nabla \cdot \rho \mathbf{v} \right] = 0 \quad (B5.7)$$

Because of continuity the rate of change in density is equal to density entering or leaving. The second vector $\mathbf{T}^1$ has the components from the second row of the T^{ik} matrix. Its divergence is

$$\nabla \cdot \mathbf{T}^1 = \frac{\partial(\rho v_x)}{\partial t} + \frac{\partial(\rho v_x^2)}{\partial x} + \frac{\partial(\rho v_x v_y)}{\partial y} + \frac{\partial(\rho v_x v_z)}{\partial z}$$
$$= \rho \left[\frac{\partial v_x}{\partial t} + \mathbf{v} \cdot \nabla v_x \right] + v_x \left[\frac{\partial \rho}{\partial t} + \nabla \cdot \rho \mathbf{v} \right]. \tag{B5.8}$$

The second term in the square brackets is zero due to the mass continuity. The first term in the square brackets is an expression of the Lagrangian (co-moving) derivative dv_x/dt. In the absence of forces $dv_x/dt = 0$. Thus in a force-free dynamic flow

$$\nabla \cdot \mathbf{T}^1 = 0. \qquad \text{(B5.9)}$$

The same holds for $\mathbf{T}^2$ and $\mathbf{T}^3$, and if we represent all the tensor components T^{ik} by $\mathbf{T}$ then

$$\nabla \cdot \mathbf{T} = 0. \qquad \text{(B5.10)}$$

Here we recognize the Newtonian stress tensor described earlier in this book as the submatrix

$$T^{ik}, i, k = 1, 2, 3.$$

The above is for the universe of dust. Now, if we consider adding pressure P to this physical system, it will appear on the diagonal of the 3×3 stress matrix. Thus it may appear that the effects of pressure could be simply taken care of by adding another matrix

$$S^{ik} = P \begin{bmatrix} 0 & 0 & 0 & 0 \\ 0 & 1 & 0 & 0 \\ 0 & 0 & 1 & 0 \\ 0 & 0 & 0 & 1 \end{bmatrix} \qquad \text{(B5.11)}$$

to the energy-momentum tensor. Going back to Equation (B5.8), the addition of S^{ik} gives on the right hand side a pressure force term

$$f_x = -\frac{\partial P}{\partial x} \qquad \text{(B5.12)}$$

The equation for conservation of momentum

$$\rho \frac{dv_x}{dt} = f_x \qquad \text{(B5.13)}$$

now leads to zero divergence as mentioned earlier. However, the matrix S^{ik} is not a tensor. In order to construct a suitable tensor, consider the tensor

$$S^{ik} = P\left(V^i V^k / c^2 - g^{ik}\right) \qquad \text{(B5.14)}$$

where g^{ik} is the metric tensor. Using the metric tensor of Equation (1.15), and $V^i V^k$ at the limit of small velocities (i.e., $v_i^2/c^2 = 0, Pv_i/c = 0, i = x, y, z$), the S^{ik} tensor can be written

$$S^{ik} = P \left[\begin{bmatrix} 1 & 0 & 0 & 0 \\ 0 & 0 & 0 & 0 \\ 0 & 0 & 0 & 0 \\ 0 & 0 & 0 & 0 \end{bmatrix} - \begin{bmatrix} 1 & 0 & 0 & 0 \\ 0 & -1 & 0 & 0 \\ 0 & 0 & -1 & 0 \\ 0 & 0 & 0 & -1 \end{bmatrix} \right]. \qquad (B5.15)$$

At this limit, the tensor S^{ik} has the correct components, and thus we may choose it as the additional part of the energy-momentum tensor. Together

$$T^{ik} = \rho_0 V^i V^k + P(V^i V^k / c^2 - g^{ik}) = (\rho_0 + P/c^2) V^i V^k - P g^{ik}. \qquad (B5.16)$$

Because the divergence of the metric tensor is zero, the newly defined energy momentum tensor also has zero divergence.

The energy-momentum tensor's zero divergence is equivalent to conservation of energy and momentum in the matter field.

7.3   Interpretations of the cosmological constant

The cosmological constant, as it is now understood, can be considered as a fundamental constant of nature. Its physical interpretation has varied from one decade to another, starting from studies by the Dutch astronomer of Willem de Sitter (Leiden University), the Belgian mathematician Georges Lemaître (Université de Louvain), the American physicist Richard C. Tolman (California Institute of Technology), and the Austrian-British astrophysicist Hermann Bondi (Cambridge University). In 1965, Russian theoretical physicist Erast B. Gliner (Leningrad) suggested a physical interpretation of the constant that is generally accepted now. According to Gliner (1965), the cosmological constant represents the cosmic vacuum, which is understood as a special state of cosmic energy.

We should clarify some terminology. Before Gliner, the notion of the vacuum was used in a different sense in General Relativity. It was taken to simply be empty space. For instance, when "vacuum solutions" to Einstein equations were mentioned (like the Schwarzschild solution for a point mass in empty space), one had in mind that these were solutions with zero energy-momentum tensor. On the contrary, the cosmic vacuum is characterized by a non-zero energy-momentum tensor with a non-zero energy density that is constant in space-time, in any reference frame.

The Einstein equations (Equation (7.8)) with the cosmological constant have the form:

$$R_k^i - \frac{1}{2} R g_k^i + \Lambda \delta_k^i = -\frac{8\pi G}{c^4} T_k^i. \qquad (7.11)$$

Einstein put Λ to the "geometry" side of the equations—indicating perhaps the geometric nature of the new constant. Indeed, if $\Lambda \neq 0$, even in absence of any matter when the matter energy-momentum tensor $T_k^i = 0$, the space-time cannot be the Minkowski space-time of Special Relativity, and the General Relativistic space-time must be curved.

In Gliner's interpretation, the cosmological constant has a "material" nature, rather than a geometric one. Thus, the space-time of General Relativity may be the Minkowski space-time in the absence of any forms of matter or energy, only if the vacuum energy is also zero. It is thus natural to move the cosmological constant to the energy side of the Einstein equations and identify $\Lambda \delta_k^i$ with a special contribution to the energy-momentum tensor. This contribution alone is described by an energy-momentum tensor of the form

$$T_k^i = \frac{c^4}{8\pi G} \Lambda \delta_k^i. \tag{7.12}$$

which is diagonal in any reference frame. Its structure is like that of a tensor for a perfect fluid whose density is ρ and whose pressure is P in a co-moving reference frame:

$$T_k^i = \mathrm{diag}\,[\rho c^2, -P, -P, -P]. \tag{7.13}$$

By comparing the last two equations, the vacuum energy density is

$$\rho_V c^2 = \frac{c^4}{8\pi G} \Lambda \tag{7.14}$$

and the vacuum pressure is

$$P_V = -\frac{c^4}{8\pi G} \Lambda. \tag{7.15}$$

From these two equations, the cosmic vacuum can be taken as a kind of perfect fluid with the equation of state

$$P_V = -\rho_V c^2. \tag{7.16}$$

This and only this equation of state can be reconciled with the definition of the cosmic vacuum as a form of energy with eternally and everywhere constant density, independently of the reference frame.

The vacuum equation of state and the properties of the cosmic vacuum to be discussed below are covariant. In other words, they are the same in any reference frame. This is simply because of the tensor form $\Lambda \delta_k^i$ as the cosmological constant appears in the Einstein equations.

As we shall see, observations indicate that the vacuum density is positive i.e., the vacuum pressure is negative. This may seem very unphysical. However, negative pressure is not something impossible in other physical contexts. Negative pressure appears, for instance, in a stretched rubber band. It is also helpful to compare negative and positive pressure in thermodynamics. Consider the increase of the volume of a container. There are three different cases depending on the sign of the pressure. For positive pressure (e.g., the usual gas) the internal energy decreases with increasing volume ($\Delta E = -P \times \Delta V$). For zero pressure (dust-like matter) the energy stays the same. For negative pressure (e.g., vacuum) the internal energy increases, for which the equation of state $P = -\rho c^2$ (where ρ = constant). This equation of state means that the internal energy increases in direct proportionally to the volume so that the density does not change ($\Delta E / \Delta V = \rho c^2$). But the cosmic vacuum negative pressure cannot be the absolute value of the energy density (rest energy included) of any ordinary medium or material.

The most important "mechanical" property of the cosmic vacuum is also implied by the vacuum equation of state: the state of rest and motion are not determined relative to the vacuum. Indeed, let there be two bodies or reference frames moving relative to each other. The vacuum is co-moving with both of them because the vacuum is described by its energy-momentum tensor, which is diagonal in any reference frame. In this sense, vacuum is a perfect fluid that is co-moving with any body, either in motion or in rest. It means, in particular, that the vacuum itself cannot be used as a reference frame—contrary to any other real fluid.

Finally, the cosmic vacuum is the only fluid, or, more generally, energy state, in nature that is not affected by any physical factors—at least in General Relativity which includes no quantum effects. The cosmic vacuum does not feel gravity of matter; instead it produces antigravity. Moreover, it does not feel the self-antigravity back effect either. It is simply seen from the fact that the cosmological vacuum with its density and pressure is characterized by the cosmological constant alone which is the same and unchanged everywhere and eternally.

Thus we have arrived at the second explanation of the cosmic acceleration as a special form of energy obtained by moving the Λ term from the geometry side of Einstein equations to the energy side of the equations. Mathematically, and thus also observationally, there is no difference between these two explanations of the cosmic acceleration. For now on we will follow the second explanation, and use the term dark energy when we refer to the cosmic effects of the Λ term. There are also other dark energy models, but we defer their discussion to later chapters (see Section 9.8).

Chapter 8

Model Universes

8.1 Friedmann equation

As we have said, Einstein's equations (Equations (7.8)) contain 10 independent equations with G_{ν}^{μ} being a complicated function of $g^{\nu\mu}$. The solutions of the Einstein equations are the components of the metric tensor which we want to know. Despite the large number of equations, the symmetry properties of the Friedmann metric simplify the Einstein equations a great deal. The universe is assumed to be homogeneous and isotropic. The Cosmological Principle implies that the metric of the universe including the scale factor must be of the same form as Equation (6.6):

$$ds^2 = c^2 dt^2 - R(t)^2 dl^2 \tag{8.1}$$

where dl^2 is a three dimensional metric that must be one of three alternatives: (a) flat space, (b) a constant positive curvature sphere or (c) a constant negative curvature hyperbolic space. Einstein's equations now relate the evolution of what is called **the scale factor** $R(t)$ as determined by the universe's pressure and energy of the contents of the universe.

For these alternatives only two independent equations remain—the G_0^0 term and the trace of G_{ν}^{μ} ($G_1^1 = G_2^2 = G_3^3$):

$$\frac{3}{R(t)^2}\left(\frac{\dot{R}(t)^2}{c^2} + k\right) = \frac{8\pi G}{c^2}(\rho + \rho_V) \tag{8.2}$$

$$\frac{1}{R(t)^2}\left(2R(t)\frac{\ddot{R}(t)}{c^2} + \frac{\dot{R}(t)^2}{c^2} + k\right) = -\frac{8\pi G}{c^2}\left(\frac{P}{c^2} - \rho_V\right). \tag{8.3}$$

The left hand side of Equation (8.2) is the G_0^0 component of the Einstein tensor with a minus sign. The right hand side of Equation (8.2) has the corresponding component of the energy-momentum tensor $T_0^0 = \rho c^2$ (with a minus sign) plus the cosmological constant Λ written using the quantity ρ_V, the **vacuum density**. As we already found out (Equation (7.14)), this density is related to the cosmological constant by

$$\Lambda = 8\pi G \rho_V / c^2. \tag{8.4}$$

In Newtonian terms, the right hand side of Equation (8.3) represents an additional vacuum density contribution to the radial acceleration between any two particles by the amount

$$\ddot{r} = \frac{8}{3}\pi G \rho_V r. \qquad (8.5)$$

The vacuum density is so small ($\rho_V \approx 0.73 \times 10^{-29}$ g cm^{-3}) that its influence is significant only in the scale much larger than the size of a galaxy. It is possible to set $\rho_V = 0$ and so remove the universal acceleration; however, recent observations indicate that $\rho_V > 0$.

The quantity k will take the values $+1$, 0, or -1.

The left hand side of Equation (8.3) is the negative of $G_1^1 = G_2^2 = G_3^3$ component of the Einstein tensor. The right hand side has the negative of the energy-momentum tensor $T_1^1 = T_2^2 = T_3^3 = P$ where P is pressure, plus Λ. The pressure of the present day universe of separated galaxies is negligible and may be set to zero in Equation (8.3). Recall that for an ideal gas, pressure and density are related by

$$\frac{P}{\rho} = \frac{\bar{v}^2}{3} \qquad (8.6)$$

where $\bar{v}$ is the root-mean-square velocity of gas molecules. Thus, dividing by the speed of light squared,

$$\frac{P/c^2}{\rho} = \frac{\bar{v}^2}{3c^2}. \qquad (8.7)$$

The observed random motions of galaxies are of the order of $10^{-3}c$ so that $P/c^2 \approx 3 \times 10^{-7}\rho$. Thus for the present day "gas" of galaxies, the pressure is not important.

Detailed steps leading from the Einstein equation to Equations (8.2) and (8.3) would require deeper understanding of General Relativity than we are prepared to present here. However, we can justify these equations is by comparison with Newtonian cosmology which we will carry out in Section 8.2.

First we will proceed to solve for the scale factor $R(t)$ in a zero pressure model, $P = 0$. To begin, multiply Equation (8.2) by $R(t)^3$:

$$\frac{3R(t)\dot{R}(t)^2}{c^2} + 3kR(t) = \frac{8\pi G}{c^2}R(t)^3(\rho + \rho_V). \qquad (8.8)$$

Differentiating and dividing both sides by 3 we have

$$\frac{2R(t)\ddot{R}(t)}{c^2}\dot{R}(t) + \frac{\dot{R}(t)^2}{c^2}\dot{R}(t) + k\dot{R}(t) - \frac{8\pi G}{c^2}\rho_V R(t)^2\dot{R}(t) = \frac{d}{dt}\left(\frac{8\pi G}{3c^2}\rho R(t)^3\right).$$
$$(8.9)$$

Multiplying Equation (8.3) by $R(t)^2\dot{R}(t)$:

$$\frac{2R(t)\ddot{R}(t)}{c^2}\dot{R}(t) + \frac{\dot{R}(t)^2}{c^2}\dot{R}(t) + k\dot{R}(t) - \frac{8\pi G}{c^2}\rho_V R(t)^2 \dot{R}(t) = 0 \qquad (8.10)$$

for the zero pressure model.

Subtracting Equation (8.10) from Equation (8.9) gives for the matter component

$$\frac{d}{dt}\left(\frac{8\pi G}{3c^2}\rho R(t)^3\right) = 0 \qquad (8.11)$$

that integrates to

$$\frac{4}{3}\pi R(t)^3 \rho = \text{constant} = M. \qquad (8.12)$$

The constant is the mass inside a co-moving volume of radius $R(t)$. Strictly speaking, since the large-scale geometry may be non-Euclidean, the constant M may be identified with the mass (volume times density) only in a small co-moving volume of the 3-space. Equation (8.12) is an example of important energy density vs. scale factor relations of the form

$$\rho \propto R^{-3(1+\gamma)}, \qquad (8.13)$$

where the factor γ appears in the equation of state for a cosmic matter component $P = \gamma c^2 \rho$. For example, for pressureless dust matter $\gamma = 0$, for radiation $\gamma = 1/3$, and for the vacuum density $\gamma = -1$.

Substituting Equation (8.12) into Equation (8.2):

$$\dot{R}(t)^2 = \frac{2GM}{R(t)} + \frac{8\pi G}{3}\rho_V R(t)^2 - kc^2 . \qquad (8.14)$$

The Equation (8.14) is known as the **Friedmann equation** which has the formal solution

$$t = \int \frac{dR}{\sqrt{\frac{2GM}{R(t)} + \frac{8\pi G}{3}\rho_V R(t)^2 - kc^2}}. \qquad (8.15)$$

Since it is not possible to solve Equation (8.15) using only elementary functions, we will look at some simple special cases.

8.2 The Einstein–de Sitter universe (critical density Friedmann case with no dark energy)

One special case of the Friedmann model is called the **Einstein–de Sitter universe** after its originators, Einstein and the Dutch astronomer Willem de Sitter (1932).

This universe is flat at all times. The density of matter in this model is such that the model universe is exactly between the spherical and hyperbolic cases.

During the following description, the reader may wish to refer to Figure 8.1 and Table 8.1, where this model is the third line from the top in the graph in Figure 8.1 and is in the third row of Table 8.1. This special case is the flat universe ($k = 0$) with zero vacuum density (dark energy). Integration of Equation (8.15) with zero k is now simple:

$$t = \frac{2}{3}\frac{1}{\sqrt{2GM}}R(t)^{3/2} \tag{8.16}$$

assuming that $R(t) = 0$ when $t = 0$. Inverting,

$$R(t) = \left(\frac{3}{2}\right)^{2/3}(2GM)^{1/3}t^{2/3}. \tag{8.17}$$

Substituting M from Equation (8.12) into Equation (8.16) gives the expression for the density's time evolution:

$$\rho = \left(6\pi Gt^2\right)^{-1}. \tag{8.18}$$

The significance of this result is clear if we use the Hubble constant from Equation (6.13):

$$H_0 = \frac{\dot{R}(t_0)}{R(t_0)} = \frac{\sqrt{\frac{2GM}{R(t_0)}}}{R(t_0)} = \frac{2}{3}t_0^{-1} \tag{8.19}$$

In terms of the Hubble constant, the age, t_0, of the flat universe is

$$t_0 = \frac{2}{3}H_0^{-1}. \tag{8.20}$$

Since the result is purely a function of the Hubble constant, converting it into years, if $H_0 = 71$ km/s/Mpc $\cong 71 \times 1.02$ pc/10^6 yr/10^6 pc $= 72.6 \times 10^{-12}$ yr^{-1}, gives the age in years for the universe,

$$t_0 = \frac{2}{3}\frac{1}{72.6}10^{12} \text{ yr} \cong 0.92 \cdot 10^{10} \text{ yr.} \tag{8.21}$$

This value is too low in comparison to the 13 billion year ages of the oldest objects in the universe and other age estimates mentioned in Section 5.5. Thus the Einstein–de Sitter model with modern observations of the Hubble constant is incompatable with modern estimates of the age of the universe (it would require $H_0 < 50$ km/s/Mpc to have $t_0 > 13 \times 10^9$ yr).

The value for the deceleration parameter q_0 is found from Equation (8.3) by putting $P = 0$ and $t = t_0$:

$$\frac{\dot{R}(t_0)^2}{R(t_0)^2}\left[2\frac{R(t_0)\ddot{R}(t_0)}{c^2\dot{R}(t_0)^2} + \frac{1}{c^2} + \frac{k}{\dot{R}(t_0)^2}\right] - \frac{8\pi G}{c^2}\rho_V = 0 \qquad (8.22)$$

Substituting for $k = \rho_V = 0$ we obtain

$$H_0^2\left[-2\frac{q_0}{c^2} + \frac{1}{c^2}\right] = 0. \qquad (8.23)$$

Solving for the deceleration parameter,

$$q_0 = \frac{1}{2}. \qquad (8.24)$$

The mean density ρ_c of the Einstein–de Sitter universe can be obtained by dividing Equation (8.2) by 3 and subtracting it from Equation (8.3):

$$\frac{\ddot{R}(t)}{R(t)} = \frac{4\pi G}{3}(2\rho_V - \rho). \qquad (8.25)$$

At the present time with $\rho_V = 0$

$$\rho = \frac{3}{4\pi G}\left(-\frac{R(t_0)\ddot{R}(t_0)}{\dot{R}(t_0)^2} \cdot \frac{\dot{R}(t_0)^2}{R(t_0)^2}\right) = \frac{3q_0 H_0^2}{4\pi G}. \qquad (8.26)$$

For $q_0 = \frac{1}{2}$ and $H_0 = 71$ km/s/Mpc the value of ρ,

$$\rho_c = \frac{3H_0^2}{8\pi G} \cong 0.95 \times 10^{-29}\text{g cm}^{-3} \qquad (8.27)$$

where we have calculated **the critical density**, the previously mentioned value of the present gravitating matter density which would give exactly $k = 0$, i.e., a flat 3-space, in Friedmann models of zero vacuum density. Although the age with the observed Hubble constant is too young, this density is a useful reference number. If the present day true matter density is $\rho_0 = \rho(t_0)$, then we define the density parameter as

$$\Omega_M = \rho_0/\rho_c = \frac{8\pi G}{3H_0^2}\rho_0. \qquad (8.28)$$

Now this definition results in $\Omega_M = 1$ for the Einstein–de Sitter universe. The observed local density of gravitating matter ρ_0 is lower than ρ_c by about a factor of 3.7. This is another reason why the Einstein–de Sitter universe cannot serve as a model for our universe.

8.3   The de Sitter universe (introducing dark energy with no matter)

Our next model is the **de Sitter universe**, an empty Friedmann case with flat geometry and dark energy. For this special case, we put $M = 0$ and $k = 0$, into Equation (8.14). Equation (8.15) is then easily integrated giving

$$\sqrt{\frac{8\pi G \rho_V}{3}}(t_0 - t) = \ln \frac{R(t_0)}{R(t)}$$

or

$$R(t) = R(t_0) \exp\left[\sqrt{\Omega_V} H_0 (t - t_0)\right]. \tag{8.29}$$

We have introduced the parameter Ω_V, the ratio of the cosmic vacuum density to the critical density:

$$\Omega_V = \frac{\rho_V}{\rho_c} = \frac{8\pi G}{3H_0^2} \rho_V. \tag{8.30}$$

Willem de Sitter originally introduced this model in 1917. It is an accelerating universe ($q_0 = -1$), a limiting case of all models with $\rho_V > 0$. This model is not listed in Table 8.1 or Figure 8.1. However, the Concordance model described in the following section approaches it at large times. Note that in the de Sitter universe with its exponential expansion the Hubble constant is a true constant everywhere and at all times: $H_0 = dR/dt/R = (8\pi G \rho_V/3)^{1/2}$.

8.4   The Concordance Model (both matter and dark energy so $k = 0$)

During the following description, the reader may wish to refer to Figure 8.1 and Table 8.1 below where this model is summarized in the top row of Table 8.1 and as the top line on the Figure 8.1.

According to recent observational evidence, we should consider models where ρ_V is different from zero, while $k = 0$. Let us determine under what conditions $k = 0$ in a model where $\Omega_V > 0$. Assume that pressure $P = 0$. We get a relation between Ω_M, q_0 and k by taking the difference of Equations (8.2) and (8.3) with $P = 0$. Consider the present time $t = t_0$ and use the definitions of H_0, q_0 and Ω_M:

$$kc^2 = H_0^2 R(t_0)^2 \left(\frac{3}{2}\Omega_M - 1 - q_0\right). \tag{8.31}$$

The condition $k = 0$ gives

$$q_0 = \frac{3}{2}\Omega_M - 1. \tag{8.32}$$

What does this require of ρ_V? Equation (8.25) at $t = t_0$ translates to

$$8\pi G\rho_V = 3H_0^2\left(\frac{1}{2}\Omega_M - q_0\right) \tag{8.33}$$

that together with Equation (8.32) for the flat 3-space gives

$$8\pi G\rho_V = 3H_0^2(1 - \Omega_M)$$

Using Equation (8.30) in the above gives

$$\Omega_M + \Omega_V = 1. \tag{8.34}$$

So we see that there exist possible flat models with $\Omega_M < 1$ (as observed) but where the vacuum density (and thus the cosmological constant) has a positive value. This is the essence of what is called the Concordance model, thus defined because it can explain essential cosmological observations and is currently the most commonly used cosmological model. One advantage in having such a common standard model is that using it in various astronomical and cosmological calculations makes comparison of different results convenient.

From Equation (8.25), ρ_V reduces net gravity (i.e., it acts as antigravity). Gravity is responsible for the slowing deceleration, $\ddot{R}(t) < 0$. When $\rho \leq 2\rho_V$, the deceleration changes into acceleration, $\ddot{R}(t) > 0$. This must happen at some particular time t_V since $\rho \propto R(t)^{-3} \propto (1 + z)^3$ while ρ_V is constant. Using the "current" values

$$\rho_V \cong 0.73\rho_c \tag{8.35}$$

$$\rho_0 \cong 0.27\rho_c \tag{8.36}$$

and the Einstein–de Sitter model the condition $2\rho_V = \rho$ takes place at redshift z_V

$$z_V \cong 0.75 \tag{8.37}$$

corresponding to the time $t_V \cong 7.1 \times 10^9$ yr.

Observationally, the ρ_V term dominates only at low values of z, and at relatively late times in the history of the universe. In the Concordance model, after enough expansion the vacuum density term always dominates on the right hand side of Equation (8.14), and the solution approaches Equation (8.29), the de Sitter universe. For early times dark energy vacuum may be ignored with gravity dominating when $z \gg z_V$, the high redshift universe.

The Concordance model, the hyperbolic de Sitter universe, and the Einstein–de Sitter universe all extend to infinite distance, and thus they are **open** universes. The Concordance and the Einstein–de Sitter universes contain an infinite amount of stars and other gravitating matter. In contrast, a closed Friedmann model has a finite volume just like Einstein's 1917 **static** model and again a finite number of stars (or rather galaxies).

To formulate the Concordance model put $k = 0$ in Equation (8.15) and have

$$t = \int_0^R \frac{dR}{\sqrt{\frac{2GM}{R} + \frac{8}{3}\pi G\rho_V R^2}}.$$

(8.38)

Using the integral

$$\int \frac{\sqrt{x}\,dx}{\sqrt{A_\Omega + x^3}} = \frac{2}{3}\ln\left(x^{3/2} + \sqrt{A_\Omega + x^3}\right)$$

along with Equation (8.30) we get

$$t = \frac{2}{3}\frac{1}{H_0}\frac{1}{\sqrt{\Omega_V}}\ln\frac{R^{3/2} + \sqrt{A_\Omega + R^3}}{\sqrt{A_\Omega}}.$$

(8.39)

The constant $A_\Omega \equiv \frac{2GM}{H_0^2 \Omega_V}$ may be evaluated at the present time t_0. Then (Equation (8.12))

$$A_\Omega = \frac{\frac{8}{3}\pi G R_0^3 \rho_0}{\frac{8}{3}\pi G \rho_c \Omega_V} = \frac{R_0^3 \Omega_M}{\Omega_V}$$

(8.40)

where $R_0 = R(t_0)$. Finally, use $\frac{R}{R_0} = (1 + z)^{-1}$. Next, the substitution of $\Omega_M = 0.27$ and $\Omega_V = 0.73$ as well as $H_0 = 71\,\text{km/s/Mpc}$ gives

$$t = 10.77 \times 10^9\,\text{yr}\left[\ln\left((1 + z)^{-3/2} + \sqrt{0.37 + (1 + z)^{-3}}\right) + 0.497\right]$$

(8.41)

which is the age of the universe t at the time when the light signal left the object of redshift z. The current age of the universe is obtained by putting $z = 0$:

$$t_0 = 1.37 \times 10^{10}\,\text{yr}.$$

(8.42)

As discussed earlier, the age of the oldest stars in our Milky Way Galaxy are about 13.2×10^9 yr. This value of course must be less than the age of the universe. It is impressive that two numbers obtained by such different means agree so well.

In the Concordance model the expansion not only goes on forever but it is also accelerating (Table 8.1 and its associated Figure 8.1). As defined above, the capital

Table 8.1. Friedmann models of the universe in Figure 8.1.

Model	Volume	Density	Geometry	Future evolution
Concordance (0.3, 0.7)	infinite	= critical	flat	accelerating expansion
Open (0.3, 0.0)	infinite	< critical	hyperbolic	decelerating expansion
Einstein–de Sitter (1.0, 0.0)	infinite	= critical	flat	decelerating expansion
Closed (5.0, 0.0)	finite	> critical	spherical	expansion-contraction

omegas in Figure 8.1 refer to relative effects of gravitating (ordinary and dark) matter (m) and accelerating dark energy (v) with 1.0 being the sum for a critical universe. The "relative size" of the universe is actually the metric scale factor. However, the relative size can also be taken as proportional to the separation between typical systems or galaxy clusters in the universe.

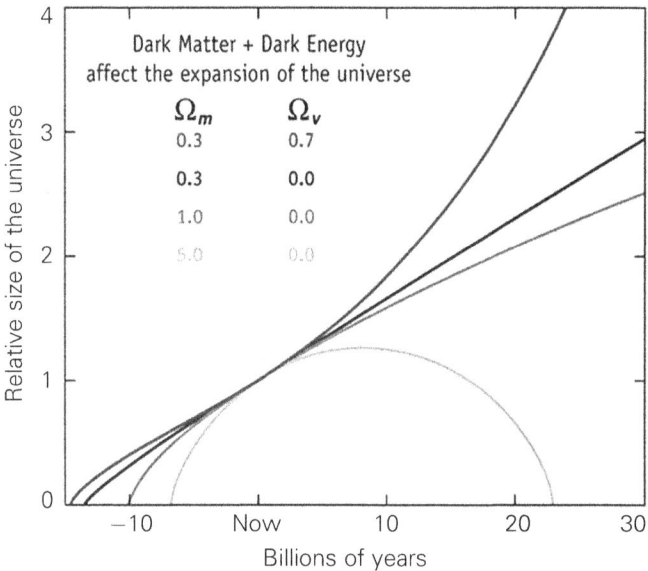

Figure 8.1. Table 8.1 "universes" graphed as a function of time. The scale length R of the vertical axis can be thought of as proportional to the average separation of galaxies. The top line is the concordance model with accelerating expansion. The three lower curves have no dark energy acceleration as indicated by zero Ω_v. The next line below the top concordance line is a "hyperbolic" model with just the deceleration due to gravity which is unable to reverse expansion. The third lowest curve is "critical" density where expansion and deceleration of gravity balance. Finally, the fourth (bottom) curve is a model whose ensity is sufficiently large so gravity eventually overcomes and actually reverses the expansion. Source: NASA http://map.gsfc.nasa.gov/m_uni/101bb2_1.html.

In the concordance model, after enough expansion the vacuum density term always dominates on the right hand side of Equation (8.14), and the solution approaches Equation (8.29) of the de Sitter universe with $\Omega_V = 1$.

8.5   Testing via the small scale Newtonian limit

On a small spatial scale (for galaxies which are cosmologically close but not orbiting around one another), one can test the above formulation by reducing it to the Newtonian case. The universe's curvature is so large that it is nearly flat or Euclidian on this scale. The starting point of most cosmological models is the cosmological principle: the universe is homogeneous and isotropic and will therefore appear the same from any viewpoint. Newtonian cosmology has an infinite Euclidean universe that satisfies the cosmological principle. In addition, the universality of physical laws remains the same in Newtonian cosmology. Newtonian gravity can be used to model the universe and non-rigorously derive the Friedmann equations that are used in the Big Bang universe. In fact, in 1934 the British astronomer Edward Milne pointed out that Big Bang expansion of the Friedmann world is described by the same formulae which determine the behavior of a uniform Newtonian "dust" universe.[1]

Consider two galaxies A and B at positions $\boldsymbol{a}$ and $\boldsymbol{b}$ relative to an arbitrary reference point O (Figure 8.2). The position of B relative to A is $\mathbf{r} = \mathbf{b} - \boldsymbol{a}$. Relative to the point O galaxies A and B move with the velocity $\mathbf{v}(\boldsymbol{a})$ and $\mathbf{v}(\mathbf{b})$, respectively; and their relative velocity is

$$\mathbf{v}(\mathbf{b} - \boldsymbol{a}) = \mathbf{v}(\boldsymbol{a}) - \mathbf{v}(\mathbf{b}) \tag{8.43}$$

A function $\mathbf{v}$ that satisfies the above Newtonian universe requirement is of the form

$$v = H(t)\mathbf{r}. \tag{8.44}$$

The coefficient $H(t)$ is called the **Hubble parameter**. In an isotropic universe this parameter can be a function only of time. Now take galaxy A to be our Galaxy, and B some distant galaxy. Then we get the **Hubble law**:

$$\mathbf{v} = H_0 \mathbf{r} \tag{8.45}$$

where the subscript 0 indicates the present time, t_0. Here $H(t_0)$ is shortened to H_0, the Hubble constant as seen at the present time. Recall that we introduced the Hubble constant in Equation (6.13).

---

1 As we noted above, Newton himself did not consider a globally non-static world structure; the cosmology he preferred in his correspondence with Bentley was made of an infinite and static distribution of uniformly distributed stars.

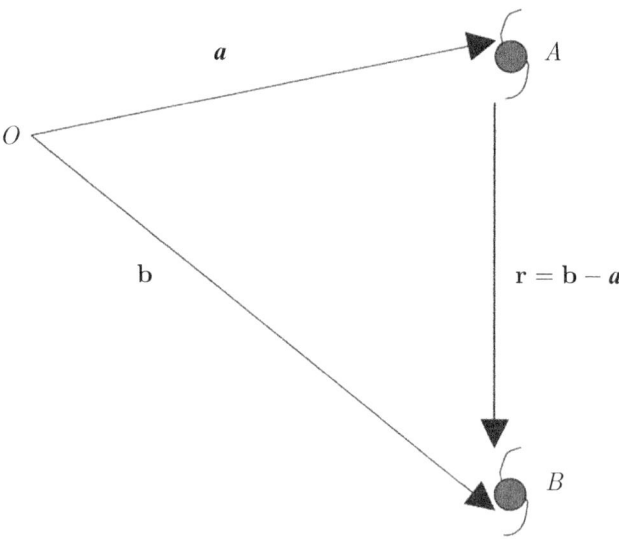

Figure 8.2. The expanding universe. galaxy B is observed from our Galaxy A. The positions of the galaxies are given relative to the reference point O.

The above Hubble law describes a uniformly expanding Euclidean space. Assume the galaxies to be stationary in a Lagrangian or co-moving expanding coordinate system. In contrast, ordinary or Eulerian coordinates $\mathbf{r}(t)$ increase with time. Designate the magnitude by $r(t) = |\mathbf{r}(t)|$. The ratio

$$R(t) = r(t)/r(t_0) \tag{8.46}$$

is the factor describing expansion between some initial time t_0 and a later or earlier time t. This corresponds to the scale factor $R(t)$ in earlier sections.

Equation (8.46) $r = r(t_0)R(t)$, and Equation (8.44) imply $\dot{r} = H(t)r$. Therefore,

$$\frac{1}{R(t)}\frac{dR(t)}{dt} = H(t). \tag{8.47}$$

The expansion factor may be obtained from this equation if $H(t)$ is known. We need an equation which governs $H(t)$.

Expansion is affected over time by gravitation. Considering a co-moving element of unit mass in the flow and assuming no effects of pressure, the acceleration of this element is given by

$$\frac{d\mathbf{v}}{dt} = -\nabla\Phi \tag{8.48}$$

where Φ is the gravitational potential which is derived from the Poisson equation

$$\nabla^2 \Phi = 4\pi G \rho \tag{8.49}$$

where ρ is the mass density. Since $\mathbf{v} = H\mathbf{r}$ and $H = H(t)$,

$$\frac{d\mathbf{v}}{dt} = \frac{dH}{dt}\mathbf{r} + H\mathbf{v} = \left(\frac{dH}{dt} + H^2\right)\mathbf{r}. \tag{8.50}$$

Thus

$$\left(\frac{dH}{dt} + H^2\right)\mathbf{r} = -\nabla\Phi. \tag{8.51}$$

Take the divergence of both sides to obtain

$$3\left(\frac{dH}{dt} + H^2\right) = -\nabla^2\Phi = -4\pi G\rho. \tag{8.52}$$

Substitute H from Equation (8.47) (writing $R = R(t)$):

$$\frac{d}{dt}\left(\frac{1}{R}\frac{dR}{dt}\right) + \frac{1}{R^2}\left(\frac{dR}{dt}\right)^2 = -\frac{4\pi G\rho}{3} \tag{8.53}$$

Simplifying, we obtain

$$\frac{1}{R}\frac{d^2R}{dt^2} + \frac{4\pi G\rho}{3} = 0. \tag{8.54}$$

Make the choice that $R(t_0) = 1$ where t_0 is the current time, and as before use the compact notation, $\rho(t_0) = \rho_0$. Mass conservation over time tells us

$$\rho R^3 = \rho_0. \tag{8.55}$$

This is substituted above and both sides are multiplied by $2R\,dR/dt$. Thus Equation (8.54) becomes

$$2\frac{dR}{dt}\frac{d^2R}{dt^2} + \frac{8\pi G\rho_0}{3}\frac{1}{R^2}\frac{dR}{dt} = 0 \tag{8.56}$$

i.e.,

$$\frac{d}{dt}\left[\left(\frac{dR}{dt}\right)^2\right] - \frac{8\pi G\rho_0}{3}\frac{d}{dt}\left(\frac{1}{R}\right) = 0. \tag{8.57}$$

Integrating gives

$$\left(\frac{dR}{dt}\right)^2 - \frac{8\pi G\rho_0}{3}\frac{1}{R} = \text{constant} = -kc^2. \tag{8.58}$$

In Equation (8.58), the constant k is arbitrary, and it does not necessarily bear any relation to the curvature k of previous sections. However, by this choice of notation and if we also choose $M = [4\pi\rho_0/3] R^3$ to be constant in Equation (8.57), we arrive at formal agreement with the Friedmann equation (Equation (8.14)):

$$\left(\frac{dR}{dt}\right)^2 = \frac{2GM}{R} - kc^2. \tag{8.59}$$

In Equation (8.59), the ρ_V term of the Friedmann equation is missing since Newtonian gravity does not have a cosmic acceleration of the type in Equation (8.5). We could formally carry it through to Equation (8.59) by adding the ρ_V term to the right hand side of Equation (8.48).

Problem. Show that adding the vacuum density (Equation (8.5)) to Equation (8.48) leads to the Friedmann equation in Newtonian cosmology.

8.6   Newtonian cosmology and the "k" parameter

Now we will deduce the significance of k in Newtonian cosmology. Consider a spherical volume sliced out of the universe, centered on an arbitrary point. Take the radius of the volume to be r. A particle at the surface of the sphere moves with speed $v = Hr$ relative to the center (Equation (8.44)). The kinetic energy per unit mass of the particle,

$$E_{KIN} = \frac{1}{2}v^2 = \frac{1}{2}H^2 r^2 = \frac{1}{2}\frac{r^2}{R^2}\dot{R}^2 = \frac{1}{2}r^2\dot{R}^2 \tag{8.60}$$

at the current time when $R = 1$.

The spherical volume's mass is $(4\pi/3)r^3\rho_0$ implying the potential energy of a particle on the surface of the sphere is, per unit mass,

$$W = -\frac{4\pi G r^2 \rho_0}{3}. \tag{8.61}$$

Substitute kinetic energy and potential energy in Equation (8.58) ($R = 1$):

$$E_{KIN} + W = -\frac{1}{2}kc^2 r^2.$$

Examining the equation above, the energy per unit mass of the particle on the surface of the sphere, $E_{KIN} + W$, is negative if $k > 0$. Because of the homogeneity and isotropy of the universe, this statement applies to any particle, in relation to any point. Thus $k > 0$ implies a universe with negative total energy. Such a universe must stop expanding at some point and re-collapse.

Problem. Obtain with $k = 0$, for Newtonian cosmology, the critical density Equation (8.27).

The opposite, $k < 0$ leads to positive total energy and expansion forever. The borderline $k = 0$ has zero energy and expansion which never quite stops but approaches standstill at very large times.

In General Relativity with $\rho_V = 0$ these same general statements of future behavior apply. Then in General Relativity, one adds that the $k > 0$ universe has a bound (closed) space of a 3-sphere, the $k < 0$ universe is an unbound (open) space of a 3-hyperboloid while the $k = 0$ case has a flat 3-space. Figure 8.2 illustrates the evolution of the scale factor R in the three cases.

One can represent the solutions of Equation (8.59) for $k = +1$ and $k = -1$ in a parametric way by defining a parameter in several ways as follows:

For $k = +1$

$$R = \frac{GM}{c^2}(1 - \cos \eta)$$

$$t = \frac{GM}{c^3}(\eta - \sin \eta). \tag{8.62}$$

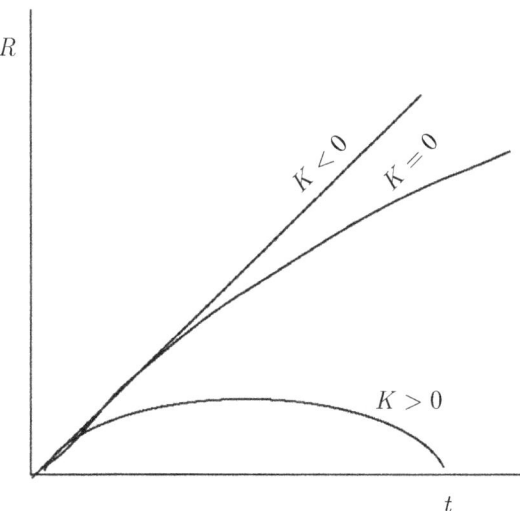

Figure 8.3. Evolution of the scale factor R in three different cosmologies.

We see this by taking

$$\frac{dR}{d\eta} = \frac{GM}{c^2}\sin\eta$$

$$\frac{dt}{d\eta} = \frac{GM}{c^3}(1-\cos\eta)$$

whereby

$$\frac{dR}{dt} = \frac{c\sin\eta}{1-\cos\eta}$$

and

$$\left(\frac{dR}{dt}\right)^2 = \frac{c^2\sin^2\eta}{(1-\cos\eta)^2} = \frac{c^2(1+\cos\eta)}{1-\cos\eta}$$

$$= \frac{GM}{R}\left(1+1-\frac{Rc^2}{GM}\right) = \frac{2GM}{R} - c^2.$$

Similarly we may prove for $k=-1$ that

$$R = \frac{GM}{c^2}(\cosh\eta - 1)$$

$$t = \frac{GM}{c^3}(\sinh\eta - \eta)$$

is the solution of Equation (8.59).

Problem. Prove the correctness of the above statement.

Finally, we should make a comment on the formal similarity between the Newtonian and Friedmann cosmologies. In Newtonian physics the velocities are not restricted by the velocity of light. Thus the mutual velocity of any two particles in the infinite expanding Newtonian particle system which are sufficiently far away from each other may well exceed the speed of light. Because the expansion of the Friedmann model (the behavior of the scale factor $R(t)$) is described by formally similar equations, then also in the infinite Friedmann world the distance between two galaxies may increase at a rate larger than the speed of light. This does not contradict modern physics because actually the galaxies (in the ideal case) are at rest relative to the space and it is the space which is expanding—the galaxies are not superluminally moving or carrying information to anywhere in space.

Chapter 9

Dark energy discovered

In the 1950s, Friedmann's universe with its origin a long time ago and the steady state cosmos, without beginning (Bondi and Gold 1948, Hoyle 1948), were competing alternative cosmological models. This was before the discovery of the cosmic microwave background radiation in 1965 which quickly made the Big Bang idea the ruling paradigm. The accurate thermal spectrum of the background radiation was naturally explained as the cooled down relict of the hot big bang. This predicted cooling as a function of redshift ($T(z) = T_0(1 + z)$), is a critical verification of the Friedmann model (and was not at all expected in the steady state model). Aside from the observation at the redshift of the CMB, the cooling has now been detected in observations from different redshifts (see, e.g., Luzzi et al. 2009).

It is interesting to note that the steady state model contained hidden in it the main property of the current cosmology. Hoyle and Sandage (1956) pointed out that an accelerating expansion would be a signature of the steady state, while a deceleration would be expected in a Friedmann cosmos containing only gravitating matter. In fact, the exponential behavior of the scale factor (which means a truly constant Hubble constant with time) is similar in the steady state model and in the dark energy filled de Sitter model. The dark energy filled model is an extreme Λ model among the Friedmann models (Section 7.3).

9.1 The era of zero-Lambda models

Two types of Friedmann models, both without dark energy (i.e., $\Lambda = 0$), were from the 1970s to 1990s the center of cosmological study:

1) an open universe, with the mean gravitating matter mass density considerably less than the critical density, and
2) the Einstein–de Sitter universe having exactly the critical density.

In both cases, one expects a decelerating expansion, which should be observable in cosmological tests extending from small to sufficiently large redshifts.

In an early study Tolman (1930) discussed angular sizes and light fluxes of distant galaxies in a non-static curved space and considered their relation to redshift as a test of expansion. Fred Hoyle (1959), when discussing the steady state model at a radio astronomy meeting, suggested studying the angular size—redshift relation to distinguish between alternative cosmologies. The idea is simple. One needs a celestial body with a constant linear size and which can be observed at various distances. Hoyle's proposal of the angular size vs. redshift test, which he hoped could support the classical steady state model, illustrates the fact that some alternative cosmological approaches can offer useful insights even when they may not explain all key observations as well as the mainstream cosmology does (Baryshev and Teerikorpi 2011). Alternative approaches may contain a part of a future world model, as a new cosmological process or law, or they may help formulate tests able to distinguish between rival theoretical ideas.

In his classic paper, *The ability of the 200-inch telescope to discriminate between selected world models*, Sandage (1961) discussed in detail several cosmological tests, including the magnitude-redshift ($m(z)$) and the angular size-redshift ($\theta(z)$) relations which might be used to decide between different Friedmann models. As Sandage said, the first task of cosmology was "the search for two numbers". The two numbers are the Hubble constant H_0 and the deceleration parameter q_0 (equivalently H_0 and the cosmic gravitating matter density ρ_0). If the dark energy Lambda term is zero (as was thought then), these two numbers are sufficient to define the Friedmann universe of pressureless matter ("dust"). We begin by discussing tests which measure the dependence on redshift of the luminosity distance and the angular-diameter distance. We may look for acceleration by dark energy as well as deceleration by gravitating matter.

9.2   Cosmological angular-diameter distance estimates

Earlier we discussed the distance estimates using **standard candles**. Here we introduce the **angular-diameter distance** method which, via the $\theta(z)$ test mentioned above, has become very important in the interpretation of cosmic microwave background observations in particular in the determination of the relative proportions of gravitating matter and dark energy. This method is based on the concept that a **standard rod** appears smaller and smaller the further it is moved from us. For a flat Euclidian space, there is an inverse dependence between the distance d_A and the angular size $d\theta_{AB}$

$$d_A = \frac{AB}{d\theta_{AB}}. \tag{9.1}$$

In Equation (9.1), AB is the length of the standard rod which is viewed perpendicular to the line of sight.

To measure the angular size, we observe the angle $d\theta_{AB}$ between two light rays from the two ends of the rod (A and B). Since the angle $d\theta_{AB}$ is small and the rod is perpendicular to the line of sight, the two light rays have departed at the same time (which we denote by t_1) so that $dt = 0$. We may also specify the θ, ϕ coordinate frame in the sky such that $d\phi = 0$. Similarly, the two radial coordinates Ψ for the two ends of the rod are also taken to be the same value, say Ψ_1, i.e., $d\Psi = 0$. The local spatial distance between the two ends is (using the metric Equations (4.18) and (6.1)):

$$AB = R(t_1)f(\Psi_1)\,d\theta_{AB}. \tag{9.2}$$

Now, because $R(t_1) = R(t_0)/(1+z)$,

$$d_A = \frac{R(t_0)f(\Psi_1)}{(1+z)}$$
$$\tag{9.3}$$
$$d\theta_{AB} = \frac{AB}{R(t_0)f(\Psi_1)}(1+z).$$

The angular size, $d\theta_{AB}$, is observed, and we take the size of AB to be known in kpc (or at least to be constant at different redshifts) making it a standard rod (rather than standard candle). The product $R(t_0)f(\Psi_1)$ is a function of space-time curvature for a given redshift (z) and for given values of Ω_M and H_0 (see Mattig's formula in Box 6). Thus Equation (9.3) can be used to determine the curvature.

One cannot express the angular-diameter distance as a function of redshift with a simple analytic form for all the cosmological models we have discussed. However, the flat **Friedmann** universe ($K = 0$) permits derivation of a relatively simple approximate expression. Calculate first $R(t_0)\Psi_1$:

$$R(t_0)\Psi_1 = cR(t_0)\int_{t_1}^{t_0}\frac{dt}{R} = cR(t_0)\int_{R_1}^{R_0}\frac{dR}{R\dot{R}} = c\int_0^z\frac{dz}{H(z)}. \tag{9.4}$$

Now $H(z)$ is the Hubble parameter at z corresponding to t and $R = R(t)$. $H(z)^2$ is obtained from Equation (8.14) (with $k = 0$) by dividing it by $R(t)^2$:

$$H(z)^2 = \frac{\dot{R}(t)^2}{R(t)^2} = H_0^2\Big[\Omega_M(1+z)^3 + (1-\Omega_M)\Big]. \tag{9.5}$$

Equations (8.12) and (8.27)–(8.30) are used. Substituting $H(z)$ in Equation (9.4), we obtain a result which can then be integrated. This integral does not have a simple analytic form, but the result can be expressed well by the following function for the standard Concordance values for matter and dark energy:

$$d_A = \frac{c}{H_0}\frac{z}{\sqrt{(1+0.6z)^4 - 0.18z^2}} \tag{9.6}$$

The function is accurate to 1.5% in the redshift range $0 < z < 10$ (Demianski et al. 2003). The approximation is a function of Ω_M; we have assumed $\Omega_M = 0.27$ and $\Omega_V = 0.73$, taking the sum to be one. This is the flat Concordance model which was explained in Chapter 8.

Box 6. Mattig's formula.

Wolfgang Mattig (1958) derived an important expression for the z-dependence of the factor $R(t_0)f(\Psi_1)$ needed in the calculation of the angular diameter and luminosity distances in the zero-pressure dust Friedmann universe. For a general case (see e.g., Carroll, Press and Turner 1992) it can be written as follows:

$$R(t_0)f(\Psi_1)$$
$$= c/H_0(|\omega|)^{-1/2}\sin n\left\{(|\omega|)^{1/2}\int_0^z \left[\Omega_M(1+x)^3 + \Omega_\Lambda + \omega(1+x)^2\right]^{-1/2}dx\right\}$$

where

$$\omega = 1 - \Omega_M - \Omega_\Lambda$$

and the function

$$\sin n(y) = \sinh(y) \quad \text{for } \omega > 0$$
$$\sin n(y) = y \quad\quad\quad \text{for } \omega = 0$$
$$\sin n(y) = \sin(y) \quad\; \text{for } \omega < 0.$$

We list a few special cases:

1) Without dark energy ($\Omega_\Lambda = 0$) Mattig's formula becomes

$$R(t_0)f(\Psi_1) = (c/H_0)(2/\Omega_M^2)\left[\Omega_M z + (\Omega_M - 2)(-1 + (1 + \Omega_M z)^{1/2})\right]/(1+z).$$

This formula is valid for all $\Omega_M > 0$. However, for very low density universes it is numerically more accurate to use another form derived by Terrell (1977) which is valid also for $\Omega_M = 0$:

$$R(t_0)f(\Psi_1) = (c/H_0)z\left\{1 + z(1 - \Omega_M/2)\left[(1 + \Omega_M z)^{1/2} + 1 + (\Omega_M/2)z\right]^{-1}\right\}.$$

2) For the flat dust matter Einstein–de Sitter universe ($\Omega_M = 1$, $\Omega_\Lambda = 0$) we have

$$R(t_0)f(\Psi_1) = 2(c/H_0)\left[1 - (1+z)^{-1/2}\right].$$

3) For the flat pure vacuum universe ($\Omega_M = 0$, $\Omega_\Lambda = 1$), the formula becomes simply

$$R(t_0)f(\Psi_1) = (c/H_0)z.$$

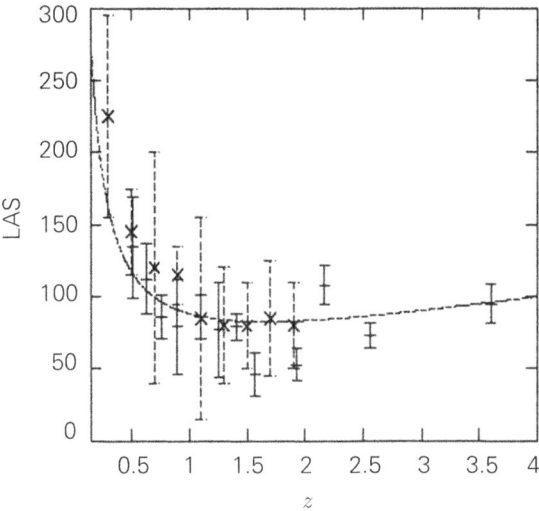

Figure 9.1. Largest angular sizes (LAS) in arcsec of the biggest double radio sources are plotted at various redshift intervals (Nilsson et al. 1993; see crosses with error bars). Also plotted are the median angular sizes of compact radio sources at similar redshift intervals in units of 45 microarcsec (Gurvits et al. 1999; horizontal tics with error bars). The solid theoretical line corresponds to Equation (9.6), with the Concordance model parameter values, for the standard rod length of 740 kpc for the double radio sources and 34 pc for the compact radio sources.

To show how the angular diameter method works, we will discuss the suggestion that the largest double radio sources could be of standard linear size. Observing these at different redshifts would help us find the correct cosmological model (Miley 1971). However, it has become obvious that there is a maximum size for every radio luminosity, and that this maximum is smaller for brighter luminosities (Hooley et al. 1978). Because from large distances we tend to select more-luminous-than-average sources (the flux limit!), the angular diameter vs. redshift relation from a raw sample of radio sources becomes distorted. Thus one must use radio sources of a given radio power, or "standardize" the observed sizes with a luminosity-dependent correction. Figure 9.1 shows how observations of radio angular diameters, after the correction, agree with Equation (9.6) (Nilsson et al. 1993). In this graph, observational points for compact radio sources are also plotted (Gurvits et al. 1999) and show equally good agreement with theory, but without the need for a luminosity correction. Although the diagram is in overall agreement with the Concordance model, uncertainty about the amount of the luminosity correction and cosmological evolution of radio source sizes have hampered the use of this method for the determination of cosmological parameters (Krauss and Schramm 1993).

There are interesting differences in the angular size versus distance relation between cosmological models. In the Concordance model there is a predicted minimum around the redshift $z \approx 1.5$ followed by an increase of angular size at higher redshifts. The existence of a minimum in the angular diameter θ (which corresponds to a maximum in the angular diameter distance d_A) is typical for Friedmann models. For the Einstein–de Sitter model (critical density, no dark energy) the minimum is expected at $z = 1.25$. In the extreme case of the flat dark energy filled de Sitter model (and also in the steady state model) there is no minimum (it is shifted to infinite redshift).

9.3 Cosmological standard candle distance estimates

Comparing the apparent brightness of a standard candle to its intrinsic brightness permits the estimate of the **luminosity distance** in cosmology. However, similar to the standard rod method, the universe's expansion and its possibly non-Euclidean spatial geometry change the familiar inverse square law of light intensity with distance for distant objects. Assume a source that has the radial coordinate Ψ_1 (Equation (4.14)) and the absolute brightness or energy/sec emitted, B. During the time interval Δt_1 it emits energy $B\Delta t_1$. By the time the radiation is received, the energy is spread over a sphere of coordinate radius $R(t_0)f(\Psi_1)$ (Equation (4.12)). The reception of the energy thus takes place during the time interval Δt_0 which is a factor $(1 + z)$ longer than Δt_1 with a corresponding factor of $1/(1 + z)$ decrease in intensity. Also note that the energy of each photon $h_P\nu$ is "redshift decreased" by the ratio

$$\frac{h_P\nu_0}{h_P\nu_1} = \frac{R(t_1)}{R(t_0)} \tag{9.7}$$

during its travel between the emission of the photon and its reception (Equation (6.7)). This is again a corresponding factor of $1/(1 + z)$ decrease in flux. Therefore the observed energy flux is

$$b = \frac{B}{4\pi R(t_0)^2 f(\Psi_1)^2} \frac{\Delta t_1}{\Delta t_0} \frac{R(t_1)}{R(t_0)} = \frac{B(1 + z)^{-2}}{4\pi R(t_0)^2 f(\Psi_1)^2}. \tag{9.8}$$

In the non-expanding static Euclidean space, the familiar inverse square law of flux gives,

$$b = \frac{B}{4\pi d_L^2} \tag{9.9}$$

where the distance is d_L. Therefore equating the two intensities we may define the luminosity distance as

$$d_L = (1+z)R(t_0)f(\Psi_1) \tag{9.10}$$

We remind the reader that the factor $R(t_0)f(\Psi_1)$ can be calculated from Mattig's formula (Box 6) for fixed parameters of the Friedmann model.

By comparing Equations (9.3) and (9.10) we see the angular diameter distance is related to the luminosity distance as follows

$$d_L = (1+z)^2 d_A. \tag{9.11}$$

The relation between d_L and z in the standard concordance theory is obtained using Equation (9.6).

The definition d_L is compatible with the inverse square law of dimming of the light flux with increasing distance, as in a static Euclidian space. Take the **distance modulus** as usually defined

$$m - M = 5 \log d_L - 5 \tag{9.12}$$

$$d_L = 10^{(m-M+5)/5}$$

where d_L is in parsecs.

We thus see that with Equation (9.12), one can use the observed distance modulus, $m - M$, to compute luminosity distances for the standard candles for which the absolute magnitude is known or assumed. These distances and their independently observed redshifts, cz, can be compared to theoretical distances and redshifts for different models of the universe.

9.4   More luminous standard candles

After Hubble's discovery of the expansion of the universe, and the other evidence for an origin in time accumulated, it became obvious that Λ could not be as big as Einstein had calculated for his static universe model. However, the effects of some smaller values of Λ were studied, even though it became common to assume that $\Lambda = 0$. Although the universe is definitely not of a static spherical geometry, today there is strong observational evidence of something like the lambda term (dark energy) acting in the universe. The ideas about the existence of lambda changed in the late 1990s when it became possible to use extremely luminous standard candles to estimate distances of cosmological galaxies whose redshifts z are comparable to unity. These large z distance estimates can detect nonlinearities in the distance versus redshift relation.

For galaxies whose Cepheids are too distant to observe (i.e., beyond about 25 Mpc or outside of the Local Supercluster of galaxies), astronomers need other objects or distance indicators to set the distance scale. We mentioned a few of them in Section 5.4.

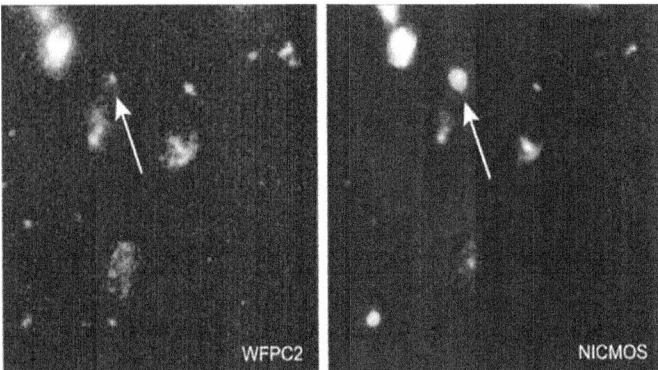

Figure 9.2. A very distant supernova at $z = 1.7$. The two pictures are taken before and after the explosion. The picture taken before shows the host galaxy. (Riess et al. 2001).

In the 1950s to 1980s much attention was paid to a special class of giant galaxies: the brightest elliptical galaxies in clusters. They have a narrow magnitude-redshift relation, which suggested that they form a good standard candle class. However, it became clear that they are standard candles in restricted redshift (or time) ranges only. Namely, the luminosity of giant ellipticals decreases with time when their stellar contents evolve. Up to now there is no sufficiently reliable theory of galaxy formation and evolution, and no definitive results have been obtained in this way (Sandage 1995, Yoshii and Takahara 1988).

The exploding stars called supernovae are much brighter than Cepheids and thus can be seen at greater distances. As an example, the supernova of 1885 in the Andromeda Galaxy was 1/10 of the brightness of the whole Galaxy. Supernovae are as bright as 10^9 Suns or more. They may outshine their host galaxy for a period of weeks (Figure 9.2). The usefulness of the supernovae is based on this great luminosity that allows observations from a large distance, and also on the discovery that a special class of supernovae can be considered a standard candle. Recently there has been a dramatic increase in the detection of distant supernovae, as well as in the ability to measure their apparent magnitudes and their behavior.

Supernovae (Sn) come in two main varieties: those whose optical spectra exhibit hydrogen lines are Type II, while hydrogen-deficient supernovae are Type I. Type I Sn are further subdivided according to the appearance of the early-time spectrum: Type Ia Sn are characterized by a strong absorption line of Si II; Type Ib lack this feature but instead show prominent He I lines; and Type Ic Sn have no Si II lines and weak or no He I lines.

Type SnIa supernovae are useful as **standard candles**. This class is distinguished first of all by the evolution of its brightness (its **light curve**), and secondly by the above special spectral features. Many supernova explosions involve only one very massive star that has come to the end of its evolution after exhausting all the nuclear fuels available. These supernovae may have different masses and different magnitudes of explosion. The SnIa type supernovae are different in that they originate from the explosion of a white dwarf of well defined mass.

A Type Ia supernova's proper luminosity (at its maximum) may be estimated from the shape of the supernova's light curve, and it is found to lie within rather narrow limits. In other words, after a suitable calibration illustrated in Figure 9.3, Type Ia supernovae all seem to have the same maximum luminosity. This allows the determination of the supernovae distances by comparing the flux of light to what the flux would be at the standard 10 parsecs (32.6 light years) distance. In addition to obtaining observations of the intensity, a spectrum of each supernova is obtained and the shifting of spectral lines used to estimate its redshift, z.

Recall that white dwarfs are the end products of evolution of lower mass stars that stay at the same radius and slowly cool. If a white dwarf star is below a critical mass value of $1.4\,M_\odot$, called the **Chandrasekhar limit**, it is stable against collapse. If for any reason more mass is dumped on it, e.g., from a nearby companion star (or in some explanations, the collision of two stars), then it becomes unstable and explodes. This explosion always takes place in a star of about the same mass (near its Chandrasekhar limit), and we have good reasons to expect that the explosions are quite similar. This is not quite true since the white dwarfs may have had somewhat different histories reflected in their chemical composition and that is expected to cause some scatter in their explosion characteristics (Höflich et al. 1998). The supernova explosion produces a large amount ($\sim 0.6\,M_\odot$) of heavy radioactive nuclei which afterward decay. After the maximum light the decay process

$$^{56}\mathrm{Ni} \to {}^{56}\mathrm{Co} \to {}^{56}\mathrm{Fe}$$

is responsible for most of the supernova light emission. Nickel as well as smaller quantities of silicon and calcium are produced in the supernova shock wave from carbon and oxygen nuclei. These fusion products as well as iron are observed in absorption in SnIa type supernovae spectra during peak brightness.

The key discovery that justified SnIa supernovae as standard candles was the observation by the Moscow University Russian astronomer Yurii P. Pskovskii (1977) that the maximum brightness is strongly related to the rate of decay subsequent to the maximum light. Usually the drop of brightness during the 15 days after the maximum is used as a quantitative measure. The supernovae which decay faster are fainter in absolute magnitude or luminosity (Figure 9.3).

Thus observations of SnIa supernovae at maximum brightness and during a subsequent period are useful as excellent standard candles. It is thus possible to do what Hubble did with Cepheids and plot redshifts versus distances for galaxies in

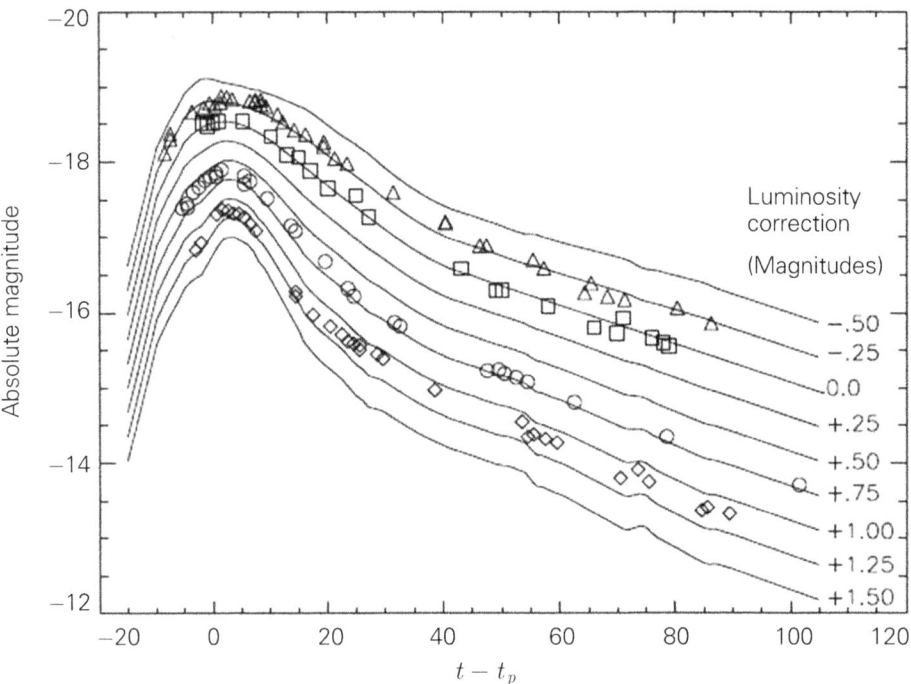

Figure 9.3. Empirically derived set of light curves which can be used to obtain the absolute brightness of a SnIa supernova at maximum brightness. Observations from four different supernovae are also shown with different symbols. (Riess et al. 1995).

which supernova brightness observations and redshifts are obtained. In an early conference devoted to the scientific possibilities of the future Hubble Space Telescope, Gustav Tammann (1979) proposed using type Ia supernovae occurring in very distant galaxies as a way to test whether Einstein's cosmological constant Λ is non-zero. In effect, this was to be the final goal of the Hubble–Sandage program of determining the correct cosmological model.

9.5   Observational discovery of dark energy

In 1990–1999 two groups discovered the cosmic vacuum (or dark energy) by studying remote supernova explosions (Riess et al. 1998a and 1998b, and Perlmutter et al. 1999). One group, the Supernova Cosmology Project, was led by Saul Perlmutter (Lawrence Berkeley National Laboratory in California). To discover supernovae, they used wide-angle cameras on large telescopes to examine many galaxies to "catch" supernovae as they exploded. The second group, the High-z Supernova Search Team, led by Robert Kirshner and his former student Brian Schmidt

(Harvard–Smithsonian Center for Astrophysics) used similar methods. To get quickly detailed brightness observations after discovery, both teams used the Hubble Space Telescope and the biggest ground based telescopes, such as the 10-meter Keck Telescope in Hawaii. The light curves permitted estimates of the apparent magnitudes, m, as well as the absolute magnitudes, M, from the rate of brightness decline as we discussed earlier. Finally, the redshifts, z, were obtained via spectra.

By 1997 the High-z Supernova Search Team had discovered 16 high redshift supernovae giving the first indication that the universe is accelerating, an amazing discovery. The supernovae appeared fainter and thus more distant than they should be in a gravitationally decelerating universe. An obvious explanation was that the universe must contain something like Einstein's antigravity lambda!

The High-z Supernova Search Team reported its results at the January 1998 meeting of American Astronomical Society. At the same meeting, the Supernova Cosmology Project also gave tentative evidence for the cosmic acceleration. Both results were immediately reviewed in *Science* magazine. Later in that year the High-z team with Adam Riess (another former student of Kirshner) of University of California at Berkeley as first author published the results in the *Astronomical Journal* (1998). Among the many parameters they were able to determine with more precision was the age of the universe, close to 14 billion years.

The work of the Supernova Cosmology project appeared in 1999 in the *Astrophysical Journal*. The Project's results were based on a larger independent set of 42 high redshift supernovae which confirmed the findings of the other group. This scientific discovery was made and verified "beyond reasonable doubt" within one year in this interesting scientific "horse race".

Even group members did not expect their results. Schmidt is quoted as saying,

"My own reaction is somewhere between amazement and horror; amazement because I just did not expect this result, and horror in knowing that it will likely be disbelieved by a majority of astronomers who, like myself, are extremely skeptical of the unexpected." (*Science* February 27, 1998)

Additional high redshift supernovae observations have confirmed the acceleration. As we will discuss later confirming evidence of an entirely different sort for the accelerating critical density universe has come from totally different observations, cosmic microwave background measurements by a satellite called the **Wilkinson Microwave Anisotropy Probe (WMAP)** in 2003, which was named after a pioneer in this field David Wilkinson (1935–2002) of Princeton University. This has led to acceptance of the "Concordance model" for the universe. It is dominated today by dark energy acceleration with minority deceleration by dark matter and ordinary (baryonic) matter like that which makes up our Earth. In 2011, Schmidt, Riess, and Perlmutter were awarded the Nobel Prize in Physics for these results. We next discuss how the supernovae indicate an accelerating universe.

9.6   Type Ia supernovae redshifts and distances vs. uniform expansion

In this section, we compare the supernova distances and redshifts with a simple (even though physically unrealistic) model called the Milne model. It is obtained from Equation (8.15) by putting $M = \rho_V = 0$, $k = -1$. This corresponds to a uniformly expanding universe with no gravitational deceleration or dark energy acceleration. This "in-between" model is useful for comparison with observations to see if they fall on the side of acceleration or deceleration. This "empty" model with no matter, radiation nor cosmic vacuum energy is simple to solve:

$$R(t) = ct. \tag{9.13}$$

For the Milne model, as one may verify from Mattig's equation (Box 6), the luminosity distance is

$$d_{LM} = \frac{c}{H_0} z \left(1 + \frac{1}{2} z \right). \tag{9.14}$$

We see that the above relation is not a straight line on the cz versus distance plots but instead curves with increasing redshift versus distance.

Figure 9.4 compares the uniform expansion Equation (9.14) luminosity distances and redshifts (shown as a line) with the luminosity distances computed from the supernovae distance moduli. The supernova data points **to the right of** the line are at greater distances than would be expected i.e., dimmer than expected even for a uniformly expanding universe. If the galaxies containing the supernovae were being decelerated, they would be **to the left of** the uniform expansion line. The supernovae are farther (dimmer) than they should be if they were in a decelerating universe. Thus the expansion of the universe of galaxies is actually being accelerated.

In Figure 9.4 we use distance modulus and redshift observations of "Gold" supernovae from Riess et al. (2004). Here parsecs are converted to light years. We see that these exceedingly luminous standard candles permit estimates of distances for distances far beyond those of Hubble's work. It is also clear that the plot at large distances is no longer a straight line; it curves. The solid curve represents a uniformly expanding "empty" Milne model universe from Equation (9.14). What is important is that observed points tend to lie to the right of the model curve as indicated by the arrow for one supernova.

A more direct comparison can be seen by comparing the Equation (9.14) $M - m$ distance modulus for the uniformly expanding universe to the observed $m - M$ for the supernovae plotted versus redshifts.

The difference between the distance moduli of the Milne model (d_{LM} from Equation (9.14)) and the observed supernova is defined as

$$\Delta(m - M) = 5 \log \frac{d_L}{d_{LM}} \tag{9.15}$$

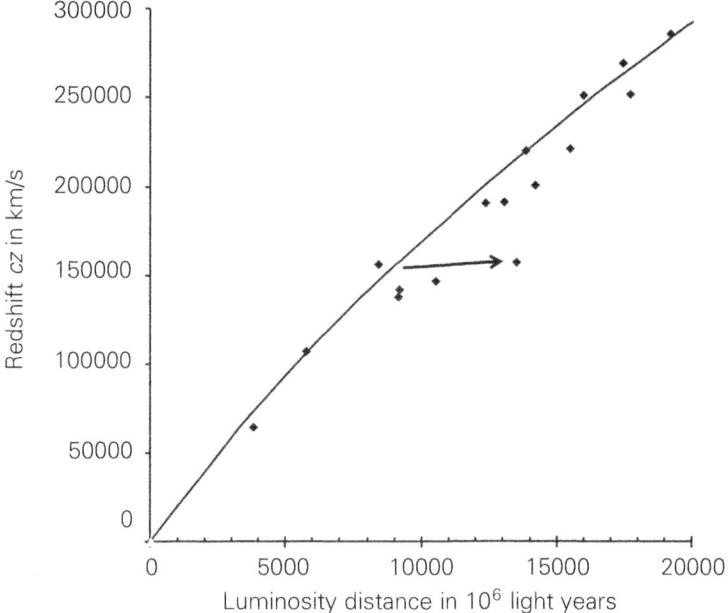

Figure 9.4. Plot of redshift times speed of light, cz, as a function of luminosity distance from Equation (9.12) for Type Ia supernovae. Data are "Gold" supernovae from Riess et al. (2004). Here pc are converted to light years. The solid curve is a uniformly expanding "empty" Milne model universe from Equation (9.14) with $H_0 = 71\,\mathrm{km/s/Mpc}$. The arrow shows how the estimated distance is larger than that expected at the observed redshift for a uniformly expanding universe due to dark energy acceleration. The much bigger sample of about 500 supernovae today gives the best fit value of the Hubble constant $73.8\,\mathrm{km/s/Mpc}$, with the uncertainty of $2.4\,\mathrm{km/s/Mpc}$.

which has been plotted in Figure 9.5 as dots with error bars relative to a horizontal line.

In Figure 9.5 positive magnitude differentials signify excess dimming of supernova light. i.e., acceleration, not deceleration. Looking at the plot, the increase of the magnitude differential above horizontal with z is a sign of an accelerated universe for redshifts up to about one. The points are above the horizontal line because the supernovae are dimmer i.e., more distant than expected due to dark energy acceleration. Note that all the binned groups of supernovae with $z < 1$ are above the horizontal line.

With this effect of acceleration, dark energy was first revealed. Its universal antigravity accelerates the cosmological expansion. The negative deceleration parameter was measured for the first time via observations of these supernovae.

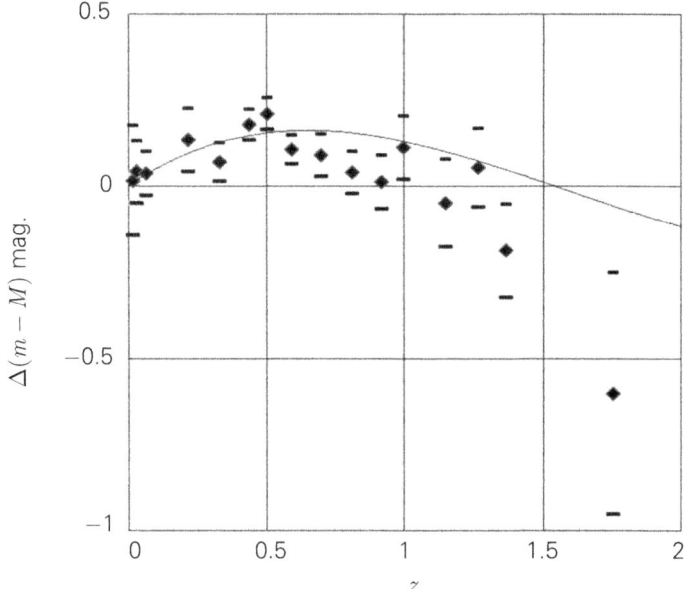

Figure 9.5. Binned supernova (type Ia) distance moduli at different redshift intervals, in comparison with those for a Milne "coasting model", i.e., a model with no deceleration or acceleration ($\Delta(m - M) = 0$ horizontal line). The expected behavior of the concordance model (solid line) is also shown together with observational points and their associated error bars. The figure is a plot adapted from Riess et al. (2007) for 206 supernovae with the most accurate distances. Also see http://www.astro.ucla.edu/~wright/sne_cosmology.html for data tables and later references.

The Concordance model because of its dark energy as well as matter produces the characteristic curved line seen in Figure 9.5 relative to the horizontal line. This curved line approximates the observations. To review the mathematics of the transition from acceleration today to deceleration in the distant past, recall the redshift velocity, cz, distance, l_0, relation prediction of the Friedmann models, Equation (6.22)

$$cz \cong H_0 l_0 + \frac{1}{2}(1 + q_0)\frac{H_0^2}{c} l_0^2. \tag{9.16}$$

The empty Milne model corresponds to $q_0 = 0$. The quantity q_0 specifies departure from a linear Hubble law (via acceleration or deceleration). As we have discussed, the supernovae distances and redshifts indicate that the current best value is $q_0 = -0.60$ which implies acceleration not deceleration of the expansion.

The relation between q_0 and the **density parameter**, Ω_M, is Equation (8.32) for a flat $k = 0$ universe

$$q_0 = \frac{3}{2}\Omega_M - 1. \tag{9.17}$$

$$\Omega_M = 0.27 \quad \text{for} \quad q_0 = -0.60. \tag{9.18}$$

For the Equation (8.34) flat $k = 0$ space condition, $\Omega_M + \Omega_V = 1$ so that $\Omega_V = 0.73$.

Finally, recall that from Equation (8.25), when $\rho = 2\rho_V$, the deceleration changes into acceleration. Let us calculate the redshift z at which this happens.

Let the universal scale factor be $R(t) = R$ at that time, in comparison with $R(t_0) = R_0$ at present. Going back to Equation (8.12),

$$\rho_0 = M/\left[(4/3)\pi R_0^3\right] \quad \text{and} \quad \rho = M\left[(4/3)\pi R^3\right]$$

where M is the same for both. Therefore, the mean density, ρ, as compared with the same quantity today ρ_0 is

$$\rho/\rho_0 = (R_0/R)^3. \tag{9.19}$$

Substituting Equation (8.28) ($\rho_0 = \Omega_M \rho_c$) gives us

$$\rho = (\Omega_M \rho_c)(R_0/R)^3. \tag{9.20}$$

Substituting Equation (9.20), along with Equation (8.30), $\rho_V = \Omega_V \rho_c$, into Equation (8.25), $\rho = 2\rho_V$ at the z of transition, we obtain

$$\Omega_M \rho_c (R_0/R)^3 = 2\Omega_V \rho_c.$$

We thus obtain

$$(R_0/R)^3 = 2\Omega_V/\Omega_M = 2(0.73/0.27) = (1.75)^3. \tag{9.21}$$

Now, from Equation (6.8), $R_0/R = 1 + z$ where z is the redshift of the observed transition. The transition from deceleration to acceleration thus happens at $z \approx 0.75$ which corresponds to the peak of the curve in Figure 9.5.

The fact that the group of high $z > 1$ supernovae in Figure 9.5 is below the horizontal line shows a change in the relative importance of dark energy versus attractive gravity about seven billion years ago in the history of the universe. The Concordance model which we have discussed earlier is a model which is dominated by gravitational deceleration early in the universe's history and later by dark energy acceleration. The expected behavior of the Concordance model (solid line) is shown in Figure 9.5 together with observational points and their associated error bars. We see that the solid line curves above the horizontal line but then curves downward so that it is below the line at redshifts greater than unity in agreement with the last binned supernova values. On a cautionary note, note the large error bars and disagreement with the Concordance model for the highest redshift points.

9.7   Could it be some problem with the standard candle method?

Even though the Λ term or dark energy is now standard in cosmological equations, it is natural that alternatives have been considered to explain why the distant supernovae appear fainter than predicted by models with $\Lambda = 0$.

Intergalactic extinction? Skeptics wondered whether it was possible that the supernovae appeared fainter than expected solely because of obscuring dust lying between us and the stars and not because of any dark energy acceleration. This dust could be genuinely intergalactic or perhaps located in the massive halos around galaxies which happen to lie close to the line of sight from the supernova to us (it is known that the halos contain small amounts of reddening dust, Teerikorpi 2002). Recapitulating, Figure 9.5 shows the supernova observations are above the horizontal line for redshifts zero up to about one. Beyond redshifts of about one, some galaxies' supernovae appear brighter (lower) than the horizontal line model. Contrary to this behavior around $z = 1$, if dust obscuration is an effective systematic source of error, the dimming should be present at high as well as low redshifts. The entire set of data in Figure 9.5 should be above the horizontal line. High redshift supernovae are seen **below the horizontal line** i.e., they are brighter than expected in a non-accelerating universe. Thus systematic effects of dust are not likely to create the excess dimming over the model seen at redshifts less than one.

Furthermore, the dust explanation can be observationally constrained since the dimming of light should be accompanied by reddening, unless the dust is unusual ("grey": extinction independent of wavelength) and suitably distributed at different redshifts (Aguirre 1999). In its original form, where the density of the dusty medium follows that of non-relativistic matter $(\sim [1 + z]^3)$, this model is contradicted by SNIa data beyond $z \approx 1.0$ where the supernovae are too bright (Riess et al. 2004), as we already noted above. In the grey dust model of Robaina and Cepa (2007) the density follows the stellar formation rate density evolution (with some delay due to the injection of the dust into intergalactic space). Such a "replenishing grey dust" model could be flexible enough for explaining the run of the Hubble relation. Independent evidence would be needed for the hypothetical intergalactic grey dust.

Selection effects? The peak luminosities of supernovae Ia have some scatter even after the correction for the dimming rate. This causes, together with the observational limiting magnitude, a selection effect (a sort of Malmquist bias): the required detection of a supernova when it is still brightening favors brighter than average objects and increasingly so at higher redshifts. Could this selection explain the "too" faint magnitudes of high-z supernovae? Actually not, because the correction for this effect would mean making the observed magnitudes still fainter by some amount. Minor effects due to this selection problem and another effect, gravitational lensing, have been discussed, e.g., by Kowalski et al. (2008) and Sarkar et al. (2008).

Problems with the candles? Could the average properties of Ia supernovae be changing over cosmological times? After all, their use is a purely empirical method to determine the luminosity distance. Hence, along with collecting SNe Ia events, it is important to continue studying their physical nature as thermonuclear explosions of carbon-oxygen white dwarfs which have in some way gained mass from their surroundings. Theoretically, these explosions are not yet fully understood.

In fact, the Ia supernovae do not seem to form a uniform class (even when the peak luminosity vs. decay rate dependence discussed above is taken into account). One speaks about two components in their population: the "prompt" one proportional to the instantaneous host galaxy star formation rate (in S galaxies) and the "delayed" component, delayed by several Gyr and proportional to the total stellar mass (in E galaxies). Thus some of the photometric properties which are used to derive their luminosity distance depend on the host galaxy type (Hamuy et al. 1995). The more luminous prompt population with broader light curves and younger progenitor stellar population dominate at high redshifts where the star formation rate was higher. However, the resulting redshift-dependent shift in the Hubble diagram is small (~ 0.05 mag), and even if not fully controlled (Howell et al. 2007), cannot change the general run of the Hubble relation.

Another sign of non-uniformity among the Ia supernovae came from the different expansion speeds of the ejected material (Branch 1987 and Benetti et al. 2005). Maeda et al. (2010) suggested that this diversity is not intrinsic, but caused by asymmetric explosions viewed from random directions. The asymmetry would result from the ignition occurring at an offset from the centre of the white dwarf progenitor. This result also suggests that the expansion speed diversity does not undermine the use of Ia supernovae as standards.

The above explanations were based on possible problems in the standard candles. Quite another kind of approach admits that the Ia supernovae are good distance indicators, but says that it is problematic to take the magnitude (luminosity distance) vs. redshift relation at its face value and to interpret it in terms of the ideal homogeneous world model. The "acceleration" might be just a consequence of the assumed homogeneity of the actually inhomogeneous universe (e.g., Mattsson 2010; Célérier et al. 2010). Another important approach considers modified gravity theories (see below). Some suggestions analyze the propagation of light and/or the origin of the cosmological redshift in a non-standard manner (e.g., Annila 2011). As usual in science, alternative attempts are worthy of study. However, in this book we concentrate on the mainstream idea behind the Concordance model. We note that local studies of dark energy (Chapter 15) can in principle exclude many models based on no dark energy.

Confirmation from other objects? Naturally, one would like to have other types of standard objects observable from low to high redshifts and based on other physical mechanisms. We saw above that double radio sources can be used to construct the angular diameter—redshift diagram; however, they are quite problematic as standard rods. Astronomers have also attempted to use more compact radio sources (with a typical size of about 50 pc, in comparison with the 500 kpc wide double radio sources). In fact, already before the supernova breakthrough, Jackson and Dogdson (1997) used 256 ultracompact radio sources in the redshift range 0.5–3.8 to construct the redshift-angular size diagram (which was first done by Kellermann (1993) using 79 sources). Their analysis excluded the Einstein–de Sitter model and it was stated that the best-fit flat model has $\Omega_V = 0.8$ (more recent results in Jackson and Jannetta 2006).

Also, it is interesting to mention that the suggested class of the optically most luminous radio quasars (Teerikorpi 2000) appeared best when the magnitudes were calculated using the flat Friedmann model with $\Omega_V \approx 0.7$, as analyzed using the so-called cosmological Malmquist bias approach (Teerikorpi 2003). Typically, the possible members of this class are optically rather non-active (with small variability) and radiate not far from their Eddington limit (Teerikorpi 2011).

It has been proposed that gamma ray bursts could serve as standard candles. Schaefer (2006) has published a sample of γ-ray burst distances. GRBs distances are probably much less accurate than supernovae but they are observable to much higher redshifts. The high redshift GRB data could help to distinguish between non-flat geometries and dark matter equations of state (which we will discuss later).

9.8   Modified gravity theories

Could it be that there is really no dark energy, but its appearance is an illusion created by using a theory of gravitation (General Relativity) which is not quite accurate but only an approximation to the correct theory? The apparent weakening of gravity in large scales which is brought about by the negative of 2Λ in $R - 2\Lambda$ of the Lagrangian in Equation (8.11), could be an indication of a need for a bigger change in the theory of gravity. If this is true, then one could talk about "dark gravity" instead of dark energy as the source of the accelerating universe.

Einstein field equations for empty space were derived from the Lagrangian in Equation (8.11):

$$L = \frac{c^4}{16\pi G}\,(R - 2\Lambda). \tag{9.22}$$

Could we use something more complicated in place of the curvature scalar R that would allow us to put $\Lambda = 0$, and still have universal acceleration? One of the ideas which has received lots of attention in recent years is the so called $f(R)$ theories. There $R - 2\Lambda$ is replaced by some function of R in the Lagrangian, i.e.,

$$L = \frac{c^4}{16\pi G} f(R).$$ (9.23)

This would mean that gravity deviates from General Relativity from place to place, depending on the value of R at that location. And also since the overall value of R decreases due to expansion of the universe, gravity becomes also a function of time. In this way it is possible to create very complex theories of gravity. The primary motivation has been to be able to put $\Lambda = 0$ without giving up the cosmic accelerated expansion.

There are obvious restrictions to the possible forms of the function $f(R)$. If R is large, as near black holes, we have strong evidence that General Relativity is correct to high level of accuracy (in particular, in the binary black hole system OJ287, see Section 3.2). Even in less extreme systems such as the binary pulsars and in the Solar System, General Relativity works well. Thus the function $f(R)$ must be such that:

1) $f(R) \rightarrow R$ when R is large. In this case $R \gg \Lambda$, and Λ can be neglected in any case.

2) $f(R) \rightarrow R - 2\Lambda$ where R is small, as in the gentle curvature of the present universe R.

An example of a function that fits the description is

$$f(R) = R - \mu R_c \left[1 - \left(1 + R^2/R_c^2\right)^{-n}\right]$$ (9.24)

where μ, n and R_c are positive constants, and n is close to 1. Condition 1) is then satisfied if $R \gg R_c$ and 2) if $R \approx R_c$ and $\mu R_c \approx 4\Lambda$.

A small difference from the metric of General Relativity can be described as a scalar field

$$\varphi \equiv \frac{\partial f}{\partial R} - 1.$$ (9.25)

It represents a gravitational wave, carried by a massive particle. This is in addition to the ordinary gravitational waves which are carried by massless particles, gravitons.

These theories may come handy one day if the parameter (Λ) driving cosmic acceleration is found to depend on cosmic time. At the present there is no such evidence, and replacing one natural constant Λ by several new constants is hardly justified.

Another modification of the standard theory is obtained by considering gravitation as a higher-dimensional phenomenon, e.g., in 5 dimensional space-time. These are referred to as Braneworld theories. The idea is that there exists more space dimensions than the usual three, and that gravity, unlike other interactions, operates in these higher dimensions. The observed phenomena are then projections to our 4 dimensional space-time. In the action (Equation (7.1)) we need to replace d^4x by d^5x and replace the 4-dimensional Ricci tensor by a 5-dimensional Ricci tensor (together with some other more subtle changes). The main advantage of this formulation is that it connects with string theories which are popular avenues in the search for underlying theory of particle physics. However, so far no other advantages over the standard theory have been uncovered.

Chapter 10

Relics: cosmic microwave background (CMB) photons and neutrinos

Just as for the expanding universe, the existence of the cosmic microwave background radiation (CMB) was predicted prior to discovery! Also, as in other areas of astronomy involving radio waves, engineers led the way in the observational discovery with progress delayed via a communication gap between the engineers and astronomers. The existence of relic electromagnetic radiation was predicted theoretically by Russian-American physicist George A. Gamow (1946). He put forward the idea of a very hot initial stage in which photons existed in thermodynamic equilibrium with matter. These photons survived to the present epoch, and their temperature today was predicted within the range $1\,\mathrm{K} < T_0 < 10\,\mathrm{K}$ by the American physicist Ralph A. Alpher and Robert C. Herman together with Gamow (1948–1953).

10.1 The prediction and discovery of the CMB

Gamow and his co-workers predicted a phenomenon which would be a consequence of the early state of expansion and thus proof of its existence. Similar to interstellar material today in our Galaxy, matter was mostly positive hydrogen nuclei and about 25% by mass positive helium nuclei plus detached negative electrons. These nuclei formed from subatomic particles earlier in the expansion. As noted earlier, the massive atoms familiar to us on Earth were formed in subsequent generations of stars. In addition, there was electromagnetic radiation (a "gas" of photons), which filled space uniformly and expanded together with the matter. Finally, unknown to Gamov at that time, there was the non-luminous but otherwise non-interactive **dark matter** which makes up the great majority of gravitating matter compared to ordinary matter (**baryonic matter**). As we have discussed earlier, dark energy has only become dominant in the universe about 7 billion years ago. In these early times it was quite unimportant.

Radiation in thermodynamic equilibrium with a few K temperature has a maximum in the microwave band. In the late 1950s, Gamow discussed the possibility of observing a microwave background radiation with radio astronomers, but their reaction was unenthusiastic. Non-astronomers first detected the radiation. In 1957, Tigran A. Shmaonov of Pulkovo Observatory, St. Petersburg, published (in a Moscow technical journal *Experimental Devices and Methods*) background observations with a horn antenna. An isotropic signal was found at the wavelength of

3 cm, and its intensity corresponded with the temperature of a few K. The measured temperature was 4 ± 3 K. We now know that this was the first observation of the cosmic microwave background radiation. But this result attracted no attention from cosmologists. In 1961, the American radio engineer E.A. Ohm reported (in the *Bell System Technical Journal*) an excess noise equivalent to a radiation temperature of about 3 K in a similar 20-foot horn antenna, somewhat more sensitive than the Pulkovo horn. This result was noticed by Soviet cosmologist Yakov B. Zel'dovich and his colleagues in Moscow, but it was misinterpreted as evidence *against* the existence of the radiation background!

Finally, engineers made observations that got astronomers' attention. Bell Laboratories scientists Arno A. Penzias and Robert W. Wilson (1965) reported an unavoidable "noise" at a wavelength of 7.35 cm in the their horn instrument. The equivalent noise temperature was the same, about 3 K. They had heard nothing about Gamow's theory, and their efforts were directed toward satellite communication; but their results were immediately recognized by Princeton cosmologists Robert H. Dicke and P. James E. Peebles as the cosmological isotropic signal of the relic radiation background. With the Nobel Prize being awarded to Penzias and Wilson, the odyssey of the 3 K cosmic microwave background (CMB) prediction and discovery came to a happy end in 1978.

10.2   The Big Bang components

Before we discuss how the CMB is so useful in establishing the relative importance of dark energy and gravitating matter in the dynamics of the universe, we first summarize the parameters of the Big Bang theory. We will refer to the WMAP (Wilkinson Microwave Anisotropy Probe) parameters frequently in this chapter. Later, we will explain how they are estimated. This is a set of cosmological parameters derived from the temperature variations in the cosmic microwave background, together with cosmological supernovae, galaxy clustering in large scale, x-ray observations of galaxy clusters and other modern cosmological observations. These constitute the best values for the **concordance dataset** (Ostriker and Steinhardt 1995, Spergel et al. 2003), the empirical basis of the current standard cosmological model. Following the concordance (Ostriker and Steinhardt, 1995) approach, we give below what are regarded as the most reliable figures for today's densities of the four major cosmic energies for the vacuum (V), dark matter (D), baryons (B) and radiation (R) from the WMAP site seventh year data release (http://lambda.gsfc.nasa.gov/product/map/current/best_params.cfm).

The current age of the universe measured in proper time, according to the most precise set of parameters for the universe is 13.75 ± 0.11 Gyr. The rate of cosmological expansion is given by the **Hubble parameter**, $H = \dot{R}/R$, where $R \equiv R(t)$ is the cosmological scale factor. According to the WMAP measurements in combination with the Hubble Space Telescope observations and other studies, the present-day value for the Hubble parameter, the **Hubble constant** is

$$H_0 = 70.4 \pm 1.3 \, \text{km/s/Mpc}. \tag{10.1}$$

Here we should say that "constant" means the slope of the local linear recession versus distance relation. This slope can vary with time in many cosmological models. Here we also use the reduced Hubble constant, h, where $H_0/100 \, \text{km/s/Mpc} = h$ whose value today is taken to be 0.704 ± 0.013. With the cosmic age and the Hubble constant, the third major cosmological parameter characterizing the present-day universe which we have discussed is the dimensionless density parameter

$$\Omega \equiv \rho/\rho_c,$$

where ρ is the density of all the cosmic energies and $\rho_c \equiv \frac{3}{4\pi G} H_0^2$ is the critical density. The value of the present-day standard density parameter (summing all the cosmic energies at the present day) is

$$\Omega_0 = 1.0023 \pm 0.0055. \tag{10.2}$$

with mean plus and minus errors. As discussed earlier, according to the Friedmann model, if the co-moving 3D space is flat, the parameter is equal to 1. In this case, the universe's space-time can be described by the simplest form of the metric interval:

$$ds^2 = c^2 dt^2 - R(t)\left(dx^2 + dy^2 + dz^2\right), \tag{10.3}$$

where synchronous proper time (here represented by t) and Cartesian coordinates are used.

The only difference of the Friedmann model metric in Equation (10.3) from the Minkowski space-time of Special Relativity is in the time dependent scale factor $R(t)$ which describes the universe's expansion. This is the scale factor that makes the 4D space-time non-trivial with non-zero 4D curvature. The WMAP value for Ω_0 indicates that the universe's 3D space is exactly flat or at least very nearly flat, at the present-day. Its normalized cosmic densities are

$$\Omega_V = 0.728 \pm 0.0155 \tag{10.4}$$

$$\Omega_D = 0.227 \pm 0.014 \tag{10.5}$$

$$\Omega_B h^2 = 0.02260 \pm 0.00053 \tag{10.6}$$

$$\Omega_B = 0.0456 \pm 0.0016.$$

The radiation is a comparatively minor but uncertain constituent today but, as we shall see, it dominated the universe in the past.

$$\Omega_R h^2 = 2.5\alpha \times 10^{-5}, \quad 1 < \alpha < 10. \tag{10.7}$$

$$\Omega_R = 5.1\alpha \times 10^{-5}.$$

The factor α defined in the formula for Ω_R accounts for non-CMB contributions (neutrinos, gravitons *etc.*) to the density of cosmic relativistic matter; these remain poorly defined. The present-day energy densities are given here in the dimensionless form, i.e., in units of the present-day critical density:

$$\rho_c(t_0) = \frac{3}{8\pi G} H_0^2 = 0.95 \times 10^{-29} \text{g/cm}^3. \tag{10.8}$$

We can see from Equations (10.6) and (10.7) multiplied by Equation (10.8) that the ordinary dimensional densities of baryonic matter (g/cm^3), $\rho_B = \Omega_B \rho_c$, and radiation $\rho_R = \Omega_R \rho_c$ do not depend on the adopted value of the Hubble constant, H_0, while the other two dimensional densities (vacuum, dark matter) do depend on H_0. Later in this book we will describe a local method to obtain Ω_V independent of H_0 observationally. Globally, the sum of the four energy densities is certainly very near the critical density.

If $\Omega_0 = 1$ exactly, it would mean that the co-moving 3D space is exactly flat and the flat Equation (10.3) metric is valid for the past and future. In this case, a question arises: Why is space flat? This 'flatness problem' was first formulated by physicist Robert H. Dicke (1970, Princeton University): To have Ω_0 even approximately 1 today requires an accurate fine tuning of Ω in the past (we shall return to this question later in the book in connection with other "cosmic coincidences"). Even if Ω_0 is not exactly equal to one today, the metric of Equation (10.3) is a good approximation not only for the current era, but also for the past and future of the universe. All the bulk of observational data indicates that the co-moving 3D space of the universe is remarkably uniform, isotropic, and nearly or exactly flat.

In contrast to the geometry of the universe, the energy content today (diagrammed in Figure 10.1) seems to be rather complex. According to the current data given above, the major cosmic components today are dark: cosmic vacuum (which is also called dark energy) and dark matter comprise together (72% + 23%) more than 95% of the total cosmic energy. "Ordinary" matter of stars in galaxies and intergalactic gas contributes less than 5%, and its physical origin is almost as unclear as the nature of the dark sector. It is embarrassing that only the cosmic microwave background radiation which contributes 0.005% to cosmic energy is well understood in its nature and origin. This multiplicity of components and the uncertainty about their origin is an unsatisfying aspect of standard cosmology, making a deep puzzle which we will discuss later in this text.

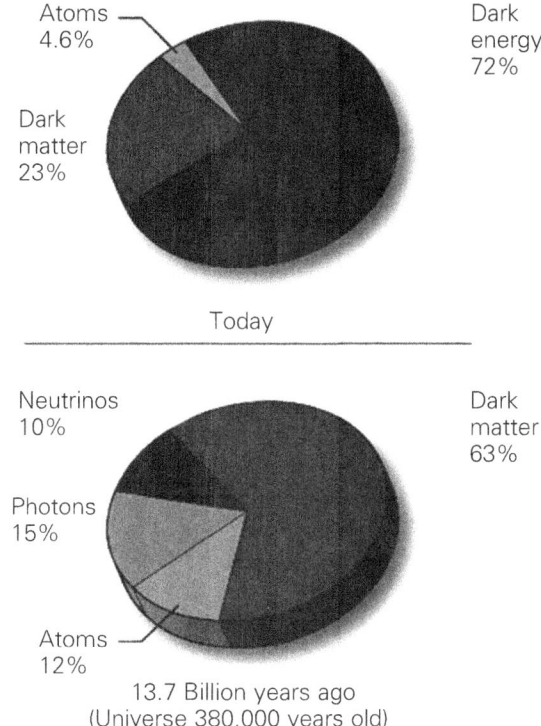

Figure 10.1. Energy components of the universe today (top). Radiation photons are also present but, at a level of ~0.01%, are too small (along with neutrinos) to show up in the chart. The bottom shows energy components at the time of the CMB emission. At that time dark energy was quite unimportant while radiation photons and neutrinos were important. See the following sections for explanation of the differences. Courtesy NASA WMAP site: http://wmap.gsfc.nasa.gov/media/080998/index.html.

10.3   The early radiation-dominated universe

The CMB today is composed of relic photons, which were in thermodynamic equilibrium with a hot cosmic plasma at early stages of the cosmological expansion. We will now study these early stages of the universe. We restrict ourselves to the evolution somewhere between the first second and the first 10^5 years. The important difference between this stage and the later stages is the dominance of radiation and radiation pressure in the early universe. Today the radiation plays a minor role as seen in Figure 10.1. Also, in this early stage, gravitation retards the expansion while dark energy is quite unimportant.

At this early stage of the universe, the pressure term on the right hand side of Equation (8.45) is significantly different from zero and the derivation of Equation (8.53) is changed. We get

$$\frac{d}{dt}\left(Mc^2\right) = -PdV.$$ (10.9)

Here we take V as the volume of the sphere of radius $R \equiv R(t)$:

$$V = 4/3 \pi R^3.$$ (10.10)

There is an alternative form for Equation (10.9) which uses the thermodynamic identity for a sample of material in the early universe,

$$|dE| = TdS - PdV.$$ (10.11)

Here $E = \rho c^2 V$ is the total energy (including the rest mass energy) in the volume V. Here T and S are the temperature and entropy of this volume, respectively. For an adiabatic expansion (which is the case here) $dS = 0$, and Equation (10.11) can be reduced to Equation (10.12) by differentiation as we will now show.

The entropy in the co-moving volume will be conserved since the expansion is adiabatic. Rearranging with TdS alone on the left side which is zero, substituting for dV and taking the time derivative:

$$0 = d\left(\rho c^2 V\right) + PdV = \rho c^2 4\pi R^2 dR + d\rho c^2 (4/3)\pi R^3 + P4\pi R^2 dR$$

$$0 = (d\rho/dt)c^2(4/3)\pi R^3 + \rho c^2 4\pi R^2 dR/dt + P4\pi R^2 dR/dt.$$

Dividing by 4π and $3c^2$, we get

$$\frac{d\rho}{dt} + 3\frac{\dot{R}}{R}\left(\rho + \frac{P}{c^2}\right) = 0$$ (10.12)

Equation (10.12) is the **Friedmann thermodynamic** (or second) **equation**. This equation also follows from the General Relativity equations, if the energy-momentum tensor describes a perfect fluid.

For non-relativistic matter, the pressure is much less than the energy density, so that $P = 0$, due to matter in the equations above. On the other hand, the radiation pressure is

$$P_R = \frac{1}{3}\rho_R c^2$$ (10.13)

where the radiation density ρ_R is related to the radiation energy density u_R by

$$u_R = \rho_R c^2.$$

Substituting P_R for ρ_R for P and ρ in Equation (10.12):

$$\frac{d}{dt}\rho_R + \frac{3\dot{R}}{R}\frac{4}{3}\rho_R = 0, \tag{10.14}$$

or

$$\frac{\dot{\rho}_R}{\rho_R} = -4\frac{\dot{R}}{R}. \tag{10.15}$$

The solution is simple

$$\rho_R \propto R^{-4}. \tag{10.16}$$

More directly, we could infer this result from the conservation of the number of photons in this co-moving volume. The frequency of a photon ν in this volume, and consequently also its energy $h_P\nu$ ($h_P = 6.626 \times 10^{-27}$ erg sec, Planck's constant) is inversely proportional to R via the cosmological redshift. Taking the product of these two, the radiation energy density should scale as $R^{-3} \times R^{-1} = R^{-4}$.

Thus the radition mass-energy density varies as the inverse fourth power of the increasing scale length while the dark and ordinary baryonic matter density decreases according to the inverse third power of the increasing scale length (as time increases). Correspondingly, going in the opposite direction of time into the past, radiation mass-energy density increases according to the inverse fourth power of the decreasing scale length. The (dark and ordinary baryonic) matter density increases according to the inverse third power of the decreasing scale length. As we have discussed earlier, looking at greater distances, or greater fractional wavelength z, the scale length is decreased by the inverse of $1 + z$. So in in the bottom frame of Figure 10.1 at the time of the CMB emission where the scale length is decreased by $1 + z \approx 1\,100$, radiation's proportion relative to matter is greatly enhanced from its insignificant portion today.

We will now go back to the Friedmann equation, Equation (8.14);

$$\dot{R}(t)^2 = \frac{2GM}{R(t)} + \frac{8\pi G}{3}\rho_V R(t)^2 - kc^2$$

and apply it to the small scale length early universe. Because $R(t)$ is small, the first term on the right hand side dominates. In practice we may put $k = \rho_V = 0$. In other words, dark energy is unimportant at this early time. So in the bottom frame of Figure 10.1, we see that the dark energy content relative to matter at the time t of the CMB emission ($R(t_0)/R(t) \approx 1\,100$) is insignificant compared to today.

Since radiation dominates the universe at this early time Equation (10.16) is valid. Thus the mass conservation relation for dust matter $\frac{4}{3}\pi R^3\rho = M$ must to be replaced by

$$\frac{4}{3}\pi R^4\rho_R = \text{constant} = E/c^2. \tag{10.17}$$

Thus the Friedmann equation becomes

$$\dot{R}^2 = \frac{2GE/c^2}{R^2} \tag{10.18}$$

This is easily solved:

$$t = \int_0^R \frac{cRdR}{\sqrt{2GE}} = \frac{cR^2}{2\sqrt{2GE}} \tag{10.19}$$

or

$$R = 2^{3/4} \left[\frac{(GE)^{1/4}}{c^{1/2}} \right] t^{1/2}. \tag{10.20}$$

Finally,

$$\rho_R = \frac{3}{32} \left(\frac{1}{\pi G t^2} \right) \tag{10.21}$$

which gives the radiation density as a function of time in the early universe.

10.4   Properties of cosmic microwave background radiation

Having discussed the discovery of the CMB, the values for the components of the universe and the radiation domination of the early universe, we now move on to the CMB's origin and spectral properties along with another relic of the early universe, neutrinos. The CMB is a collection of electromagnetic waves, or photons, which were in thermodynamic equilibrium with hot cosmic plasma during the early stages of the cosmological expansion. At these stages, plasma and radiation interacted with each other very effectively as will be discussed in Section 10.5. However, when the temperature of the medium fell down to $T \cong 3\,000$ K due to expansion cooling, recombination of electrons with protons occurred. Afterward, because of this, the interaction between radiation and atomic gas greatly diminished, and matter and radiation decoupled from each other. Since then, this relic photon gas has continued to expand retaining its homogeneous, isotropic, and thermodynamically equilibrium distribution.

The present day CMB temperature $T_0 = T(t_0)$ can be measured to a rather high accuracy:

$$T(t_0) = 2.725 \pm 0.001 \text{ K}. \tag{10.22}$$

Measurements of the CMB over various frequencies indicate that it is thermodynamic equilibrium blackbody radiation with the Planck spectral distribution:

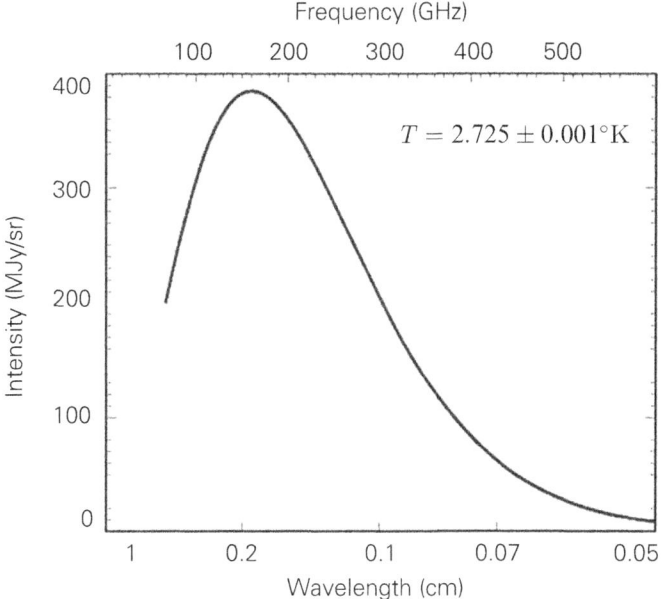

Figure 10.2. WMAP radiation intensity versus frequency (in a linear scale) and corresponding wavelengths. Here, the unit is the millijansky (symbol mJy), which is equivalent to 10^{-26} watts per square metre per hertz. The steradian (sr) is a measure of solid angle on the sky. For an area A on a sphere of radius r, the solid angle is $\theta = A/r^2$. The entire sky thus has a solid angle of 4π steradians. Courtesy NASA (http://wmap.gsfc.nasa.gov/media/ContentMedia/ 990015b.jpg).

$$n_R(\omega) = \frac{1}{\pi^2 c^3} \frac{\omega^2}{e^{\hbar\omega/k_B T} - 1}. \tag{10.23}$$

In Equation (10.23), $\hbar = h_P/2\pi = 1.055 \times 10^{-27}$ erg sec is the reduced Planck constant and $n_R(\omega)$ is the number of blackbody photons per unit volume and per unit frequency band (see Figure 10.2). Here the angular frequency $\omega = 2\pi\nu$.

The density of the radiation per unit frequency is the number density times the energy of one photon $\hbar\omega$:

$$\varepsilon_R(\omega) = \frac{1}{\pi^2 c^3} \frac{\hbar\omega^3}{e^{\hbar\omega/k_B T} - 1}. \tag{10.24}$$

The frequency of the spectral distribution maximum of the blackbody radiation energy is

$$\omega_{max} = 2.82 \, k_B T/\hbar. \tag{10.25}$$

As a function of wavelength, $\lambda = 2\pi c/\omega$, the Planck energy spectrum is:

$$\varepsilon_R(\lambda) = \frac{8\pi h_P c}{\lambda^5} \frac{1}{e^{h_P c/\lambda k_B T} - 1} \tag{10.26}$$

The energy maximum is at the wavelength (which is not equal to $c2\pi/\omega_{max}$!)

$$\lambda_{\max} = 1.26 \, \hbar c/k_B T. \tag{10.27}$$

For the present-day CMB temperature, the frequency and wavelength maxima are:

$$\omega_{max}(t_0) = 1 \times 10^{12} \sec^{-1},$$
$$\lambda_{max}(t_0) = 0.2 \, \mathrm{cm}. \tag{10.28}$$

The total energy density can be obtained by integrating the energy spectrum over all frequencies to give

$$\varepsilon_R = (4\sigma_{SB}/c) \, T^4, \tag{10.29}$$

where $\sigma_{SB} = 2\pi^5 k_B^4/(15 \, h_P^3 \, c^2) = 5.67 \times 10^{-5} \, \mathrm{erg/cm^2/K^4/sec}$, the Stefan-Boltzmann constant. Calculated with the above formula for the temperature $T = T_0$ given above, the present-day CMB energy density is

$$\varepsilon_R(t_0) = 5.2 \times 10^{-13} \, \mathrm{erg/cm^3}. \tag{10.30}$$

the present-day radiation mass density corresponding to this energy density is

$$\rho_R(t_0) = \varepsilon_R(t_0)/c^2 = 4.7 \times 10^{-34} \, \mathrm{g/cm^3}. \tag{10.31}$$

This photon energy mass density is quite insignificant compared to other components and also to the critical density ($\sim 10^{-29} \, \mathrm{g/cm^3}$). The number density of photons is equal to the integral of the spectrum $n(\omega)$ over all frequencies:

$$n_R = 0.24(k_B T_0/\hbar c)^3. \tag{10.32}$$

For the observed CMB temperature today,

$$n_R(t_0) = 0.24(k_B T_0/\hbar c)^3 = 410 \, \mathrm{cm}^{-3}. \tag{10.33}$$

The energy of the most abundant photons of the blackbody radiation is $\hbar\omega_{max} \cong k_B T \times 2.82$, see Equation (10.25). Using Equation (10.32), the radiation mass density is

$$\rho_R \cong 2.82 \, \frac{k_B T}{c^2} \, n_R \cong 0.7 \, \frac{k_B T}{c^2} \left(\frac{k_B T}{\hbar c}\right)^3. \tag{10.34}$$

For a given present day unit volume, the entropy (predominantly in the CMB) is

$$S(t_0) = \frac{4}{3}(4\sigma_{SB}/c)\,T_0^3 = 2.2 \times 10^{-13} \text{ erg/cm}^3/\text{K}. \tag{10.35}$$

with σ_{SB} given earlier. See Leff (2002) for a derivation from first principles of Equation (10.35) plus a general discussion of the properties of a photon gas. The entropy in the co-moving volume is conserved under the adiabatic expansion:

$$S(t)R(t)^3 \propto T(t)^3 R(t)^3 = \text{Const}. \tag{10.36}$$

Thus the radiation temperature scales with time as

$$T(t) \propto R(t)^{-1}, \tag{10.37}$$

so

$$T(t) = T_0 R(t_0)/R(t) = T_0(1 + z), \tag{10.38}$$

where z is the cosmic redshift.

With this relation and those above, the density of radiation and number density of photons decrease with the cosmic expansion can be found as follows:

$$\rho_R(t) = \rho_R(t_0)(1 + z)^4. \tag{10.39}$$

$$n_R(t) = n_R(t_0)(1 + z)^3. \tag{10.40}$$

Conservation of entropy in the co-moving volume leads to conservation of the total number of photons within that same volume. Now the number of baryons, $\propto n_B R(t)^3$, is also conserved in the co-moving volume. Thus there is a constant ratio of the number density of photons to the number density of baryons:

$$B \equiv \frac{n_R(t)}{n_B(t)} = \frac{n_R(t_0)m_H}{\rho_B(t_0)} = 1.5 \times 10^9, \tag{10.41}$$

where m_H is the hydrogen atom mass. The origin of this **big baryonic number** B remains unknown—mostly because the nature of baryons is uncertain (see below).

Recall that we found earlier that non-relativistic matter (dark matter and baryons) density scales as $\rho_M \propto R(t)^{-3} \propto (1 + z)^3$. Thus the ratio of radiation to non-relativistic matter densities is

$$\frac{\rho_R(t)}{\rho_M(t)} = \frac{\rho_R(t_0)}{\rho_M(t_0)}(1 + z), \tag{10.42}$$

where $\rho_M = \rho_D + \rho_B$.

From Equation (10.42), at very large z, the radiation density dominates in the early stages of the cosmic expansion. The two densities are equal at at a redshift which we denote by z_*.

By Equation (10.42):

$$1 + z_* = \rho_M(t_0)/\rho_R(t_0). \tag{10.43}$$

Using the current data on densities in Equations (10.5)–(10.7),

$$1 + z_* = \rho_M(t_0)/\rho_R(t_0) = 5\,500\,\alpha^{-1} \tag{10.44}$$

where $\rho_M(t_0) = 2.6 \times 10^{-30}$ g/cm^3, and non-CMB contributions to the radiation density are accounted for by setting $\alpha = 3/2 > 1$.

Using $\rho_R(t_*) = 4.7 \times 10^{-34}$ g/cm$^3 \times (1 + z_*)^4$ (Equation (10.31)), $1 + z_* = 5\,500\,\alpha^{-1}$ (Equation (10.44)) and $\rho(t_*) = (3/32)/[\pi G t_*^2$ (Equation (10.21)), we can make a rough calculation of the time at which the energies become equal. This happened at $t_* = 76\,000$ years $(= 2.4 \times 10^{12}$ sec). This calculation is only approximate. A more thorough calculation (e.g. Chernin 1965) gives a slightly lower value $t_* \sim 70\,000$ yr, i.e. $t_* \sim 2 \times 10^{12}$ s (Spergel et al. 2003). We should emphasize that the z_* is the equal energy z which is larger and earlier than the $z_{rec} \approx 1\,100$ of the CMB emission at which recombination of electrons and nuclei happened and space became transparent.

10.5   Why a CMB thermal spectrum?

The creation of the background radiation is the result of the recombination of the ionized hot gas. Because of its high temperature, this collection of electromagnetic waves, or photons, was in thermodynamic equilibrium with the hot cosmic plasma (positive nuclei and electrons) at early stages of the cosmological expansion. At these stages, the plasma and photons interacted with each other very effectively. As we shall see in the paragraphs immediately following, it can be estimated that when the temperature of the medium fell due to expansion down to a fairly well-defined temperature $T \cong 3\,000$ K, recombination of electrons with protons occurred. After that, the interaction between radiation and atomic gas vanished, and matter and radiation decoupled from each other. The emitted energy versus wavelength distribution of the photons was like that of a 3 000 K continuous black body thermal spectrum. The large redshift between that time and today shifts the wavelengths to the microwave region.

After the formation of protons, neutrons, and electrons, and then the nuclei of the light elements, the nucleons, nuclei, and even electrons are non-relativistic. In other words, the particle thermal energy is less than their rest energy ever since the cosmic age of one second. The baryonic plasma (nucleons and nuclei) is completely ionized during the whole epoch of radiation domination. We show below how the plasma and radiation interact strongly with each other—via electromagnetic interaction. Because of this, they have the same temperature and behave like one fluid in thermodynamic equilibrium. Using z as a reverse indicator of the time span ($z = 0$ at present and greater z farther in the past), we show that the thermodynamic equilibrium extends well back through the radiation dominated time from the z_{rec} of CMB emission.

Equilibrium is possible via scattering of photons by non-relativistic free electrons of the plasma. When photon energy can be neglected in comparison with the electron rest energy, the momentum of the scattered photon differs from the initial only in its direction, and the photon energy does not change (an analogy is elastic bouncing of billiard balls off the sides of a massive table). This is **Thomson scattering** whose cross-section does not depend on the photon frequency and is determined by the electromagnetic radius of the electron which is

$$r_e = \frac{e^2}{m_e c^2} \tag{10.45}$$

Here e represents the electron charge. Using this radius, the Thomson cross-section is

$$\sigma_T = \frac{8\pi}{3} r_e^2 \cong 0.67 \times 10^{-24} \, \text{cm}^2. \tag{10.46}$$

The path length between Thomson scattering of electrons is the mean free-path of the photon:

$$L_R = (n_B \sigma_T)^{-1} \tag{10.47}$$

Here $n_B = \rho_B / m_H$ and m_H is the mass of hydrogen nuclei which make up the great majority of the density and the number density of the nuclei. The time-rate of the process is

$$\Gamma_R = (L_R/c)^{-1} = n_B \sigma_T c \tag{10.48}$$

which is the reciprocal of the mean free path divided by the speed of light.

This time-rate is greater than the time-rate of the cosmological expansion, $\cong 1/t$, if

$$n_B > (\sigma_T ct)^{-1}. \tag{10.49}$$

If this inequality is true, the expansion of the universe is slow enough so the photons can overcome electrons' recession with expanding space.

Using Equations (6.8) and (10.19),

$$t = t_*(1 + z_*)^2/(1 + z)^2$$

and also Equation (8.55)

$$\rho_B = m_H n_B = \rho_B(t_0)(1 + z)^3,$$

Equation (10.49) becomes

$$1 + z > m_H \left[\sigma_T c \rho_B(t_0) t_*(1 + z_*)^2\right]^{-1}. \tag{10.50}$$

The size of $1 + z_*$ is given by Equation (10.44) while $\rho_B(t_0)$ may be taken from Equations (10.6)–(10.8). The time when radiation and matter densities became equal was estimated earlier as $t_*(0.7 \times 10^{12}$ s) while σ_T is given by Equation (10.46). Using $m_H = 1.67 \times 10^{-24}$ g, the right hand side is

$$\frac{1.67 \cdot 10^{-24}\text{g}\,\alpha^{1/2}}{0.67 \cdot 10^{-24}\,\text{cm}^{-2}\,3 \cdot 10^{10}\,\text{cm/s}\,0.042 \cdot 10^{-29}\text{g/cm}^3\,0.7 \cdot 10^{12}\,\text{s}\,5\,500^2} \cong 10\alpha^{1/2}.$$

For a reasonable value of α (e.g., 3/2), Equation (10.50) is satisfied by a wide margin when $z \geq z_*$. Scattering (and thermal equilibrium) is thus very fast and effective from the time when the CMB emission arises back through all the radiation domination epoch. There is said to be "close coupling" of photons and electrons through this time interval.

The cross-section of the photon scattering on ions (primarily protons at this time) is less that σ_T by the factor $(m_e/m_H)^2 \cong 3 \times 10^{-7}$. Therefore photon scattering on ions is not very effective. But ions are closely connected to electrons by the **Coulomb process**. In this process, electrons and ions of plasma interact with each other due to the Coulomb electrostatic forces that provide the exchange of energy and momentum between them. The **Coulomb radius** determines the cross-section of the process

$$r_C = \frac{e^2}{k_B T}. \tag{10.51}$$

This radius is the distance at which the energy of the electrostatic interaction potential between an electron and a proton e^2/r is equal to the mean thermal energy of the particles $\approx k_B T$. Using the radius and mean thermal energy, the Coulomb cross-section can be shown to be

$$\sigma_C = \left(e^2/k_B T\right)^2 C \tag{10.52}$$

The quantity in the parenthesis is clearly the Coulomb radius squared. On the other hand, the far less obvious quantity

$$C = 2\pi \ln(r_D/r_C) \tag{10.53}$$

is a dimensionless factor (the "Coulomb logarithm") which accounts for the fact that the electrostatic potential decreases slowly ($\propto 1/r$) with distance; because of this, increasingly more numerous distant encounters contribute also to the net result of the interaction. However, at large distance the positive and negative charges of nuclei and electrons cancel. The logarithm includes this **Debye radius**,

$$r_D = \frac{k_B T}{4\pi n e^2} \tag{10.54}$$

At larger scales than this plasma can be considered as a quasi-neutral fluid. Under typical conditions, $C \approx 10$. The time-rate of the Coulomb process is

$$\Gamma_C = \sigma_C n_B c_e \frac{m_e}{m_H} \tag{10.55}$$

where $c_e \cong (k_B T/m_e)^{1/2}$ is the mean thermal velocity of the electron. Because of the large mass difference, the electron's thermal velocity is larger than the ion's. The ratio m_e/m_H means here that the energy portion $\approx k_B T$ is transferred not in a single collision, but in m_H/m_e collisions. A similar estimate shows that $\Gamma_C \gg 1/t$, and the energy-momentum exchange between non-relativistic electrons and ions of cosmic plasma is very effective during the radiation domination epoch. Consequently, electrons and ions are strongly coupled with each other and with photons. This photon-plasma mix is thus in thermodynamic equilibrium and moves as one fluid as the universe expands.

Now, we show that the recombination of the electrons and the nuclei and the release of the future CMB thermal radiation happens at a rather definite temperature. At the end of radiation domination, the temperature $T_* = T_R(t_0)(1 + z_*) \cong 10^4$ K. At this temperature, the hydrogen gas is still completely ionized. Indeed, the equilibrium ionization is given by the general relation

$$\frac{x^2}{1-x} = \frac{1}{n}(k_B T/b)^{3/2} \exp[-E_H/k_B T] \tag{10.56}$$

where the ionization fraction is $x = n_e/n_B = n_i/n_B$ is, n_e is the number density of electrons, the number density of ions is $n_i = n_e$, the total density of atoms, both neutral and ionized is $n_B \cong \rho_B/m_H$, a combination of physical constants is represented by $b = \frac{2\pi\hbar^2}{m_e}$, the electron mass is m_e, and, finally, the hydrogen atom electron binding energy is $E_H = 13.6$ eV. The quantities m_e, m_p, m_H are masses of the electron, proton, and the hydrogen atom. With Equation (10.56), at $T \cong 10^4$ K,

$$n\frac{x^2}{1-x} \cong 10^{14}\text{cm}^{-3}. \tag{10.57}$$

Equation (10.57) means that ionization is practically complete i.e., x is close to one for number densities $n < 10^{14}$ cm^{-3}. The particle density $n(t_*) \cong \rho_B(t_*)/$

$m_H \cong 10^5 \, \text{cm}^{-3}$, so the cosmic plasma was indeed ionized at this moment in the past.

However, note that the dependence on temperature is fairly strong (exponential) for ionization equilibrium, and at a slightly lower temperature $T = 3 \times 10^3$ K, one finds

$$n \frac{x^2}{1-x} \cong 10^{-3} \, \text{cm}^{-3}. \tag{10.58}$$

For 3 000 K, we see that ionization is possible only if hydrogen gas was of very low density, $n < 10^{-3} \, \text{cm}^{-3}$ which, as we shall see, is certainly not expected at that epoch.

Using the observed background spectrum wavelength dependence, the present day radiation temperature is 2.725 K. Since we have the simple reciprocal Wein's law dependence of the peak wavelength on temperature, we can see that the temperature drops to 3 000 K at the fractional wavelength redshift of about 1 100. The change in the scale length is thus

$$1 + z_{rec} \cong 3 \times 10^3 \, \text{K} / T_R(t_0) \cong 1\,100. \tag{10.59}$$

At this redshift, the baryonic density is estimated to be

$$\rho_B(t_{rec}) \cong 4 \times 10^{-22} \, \text{g/cm}^3, \tag{10.60}$$

(about $1\,100^3$ times the present day baryonic density).

The baryonic number density is thus estimated as

$$n_B(t_{rec}) \cong 200 \, \text{cm}^{-3}. \tag{10.61}$$

This mass density and number density are far greater than the corresponding values today.

Thus the plasma recombines and is mostly neutral by the time $t = t_{rec}$ when $T(t_{rec}) = T_{rec} \cong 3 \times 10^3$ K.

The time of CMB emission can be roughly estimated in a similar way as was done earlier for the radiation vs. matter equality time where we obtained $t_* \approx 76\,000$ years. Between t_* and t_{rec} it is appropriate to use the Einstein–De Sitter model for which

$$R(t) \propto t^{2/3},$$

(Equation (8.59)). Thus

$$t_{rec} \approx t_* \left(\frac{1 + z_*}{1 + z_{rec}} \right)^{3/2} \approx 450\,000 \, \text{yr}.$$

This is similar to the more precise 380 000 year time since the origin of the universe given in the lower frame of Figure 10.1. The lower frame of Figure 10.1 shows that by the CMB emission time radiation domination is no longer the case. One should again use the Chernin (1965) method for a more general approach.

At recombination, cosmic radiation decouples from matter so photons with temperature $< T_{rec}$ can no longer interact with neutral atoms. Thermodynamic equilibrium of matter and radiation becomes impossible so they expand independently. The photons (or, to be exact, most of them) do not scatter off matter any more and they propagate freely. Via the cosmological redshift, they cool with the cosmological expansion as a relic of the early hot state of the universe constituting todays's CMB radiation background. Meanwhile baryons together with dark matter undergo quite a different evolution controlled by gravitational instability to form the variety of the systems observed today. At this point, for those readers who need a bit of review or have limited background in this area for the section at the end of this chapter and other parts of this text, we define and explain basic notions of particle physics in Boxes 7–10. We mostly follow here Okun's (1985, *Particle Physics: The Quest for the Substance of Substance*) which is an excellent introduction to the field.

Box 7. Fundamental physical interactions.

Elementary particles are usually defined as particles not decomposed into constituents in nuclear reactions. These are electrons, protons, and neutrons (from which "ordinary" matter is made), but excludes atoms and atomic nuclei. Protons and neutrons form nuclei and are referred to as nucleons.

Another elementary particle comprising light, is the photon, which is abundant in the universe today in the CMB. Only slightly less abundant are neutrinos. Neutrinos' interaction with nucleons and electrons is very weak which makes them difficult to observe.

Neutrinos, photons, electrons, and protons are stable, not decaying at all or decaying at a very low rate. The experimental lower limit on the lifetime of electrons is about 10^{22} years and on the protons 10^{33} years; both are $\gg$ the 10^{10} year age of the universe. A free neutron is unstable and decays in about 10^3 seconds; but neutrons bound in nuclei are stable.

In addition to these, several hundred unstable particles are known, with lifetimes of 10^{-24} to 10^{-6} seconds. Most are less than 10^{-20} seconds; these are called resonances.

There are only four types of fundamental interactions underlying all the large number of processes involving elementary particles: gravitational, electromagnetic, weak and strong. The gravitational interaction is universal involving all elementary particles. Beyond Newton's theory, we have seen how the General Relativistic theory of gravitation developed by Einstein is extremely successful. Any attempts to construct a quantum theory of gravity have so far been unsuccessful.

The classical (non-quantum) theory of electromagnetic interaction, including the interaction of electric charges with electromagnetic fields, was created by Maxwell. This is a relativistic theory. Quantum relativistic electrodynamics was constructed by Dirac, Feynman, Schwinger, Tomonago, Dyson and others in the 1920s through the 1950s. The electromagnetic interaction is much better studied than other fundamental interactions.

The weak interaction was first revealed in 1896 when Becquerel discovered radioactivity of uranium salts which emitted β rays, streams of electrons in a process called β decay. In the 1930s, Pauli realized that these nuclei also emit, with the electrons, light neutral particles called neutrinos. The decay of the free neutron and many other unstable particles is of the same physical nature. Contrary to gravity and electromagnetism, the weak interaction is a short-distance interaction, not taking place at distances larger than $\approx 10^{-15}$ cm. A quantum description of the weak interaction was developed by Fermi in the 1930s.

The nuclei may emit γ rays via gamma decay. Also nuclei may emit α-particles, helium nuclei composed of two protons and two neutrons. The α-decay is due to the quantum penetration of α particles through the potential barrier of the nuclear (strong) force while γ-decay is due to an energy release in transitions between the quantum energy levels of nuclei. Note that α-decay and γ-decay are not weak interaction phenomena.

Like the weak interaction, the strong interaction is a short-distance interaction but with a larger characteristic length about 10^{-13} cm. The strong interaction confines protons and neutrons inside nuclei. Protons, neutrons, and other particles participating in the strong interaction are called hadrons. With the significant exceptions are photons, electrons, and neutrinos, the majority of elementary particles are hadrons. Mesons are hadrons with integral spin. Baryons are hadrons with half-integer spin. Protons and neutrons are baryons. Yang and Mills, Gell-Mann, Nambu, and others in the 1950s–1970s developed the modern theory of strong interaction called quantum chromodynamics. The "chrom"—color—in its name is explained in the following box.

The strengths of the strong, weak, and electromagnetic interactions are specified by the corresponding so-called coupling constants: $\alpha_{strong} \approx 0.1$; $\alpha_{weak} \approx 0.04$; $\alpha_{em} \approx 0.008$. The strong interaction has the largest coupling constant and the weak interaction constant is bigger than the constant for the electromagnetic interaction. The gravitational interaction is much weaker than the weak and even the electromagnetic interaction. As an example, calculate the ratio of gravitational to electrostatic forces between two otherwise isolated electrons. At separation R, the gravitational attraction force is Gm_e^2/R^2 (where $m_e \cong 0.91 \times 10^{-27}$ g is the electron mass) while the electrostatic repulsion force is e^2/R^2 (where

$e \approx 4.8 \times 10^{-10}$ esu (cgs units) is the electron electric charge). We see that the ratio is 10^{-43}. The nature of the extreme weakness of gravity in comparison to the three other fundamental interactions is one of the greatest problems of fundamental physics.

Box 8. Three generations.

At present, electrons are considered as structureless particles, and are thus truly elementary. Electrons together with muons, tau-particles and neutrinos are called leptons—these are spin 1/2 electrically charged particles that do not participate in the strong interaction. There are also three neutral spin 1/2 leptons: electron neutrinos, muon neutrinos, and tau neutrinos. The leptons (together with their corresponding anti-particles) are all true elementary particles.

In contrast, protons and neutrons, as well as all other hadrons, are elementary with certain reservations. All the numerous hadrons are elementary, in the sense that none of them can be broken into constituents. But hadrons have an internal structure consisting of quarks. The quarks appear to be structureless, i.e., truly elementary, fundamental particles.

There are six kinds of quarks and six corresponding antiquarks. Protons and neutrons are made of **u** (up) and **d** (down) quarks. As with other quarks, their spin is ½. Their electric charges are non-integral values of the -1 electron charge: $+2/3$ for **u** and $-1/3$ for **d**. Their masses are 5 and 7 GeV, respectively. The proton is built of two **u** and one **d** quarks: **p** $=$ **u u d**. The formula for the neutron is **n** $=$ **d d u**. In both, the spins of two of the quarks are anti-parallel, so the total spin is 1/2 for each of the nucleons. When masses are expressed in energy units (e.g., GeV), we actually speak of the quantity mass $\times c^2$.

With photons, the truly elementary particles of what we call baryonic matter are quarks **u** and **d**, electrons **e** and electron neutrinos ν_e. These two quarks and two leptons form a set that is called the first family, or the first generation, of elementary particles.

Two more such families or generations are known each with two quarks and two leptons. The families are usually presented as three columns:

$$
\begin{array}{ccc}
\mathbf{u} & \mathbf{s} & \mathbf{t} \\
\mathbf{d} & \mathbf{c} & \mathbf{b} \\
\mathbf{e} & \mu & \tau \\
\nu_e & \nu_\mu & \nu_\tau
\end{array}
$$

In the second generation, the **s** and **c** indicate strange and charm quarks, respectively; μ is the muon, and ν_μ is the muon neutrino in the second generation.

In the third generation, **t** and **b** are top (or true) and bottom (or beauty) quarks, τ is the tau-lepton, and ν_τ is the tau-neutrino.

All numerous hadrons in nature are composed of only 12 truly elementary particles—six quarks and six anti-quarks. Baryons consist of three quarks each; each meson consist of two quarks, one of which is an anti-quark.

The particle generation is specified by a new quantum number called *flavor*. Speaking of the same flavor, one has in mind particles of one family.

To describe the interactions of quarks, one more quantum number is introduced. Quarks of a given flavor exist in three different species called "colors", and the three colors are—quite arbitrarily—yellow, blue, and red. Colors are "charges" of quarks in their strong interactions with one other. The term "chromodynamics" is used for the theory of this interaction.

The first generation and later the second and third generations of particles were found observationally. Does a fourth or further generations exist? Cosmology gives a clear answer: three and only three families are consistent with the Big Bang Nucleosynthesis (BBN), as we explain in Chapter 11.

Twenty-four truly elementary particles make up three generations. Both particles and antiparticles are all fermions since their spin is half-integral. These fermions interact with spin 1 bosons that mediate three of the four interactions of nature. These bosons are the photon γ (for the electromagnetic interaction), $\mathbf{W}^+$, $\mathbf{W}^-$, and $\mathbf{Z}$ bosons (for the weak interactions), and eight color gluons (for the strong interactions).

There are $24 + 12 = 36$ truly elementary particles in the so-called Standard Model of particle physics. Despite its success, this is hardly the "final" theory: it contains two dozen parameters that have to be adjusted "by hand" to fit the experimental data. Moreover, it does not include gravity. A boson which is assumed to be responsible for gravitational interaction (beyond the Standard Model) is a hypothetic particle of spin 2 called the graviton.

Box 9. Grand Unification.

Starting from the known values of the coupling constants (see Box 7), one may extrapolate the trends on the basis of their known dependence on energy. At energies $M_{EW} \geq 1$ TeV ($= 10^3$ GeV), the weak and electromagnetic interactions unite as a common electroweak interaction. The electroweak interaction is described well by a theory developed by Glashow, Weinberg, Salam and other in the 1960s–1970s. It may be seen that, at a still higher energy, other coupling "constants" α can turn out to be equal to each other or at least be very near to each other. At even higher energies, the strong interaction might join the electroweak interaction, so that the three of the four fundamental interaction of nature might become one common interaction. Starting from the known values of the coupling constants (see Box 7), one may extrapolate the trends of their known dependence on energy up to energies about $M_{GU} \approx 10^{15}$–10^{16} GeV.

A single "Grand Unified Theory" describing physics at M_{GU} energies is an ambitious goal of particle physics. The Grand Unification energy M_{GU} is much larger than the electroweak energy M_{EW}, and close, in order of magnitude, to the Planck mass, $M_{Pl} \approx 10^{19}$ GeV, the energy at which gravity, relativity and quantum effects are similar. M_{Pl} is the energy at which a particle's Compton wavelength is equal to its Schwarzschild radius. Speculating, all the four fundamental interactions might unite at the M_{Pl} energy.

Energies above ≈ 1 TeV are not yet available in accelerators so the interaction unification has not yet been verified in experiments. The so-called "desert" hypothesis assumes there is no new physics between the energy scales M_{EW} and the Planck energy. A recent alternative, suggested by Arkani-Hamed et al. (1998), is based on the idea that there is no desert at all, and the M_{EW} energy scale is the only truly fundamental energy scale in nature. It hypothesizes that electroweak, strong and even gravity forces would become unified long before the Grand Unification energy scale or the Planck scale. This "super-unification" assumes that there are some extra dimensions in space.

If $M_{EW} \approx 1$ TeV is really a very special or even the only fundamental energy constant of physics, one may expect a new arena for particle physics at the energy ≈ 1 TeV: a plethora of new particles, including microscopic black holes, gravitons, *etc.*, would be seen in experiments at an energy of $\approx M_{EW}$.

Which version of the unification—with an energy desert or with the magic energy at the TeV scale—is real? Expectations are connected with the Large Hadron Collider (LHC) currently operating at the European Organization for Nuclear Research (CERN) near Geneva on the France-Swiss border. This proton-proton collider is the first accelerator to access the crucial TeV region. The total collision energy of the LHC is 14 TeV; the energy available in the collisions between quarks as constituents of the proton is 1–2 TeV. Experiments at the LHC will gain higher energies in coming years. The next step beyond may be a Very Large Hadron Collider (VLHC) that would take experiments into the 100–200 TeV region.

Box 10. Symmetries.

We include this box as background for Chapter 11. An object or set of objects has symmetry if they retain their shape or relative positions under certain transformations, a definition appropriate for the symmetries considered in fundamental physics.

Symmetries are classified into geometrical and internal. Transformations corresponding to geometrical symmetries include spatial and temporal translations, spatial rotations, space-time rotations, and mirror reflections of coordinate axes (three spatial axes and one temporal).

The symmetries under each the transformations imply the conservation of the appropriate physical quantity: they are, respectively, momentum, energy, angular momentum, the Lorentz momentum, spatial parity, or **P** parity. Symmetry under reversal of the time axis corresponds to reversibility of physical processes; this is **T** parity.

Internal symmetry transformations relate different objects "of like kind". For example, "charge conjugation" transforms particles into the corresponding anti-particles which implies conservation of charge conjugation parity, or **C** parity. Parity is a quantum number that characterizes the symmetry of the wave function (of particles or their systems) under a discrete transformation. **P** parity of fermions is postulated to be +1; the parity of the anti-fermions is –1. For bosons, **P** parity of a particle and that of its antiparticle are identical, and their product is +1. **CP** parity (combined parity) corresponds to the mirror reflection together with charge conjugation.

CPT parity combines **CP** and **T** parity. A fundamental theorem of quantum theory states that the equations of the theory are invariant under the product of the three transformations: charge conjugation **C**, spatial inversion **P**, and time reversal **T**. If some process takes place in nature, there is a **CPT**-conjugate process which is equally possible. The **CPT** Theorem implies that a particle and its anti-particle have identical masses and lifetimes, and they interact identically with the gravitational field. The **CPT** Theorem shows that geometrical and internal symmetries are not isolated from each other, but rather closely connected (see more about this below).

In his book, Okun mentions that

"the notion of symmetry is inseparable from the notion of beauty. Furthermore, true beauty, in its highest forms, requires a slight departure from symmetry, imparting to beauty the mysterious and alluring element of incompleteness".

Many examples of symmetry violations exist in particle physics. In 1956–1957, weak processes were discovered not to be invariant under both mirror reflection and charge conjugation. Very small effects of **CP** violation were observed in decays of long-lived neutral K mesons. If the **CPT** Theorem is valid, **CP** violation entrails violation of time reversibility.

A special type of symmetry, the so-called supersymmetry, was introduced by Golfand and Likhtman in 1971 and later by Wess and Zumino. Supersymmetry relates fermions to bosons and predicts that every particle has its supersymmetric partner that is much heavier and differs in spin by a half unit. Each electron thus has a "selectron" partner which is a heavier boson. Each quark has a "squark".

Each photon has a "photino" which is a neutral spin 1/2 fermion similar to a neutrino. The graviton (spin 2) has a supersymmetric partner "gravitino" with spin 3/2, *etc*. No particle of the Standard Model is the superpartner of another. So supersymmetry is beyond the Standard Model; it is an extension of it.

If supersymmetry really exists, it is not exact, but strongly broken: the masses of two supersymmetric partners are not equal, but differ significantly. Indeed, no supersymmetric partners of our known particles have been observed because their masses are very large. Powerful accelerators would be required.

Heavy supersymmetric particles (for instance, the neutralino, a spin 3/2 superpartner of the photon) might be the particles of dark matter. Dark matter particles might be, for example, non-baryonic weakly interacting particles with masses near 1 TeV. The largest colliders, the LHC in particular, should find such superpartner particles, if they really exist.

According to Zeldovich (1968), the cosmic vacuum is composed of the vacua of all particles of nature. The net density of the vacuum is due to a near cancellation of fermionic vacua and bosonic vacua. If so, the major energy content of the observed Universe is due to the supersymmetry and its violation.

10.6   Relic neutrinos and Ω

Aside from ordinary and dark matter, the relativistic component of cosmic energy includes, besides the CMB radiation, relic neutrinos and antineutrinos (in equal numbers). There are three known species, or flavors, of neutrinos (and antineutrinos). These are electron, muon, and tau neutrinos. Ralph A. Alpher, James W. Follin and Robert C. Herman, of Gamow's group at George Washington University, first proposed in 1953 that neutrinos existed in the early universe and were initially in thermodynamic equilibrium with radiation and matter.

Neutrinos participate in just weak and gravitational interactions. The small cross-section of their interaction with electrons cause neutrinos to decouple from the plasma much earlier than photons. Recall that the photons decoupled at around 380 000 years. Neutrinos decoupled at about a few seconds when neutrinos and antineutrinos were still relativistic. Because neutrinos interact so weakly with each other, neutrinos and antineutrinos have not annihilated each other and have been conserved up to now as a relic of the initial hot state of the universe. Their total particle density now is about 300 cm^{-3}, close to the present particle density of the CMB photons.

Since neutrinos are very light, they easily move at relativistic speeds close to the speed of light. The relic neutrinos and antineutrinos are relativistic now, if their rest masses are less than

$$m_\nu < k_B T_0/c^2 \cong 4 \times 10^{-37} \, g \cong 2 \times 10^{-4} \, eV/c^2 \qquad (10.62)$$

Here T_0 is the CMB temperature at present and the mass is given in both grams and electron-Volts (eV) in the two alternative versions of the equation. When we quote the mass m of a particle in eV, we are actually refering to its rest mass energy mc^2. For example, for an electron the rest mass energy $m_e c^2 = 0.511$ MeV and for a proton $m_H c^2 = 938.3$ MeV (1 eV $= 1.602 \times 10^{-12}$ erg). If these are accelerated to a high speed their mass is increased beyond the rest mass with the energy being corresponding increased. By way of contrast, the rest mass energy of a photon is zero. If relic neutrinos are relativistic, their contribution to the total relativistic density ρ_R must be comparable to the CMB photon contribution. More accurate estimates show that the contribution would be near to one half of the CMB contribution. For this case, the factor $\alpha \cong 3/2$ in Equation (10.7) due to relativistic neutrinos for the present-day relativistic density.

Indeed, the observed fact neutrino oscillation requires neutrinos to have nonzero masses Karagiorgi et al. (2007). The experimental results constrain the differences in the masses of the neutrinos of various flavors; the absolute value of neutrino masses is uncertain. The experimental upper bounds on the electron neutrino mass are rather stringent: $m_{\nu_e} < 0.5$–10 eV. They do not exclude, however, the possibility that the electron neutrino may be non-relativistic. Bounds for neutrino masses of the two other flavors are less constrained: 170 Kev for muon neutrinos and 20 Mev for tau neutrinos. Indirect tentative data indicates an upper limit for the present-day non-relativistic neutrino density,

$$\rho_\nu(t_0) \cong (0.01 - 5) \times 10^{-2} \rho_c(t_0). \qquad (10.63)$$

Clustering data of galaxies and gas clouds gives an upper limit for the sum of the masses of the three neutrino species (Seljak et al. 2005). From this estimate, the 95% confidence upper limit of 0.42 eV is so low that the neutrinos contribute less than 1% of today's critical mass density of the universe. Aother estimate (Goobar et al. 2006) from WMAP 3 year data, the baryon acoustic peak, the SNLS supernovae and the Lyman-α forest gives an upper limit of 0.3 eV, similar to the 2005 results.

Accepting this 0.42 eV value, the neutrino contribution to the non-relativistic matter density now is orders of magnitude smaller than dark matter's contribution. These neutrinos and antineutrinos were relativistic at earlier higher temperature epochs. Their contribution to the radiation density was about 50% of the photon contribution at the radiation domination epoch and was still significant at the time of the CMB emission (see Figure 10.1 bottom).

Radiation dominates in the early universe. The equations above show also that the radiation temperature and the radiation density tend to infinity when the proper time goes to zero and the observed redshift goes to infinity, characterizing a special state of cosmic energy in the beginning of cosmological expansion, the **cosmological**

singularity. However, the relations of this chapter are based on standard laws of physics: the laws of quantum physics, relativity, and gravity. The laws have a limited area of applicability within which they have directly or indirectly been verified in experiments and observations. The back-extrapolation to the creation of the CMB observed today and the still earlier creation of today's neutrinos is amazing but reasonable.

Chapter 11

Baryonic matter

The **baryonic matter** of which the Earth, other planets, and stars are made consists of atoms and their parts—protons, neutrons, and electrons. Today, most—perhaps as much as 80%—of protons, neutrons, and electrons are in the form of a hot 1–100 million K intergalactic gas, as revealed by x-ray emission. Data and physical arguments (see e.g., Masso and Rota 2002) show that the number of electrons is equal to that of protons, and therefore the universe is evidently symmetrical with respect to electric charge: the total electric charge on cosmic scales is zero.

11.1   Why matter and not also anti-matter?

There are four fundamental interactions in nature: the strong interaction (the nuclear force), the weak interaction, the electromagnetic interaction, and the gravitational interaction. The strong force acts inside the nucleus confining protons and neutrons. The weak force also acts in nuclei and is responsible, for example, for β-decay. At high energies ≥ 1 TeV, (which were common before $t = 10^{-12}$ sec of cosmic time) weak and electromagnetic interactions unify and become the electroweak interaction.

In particle physics, baryons are known as strongly interacting particles with semi-integer spin. They are composed of quarks and participate in the strong force. Protons and neutrons are baryons; other baryons are practically absent and may be observed in very small numbers only in special laboratory experiments or in cosmic rays. Leptons are elementary particles that do not participate in the strong force and do carry a charge. The electron is a lepton. There are six observed leptons each with its antiparticle. See Box 8 and the more detailed Table 11.1 for a summary of different elementary particles and their properties.

As we have seen from the earlier discussion of the latest WMAP results, the baryonic component of the universe contains only a tiny amount of baryonic anti-matter (anti-protons, anti-neutrons). All of the anti-baryons that are present are due to secondary processes of particle and antiparticle birth in high-energy particle collisions—for example, in cosmic rays. Such a high asymmetry between baryons and anti-baryons may look strange: the fundamental symmetry between particles and antiparticles has been firmly established in collider experiments.

However, in the very early universe, there was a different balance between particles and antiparticles. At high temperatures when the particle thermal energies greatly exceeded their rest mass energy, baryons and anti-baryons or more precisely, the quarks and anti-quarks they consist of (see Table 11.1) must have had almost equal numbers. Particle-antiparticle pairs appear and disappear in great numbers under

Table 11.1. The elementary particles, according to quantum field theory. Protons, neutrons, and other **baryons** are not elementary particles, but are instead made of quarks and gluons. Rest mass is given in energy units, $1\ \text{eV} = 1.6 \times 10^{-12}$ erg.

Particles			Rest masses	Charges	Spins
Fermions	leptons	electron e	0.511 MeV	-1	1/2
		muon μ	0.106 GeV	-1	1/2
		tau τ	1.78 GeV	-1	1/2
		e-neutrino ν_e	<2.5 eV	0	1/2
		μ-neutrino ν_μ	<2.5 eV	0	1/2
		τ-neutrino ν_τ	<2.5 eV	0	1/2
	quarks	up u	6 MeV	2/3	1/2
		down d	10 MeV	$-1/3$	1/2
		strange s	0.25 GeV	$-1/3$	1/2
		charm c	1.2 GeV	2/3	1/2
		bottom b	4.3 GeV	$-1/3$	1/2
		top t	174 GeV	2/3	1/2
Bosons	scalar	Higgs H	<220 GeV	0	0
	vector	photon γ	0	0	1
		W boson $W^\pm$	80 GeV	±1	1
		Z boson Z^0	91 GeV	0	1
		gluon gl	0	0	1
	tensor	graviton g	0	0	2

such conditions. The number density of such pairs was almost exactly the same as the number density of photons in thermodynamic equilibrium. Later, after baryons and anti-baryons annihilated one another, the tiny excess of particles over anti-particles survived and yielded the baryonic matter observed today.

As the number of the relic 3 K photons in the co-moving volume was practically conserved as the universe expanded, the present ratio of the baryonic number density, $n_B(t_0) = \rho_B(t_0)/m_H$, to the CMB photon number density, $n_R(t_0) = 0.24(k_B T_0/\hbar c)^3$, gives the early baryon excess over anti-baryons.

$$b \equiv \frac{n_B(t_0)}{n_R(t_0)} = B^{-1} \cong 6.7 \times 10^{-10}. \qquad (11.1)$$

The big number B was introduced in Chapter 10 in Equation (10.41). The reciprocal of B, b, measures the original asymmetry with respect to baryons and anti-baryons. This small dimensionless cosmic number has ensured the survival of ordinary matter early in the universe and its existence today. The present strong baryonic asymmetry results from a very weak asymmetry given by the early small number b.

The physics initially responsible for this number could be researched in at least two ways. According to one way, the baryonic excess existed "from the very beginning", ab initio, which would mean that the universe was born asymmetric relative to particles and antiparticles. This possibility was actively debated in the beginning of the 1960s, and sometimes it was assumed that the number b that gives the cosmic **baryonic charge** as the primordial excess of particles over antiparticles could be as fundamental as, say, the Planck constant.

In a second approach, one can assume that b's value is not fundamental and that it must be "deduced" from general physical laws. In this case, the universe might be strictly symmetric in baryons and anti-baryons "at the beginning". The baryonic charge would have thus been initially zero, with the baryonic excess appearing in an evolutionary way via physical laws. This would be considered better than introducing a new constant.

This second approach makes predictions. According to Russian physicists Andrei D. Sakharov, the 1975 Nobel Peace Prize winner, (1966) and Vadim A. Kuzmin (1970), the appearance of the baryonic charge—referred to as **baryogenesis**—requires that certain conditions must be fulfilled. The most important condition is the instability of the proton. Also certain symmetries must be broken. The symmetry between particles and antiparticles (known as C-invariance) and the combined charge and parity mirror symmetry (known as CP-invariance) should also be broken. In addition, a very early rapid cosmological expansion is needed; it should be so rapid that the interactions of orginial particles of the cosmic medium occurred in a non-equilibrium way.

While the latter condition seems to be met naturally in the very early universe, the proton instability was a brave hypothesis at that time. But it was this hypothesis that made the evolutionary approach to the problem possible and fruitful. By the end of the 1970s, it became known that the proton decay is one of the consequences of the theory of Grand Unification, which assumes the unified nature of three of the four forces: strong, weak, and electromagnetic interactions. It was already Einstein's dream to elaborate a unified theory of all the four interactions, including gravitation.

In contrast to radioactive nuclei, the proton decay time must be extremely long, not less than 10^{32} years, according to current measurements. As to C- and CP-invariance violation, it is directly observed in decays of K- and anti-K mesons. Thus, both the theoretical and experimental particle physics combined with cosmology definitely indicates that all the necessary conditions favoring baryogenesis of matter in the early universe might actually have been satisfied.

Versions of the cosmological baryogenesis discussed in the last decades assign the baryogenesis to a very early epoch. At this time, the cosmic matter temperature was close to the characteristic Grand Unification temperature $T \approx 10^{16}$ GeV k_B^{-1} and the cosmic age was about 10^{-32} sec. In another version, the baryogenesis took place at much lower temperatures, typical for electroweak processes $T \approx 1$ TeV k_B^{-1}, at a cosmic age of about 10^{-12} sec.

It has been shown that the baryon excess number $b \cong 6.7 \times 10^{-10}$ can, in principle, be estimated using in both versions of the process: the number lies within the natural limits for the theories. However, it can be obtained only by using special assumptions and not from the basic principles of physics. In addition, the proton decay, which is the critical point of the Sakharov-Kuzmin concept, has not yet been reliably observed in experiments, only the lower time limit is found by the failure to observe it.

Now the number of electrons is equal to the number of protons in the real universe, the initial excess of electrons over their antiparticles—positrons—should be quantified by the same number b. The process responsible for the electron excess over positrons is called **leptogenesis**. This process is even less understood than the process of baryogenesis. Electrons and positrons are leptons, and leptons include also muons, tauons, and three types of neutrinos together with the corresponding antiparticles as seen in Box 8 and Table 11.1. As opposed to baryons, leptons do not participate in the strong interaction; instead they are characterized by the weak or electroweak interaction, an additional argument in favor of electroweak energy scale $\cong 1$ TeV as preferable for both baryogenesis and leptogenesis.

11.2   Big Bang Nucleosynthesis prediction and processes

The history of the universe can be traced back to the era of the **Big Bang Nucleosynthesis** (BBN). In this era, light chemical elements were produced by thermonuclear reactions. BBN occurred at a cosmic age of about a few minutes. This era is documented by data on the primordial abundances of primordial helium, deuterium, and lithium generated in the BBN.

Gamow first suggested in 1946 that the early universe acted as a huge nuclear reaction "factory". Systematic studies of BBN over more than a half-century revealed new and interesting details. Now BBN is a basic component of the standard cosmological model. Importantly, BBN studies in combination with the observational data are able to provide clear answers to key problems in modern cosmology and in fundamental physics. We shall see, one impressive result of this kind is that cosmic dark matter is non-baryonic.

Clearly, there were no atomic nuclei in the high temperature early universe where $T > T_N \approx 3 \times 10^{10}$ K. This was because the thermal energy of particles exceeded the binding energy of atomic nuclei (typically several MeV per nucleon). When temperature dropped below the nucleus binding energy, protons and neutrons combined via the attractive nuclear force which was then able to resist thermal motion. However, at this point, the thermal motions must still be strong enough in order that protons collide at speeds sufficient to overcome their Coulomb charge repulsion. At still lower temperatures, speeds are not sufficient. Because of this, BBN can proceed in a rather limited temperature or energy interval. Measuring temperature in MeV (convenient in nuclear physics, actually we talk about MeV k_B^{-1} where k_B is the Boltzmann constant) the "just right" temperature interval is approximately from 1 MeV to 0.05 MeV. This interval is calculated by the complex kinetics of nuclear transformations. Corresponding to this temperature range, the cosmic age is only between 1 and 300 seconds.

A 1 MeV temperature corresponds to the "freeze-out" of neutrons. Earlier, at higher temperatures, neutrons were in thermodynamic equilibrium with protons, transforming back and forth from one to the other. The neutron particle density, n_n, was related to the particle density of protons, n_p, by the thermodynamic equation

$$n_n/n_p = \exp\left[-\Delta mc^2/k_B T\right] \tag{11.2}$$

where $\Delta m = m_n - m_p = 1.29$ MeV/c^2 is the mass difference between the neutron and the proton with the neutron being a bit heavier. The equilibrium was provided by the reactions (via weak interactions)

$$n + \nu_e \leftrightarrow p + e^- \tag{11.3}$$

$$p + \bar{\nu}_e \leftrightarrow n + e^+ \tag{11.4}$$

Here n, p, e^-, e^+, ν_e, $\bar{\nu}_e$ denote, respectively, the neutron, the proton, the electron, the positron, the electron neutrino, and the electron antineutrino. These reactions happened back and forth very rapidly. In other words, the reaction time was much shorter than the age of the universe. At a later time when particle density and temperature dropped, the two way reactions slowed down and became rare. The time rate of these reactions became less than the rate of the cosmological expansion, $1/t$, at the temperature $k_B T_n \cong .66$ MeV. At that moment, the neutron-to-proton ratio is calculated to be approximately 1/7, as is seen from the Equation (11.2) for n_n/n_p.

The light elements are produced via the chain of nuclear transformations:

$$p(n,\gamma)D, \ D(p,\gamma)^3He, \ D(D,n)^3He, \ D(D,p)T, \ T(D,n)^4He, \ etc. \tag{11.5}$$

In Equation (11.5), D, T, ^3He, ^4He, and γ stand for deuterium, tritium, two isotopes of helium, and a quantum of gamma radiation, respectively.

In the last reaction on the right hand side above, practically all the neutrons which existed at this moment soon produced ^4He nuclei. From the estimate of $n_n/n_p = 1/7$, the mass fraction of helium follows. Every ^4He nucleus takes 2 neutrons; thus their number density becomes $n_n/2$. A ^4He nucleus weighs 4 times as much as a proton. Thus in terms of the mass fraction ^4He nuclei contribute

$$Y_{BBN} = \frac{4n_n/2}{n_n + n_p} = 2\frac{n_n/n_p}{1 + n_n/n_p} \cong 0.25. \qquad (11.6)$$

This is a key result. Below T_n, weak reactions above are effectively terminated. After that point, the number of neutrons decreased only due to their decay (via β-process) with a characteristic lifetime of 890 seconds. Compared to expansion, this is a relatively slow process, so that the neutron-to-proton ratio is almost unchanged until the time when the temperature has dropped to the range $T \cong 0.1$–0.3 MeV k_B^{-1}. As we have discussed, at that range, the reactions via strong interaction started to take place. These are nuclear fusion reactions, and the neutrons that have not decayed are captured into the nuclei of the light elements.

11.3   Baryon nucleosynthesis abundances and cosmological implications

Of the light nuclei, because the binding energy per nucleon is about 7 MeV in ^4He, it is energetically favorable for the nucleons to reside in ^4He. This binding energy is considerably larger than the values between 1 and 2.6 MeV for deuterium, tritium, and ^3He. Actually, most of the free neutrons are converted into ^4He, as calculations show. Doing some arithmetic, every two neutrons and fourteen protons (of the ratio 1/7), form one ^4He nucleus with twelve protons left over. From this arithmetic, about 25% of the baryons by mass were confined in ^4He, as calculated above (Equation (11.7)). A small minority (10^{-3}) of neutrons go to deuterium, the same amount to ^3He and still less, by five orders of magnitude, to ^7Li. Still heavier elements are not synthesized in the BBN, instead requiring supernova explosions later in the evolution of the universe as we have mentioned before.

Calculated light element figures are in good agreement with the observed abundances of the light elements as shown in Figure 11.1 (Kirkman et al. 2003). The abundance of the element in the nucleosynthesis model relative to hydrogen is plotted as a function of total baryonic density relative to the critical density $\Omega_B \, h^2$ or alternatively as a function of b (Equation (11.1)). Lithium is not a good choice because its surface abundance is depleted in stars as they evolve (Korn et al. 2006;) hence the abundance in Figure 11.1 is a kind of lower limit. Note that the He^4 curve is fairly flat and thus helium-4 is not as good a BBN gauge as deuterium. He^3 is also not so good for the same reason. One can see from the slope of the line versus the

Baryon to photon ratio $\times 10^{-10}$

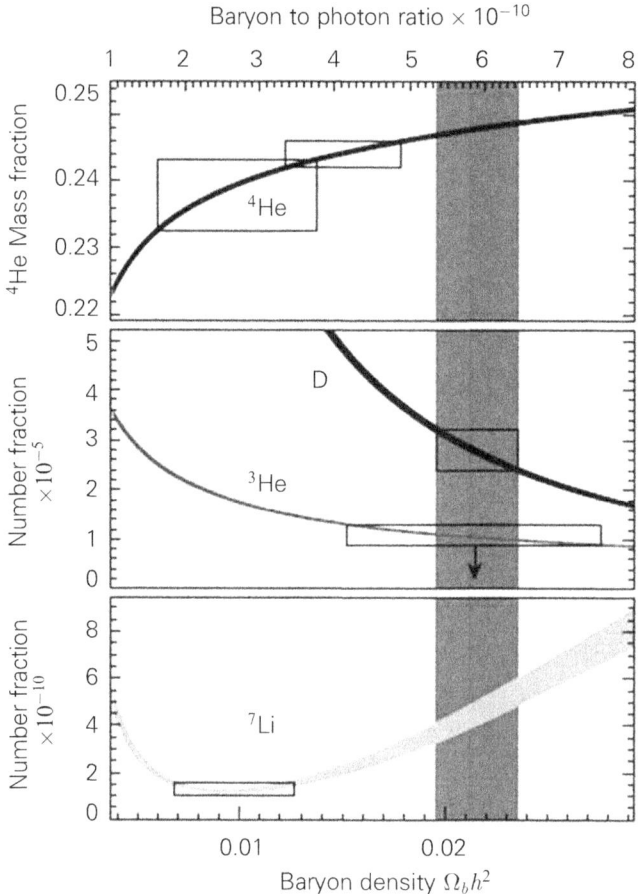

Figure 11.1. Predicted (lines) versus measured abundances (shaded region) of four light nuclei as a function of the baryon density. The figure has three vertical panels with a different linear scale for each. The top curve gives the theoretical ^4He mass as a fraction of the mass of all baryons, while the three lower curves are the number fractions D/H, $^3He/H$, and $^7Li/H$. (Kirkman et al. 2003).

horizontal axis that deuterium is a good element to determine b or $\Omega_B\, h^2$. There are no post Big Bang processes that can create significant deuterium, so what is observed is primordial in stars deuterium can be depleted, but outside of stars, interstellar clouds in our Galaxy should have deuterium abundances close to those created in the Big Bang. These intervening clouds create certain absorption lines in the spectra of stars. When the Hubble Space Telescope observed interstellar deuterium in the spectrum of the bright star Capella, astronomers found there to be less than 30 deuterium atoms for every million hydrogen atoms. Reading that number on the vertical axis of the box in Figure 11.1 for the Deuterium (D) curve, we can then read down to the corresponding middle value on the horizontal axis. We determine that the present-day observed ordinary matter density of about 4×10^{-31} grams/cm^3 (note that the horizontal axis includes the value of the Hubble constant, with its current value we get $0.0215/0.071^2 = 0.044$ of the critical density). This baryonic density would have created the observed fraction of deuterium in the universe when extrapolated back to the Big Bang. This density is consistent with other data (masses of galaxies etc.). Along with the other observational foundations of the Big Bang model, the agreement between the observed light element abundances and the predictions of this model have verified that the Big Bang is an excellent description of the evolution of the universe. We thus see that if there were only baryonic matter in the universe, it would be far below the value for a flat critical density geometry.

Relative abundances of light elements from BBN depend sensitively on parameters. The first one is the number of the neutrino species of which three are known: the electron, muon, and tau neutrinos (Box 8). If the number is more than three, this would increase the radiation density during the nucleosynthesis era, which in turn would increase the rate of expansion ($1/t \propto \rho_R^{1/2}$). If the rate were faster, nucleosynthesis would happen somewhat earlier, resulting in a higher neutron-to-proton ratio, and hence indicating a larger abundance of ^4He. But the larger amount of helium contradicts observations. Thus, BBN actually provides strong evidence that the number of neutrino species (and also the number of families in particle physics) is actually three (Shvartsman 1969).

11.4   The baryon content of cosmic systems

The second key parameter of BBN is the Big Baryonic Number $B = n_R/n_B$, the ratio of the CMB number of photons to the baryons. This ratio is one of the basic cosmological constants (see Chapter 10) related to the fundamental physics. As we have seen, BBN abundances of deuterium and ^3He are very sensitive to this ratio. Recent results give $B \cong 1.5 \times 10^9$ (Kirkman et al. 2003, Coc et al. 2004). Because the CMB temperature and thus $n_R(t_0)$ are observed precisely, the present-day baryonic density is:

$$\rho_B(t_0) = m_H n_R(t_0)/B \cong 4 \times 10^{31} \text{ g/cm}^3. \tag{11.7}$$

This is about 10 times less than the dark matter density $\rho_D(t_0) \cong 3 \times 10^{-30}$ g/cm^3. The corresponding density parameters are $\Omega_B \cong 0.04$ and $\Omega_D \cong 0.3$. Thus, dark matter only has to be almost totally non-baryonic, a significant cosmological conclusion. This value for B from light element abundances is in very good agreement with the WMAP result $B \cong 1.5 \times 10^9$ (see Chapter 10).

However, it should be noted that all those baryons have yet not been observed in the universe of galaxies. The actually observed luminous matter contribution is $\Omega_B \cong 0.005$. It may be roughly calculated from the observed optical luminosity density together with the mass-to-luminosity (M/L) ratio typical for stellar populations in galaxies. The luminosity density is obtained by summing up all the light from a representative unit volume of space, using the derived luminosity function of galaxies. There is a useful formula connecting the average M/L ratio for a volume of the galaxy universe and the implied cosmic density parameter Ω (see e.g., Efstathiou, Ellis and Peterson 1988)

$$\Omega \approx \langle M/L \rangle / 1500h \approx \langle M/L \rangle / 1000.$$

Here M/L is expressed in Solar units ($M/L = 1$ for the Sun). For instance, we see that for a universe entirely made of Sun-like stars, the detected light would imply a very low density $\Omega \cong 0.001$. Actually, galaxies contain mixtures of different stars which have typically $\langle M/L \rangle \sim 5$, so the baryonic stellar matter makes $\Omega_B \cong 0.005$, still significantly less than the predicted 0.04. One speaks about the missing baryon problem. Where are the remaining predicted baryons?

McGaugh et al. (2010) have made an inventory of the baryonic and gravitating mass in systems ranging from dwarf galaxies to rich clusters. They find that the fraction of baryons detected (stars, gas, and dust) falls short of the cosmic fraction at all scales, though the deficit depends on the mass of the system. For big systems the fraction almost approaches the predicted cosmic value, while going towards smaller systems it decreases. For spiral galaxies the range is 0.1–0.4, and for smallest galaxies the fraction may be less than 0.01. It is not known where the baryons could reside (e.g., in some form in the dark halos of galaxies). The problem is especially difficult for galaxies. McGaugh et al. (2010) write:

"The notion that we are simply not seeing many or even most of the baryons in individual galaxies is profoundly unsatisfactory"

and they conclude:

"We do not see a satisfactory solution to this missing baryon problem at present. Considerable work remains to be done to obtain a complete understanding of the universe and its contents".

The missing baryon problem can be seen as a part of the more general problem of dark matter which we shall discuss in Chapters 12 and 13.

11.5   The Lyman alpha forest

Among the possible hiding places of missing baryons are gas clouds of neutral hydrogen. Their detection is possible by the absorption lines that the gas clouds cause in the light coming from behind the clouds. Suitable background sources of light are high redshift quasars.

Let us make a brief diversion here to explain the nature of quasars. Quasars are active galactic nuclei, powered by supermassive black holes which are $3 \times 10^7 - 3 \times 10^{10}$ times more massive than the Sun. The main source of light arising from quasars is a disk of gas called accretion disk which is energized in the strong gravitational field of the supermassive black hole.

Quasars were very common in the early universe, more than a billion years after its birth, and they are so luminous that they are observed even at such great distances. We expect that clouds between the quasar and us each cause their own absorption lines within the quasar spectrum. Because the redshifts of the clouds are smaller than the redshift of the quasar, the cloud absorption lines are bluewards of the corresponding value in the quasar spectrum. There are so many absorption lines from the intervening gas clouds that they are called a "forest". Most important are Lyman α lines known as the **Lyman α forest**. The forest was predicted in 1965 by John N. Bahcall and Edwin E. Salpeter (1964) at the California Institute of Technology and Cornell University, respectively. Its discovery was made by J. N. Bahcall et al. (1968) and Roger Lynds at Kitt Peak National Observatory (1971). An example is given in Figure 11.2.

The nature of the Lyman α forest clouds is not known. Earlier they were assumed to be associated with neutral gas inside dark matter halos (e.g., by Rees 1986). However, more recently, observations indicate that an intergalactic neutral gaseous medium produces the Lyman α absorption, as well as other weaker absorption

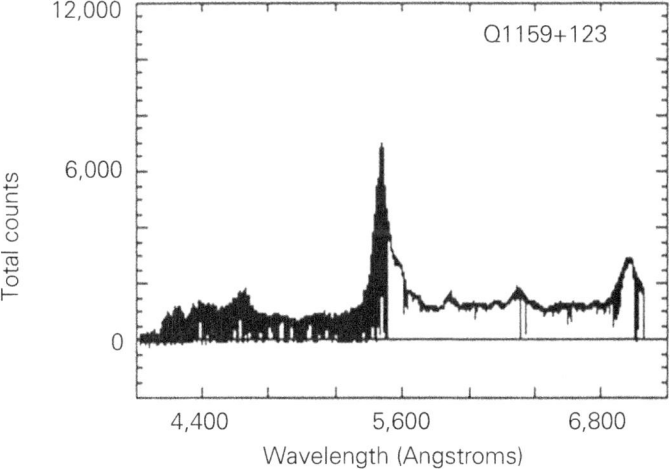

Figure 11.2. Spectrum of the quasar $1159 + 123$ obtained by the Keck telescope. The most prominent quasar emission line near 5 500 Å is Lyα at the redshift $z_e = 3.502$. Shortward of it, the Lyα forest can be seen while the few absorption lines longward of 5 500 Å are from heavier elements. (From Songaila 1998).

lines. This gas is in long filaments rather than spherical clouds, and is so tenuous that it consists of a continuous medium of variable density rather than of separate clouds. The medium is apparently closely associated with galaxies (Weinberg et al. 2003). The density at any point along the line of sight is connected with the local dark matter density. The Lyman α forest thus provides quite a direct link between cosmological observations and theory (Qian and Wasserburg 2002 and 2003, Qian et al. 2002).

Hydrogen absorption causes a reduction in flux by the relative amount

$$F_{abs} = e^{-\tau}$$

where the optical depth τ is

$$\tau \cong A(\rho/\bar{\rho})^{\beta}.$$

Here ρ is the local baryon density and $\bar{\rho}$ is its average value. The parameter β is close to or somewhat below 2. The coefficient A depends on several parameters, among them Ω_B. These relations arise from opposite effects of photoionization heating and adiabatic cooling.

Most importantly, the Lyman α forest F_{abs} as a function of wavelength translates into a record of baryon over-density as a function of distance since every distance has its associated cosmological redshift. This ignores systematic motions of the gas in the local frames of reference of each gas cloud. Obviously, the cosmological redshift correlation does not work very well close to galaxies where the gas may show large scale inflows or outflows. Actually, within roughly 1 Mpc of galaxies, it appears that there is a deficit of neutral hydrogen gas.

Because of the large redshifts of quasars and considerable amount of absorption at each redshift over the line of sight, the quasar emission can be completely absorbed over a range of frequencies. A gap in the spectrum like this is called a **Gunn–Peterson Trough**. This was proposed by James E. Gunn and Bruce A. Peterson working at Mount Wilson and Palomar Observatories in 1965, and first discussed in 1964 by Iosif S. Shklovskii of Moscow State University. In a trough the transparency is completely lost. Complete Gunn–Peterson Troughs have so far been observed only in quasars with the highest redshifts (Becker et al. 2001). Contrast Figure 11.2 to the higher redshift Figure 11.3 quasars.

The estimates of the contribution of Lyman α clouds to Ω_B have been as high as 90%, i.e., there is such a large amount of gas in intergalactic space that the amount is as high as 90% of that expected from other lines of evidence such as BBN discussed earlier. If true, there are no missing baryons. However, the amount is highly uncertain as it depends on the geometry of the clouds, whether they are somewhat spherical, sheets, strings or something more complicated. After all, we do not see images of these clouds directly, but we have to deduce their properties indirectly from the positions of absorption lines in the spectra of background quasars.

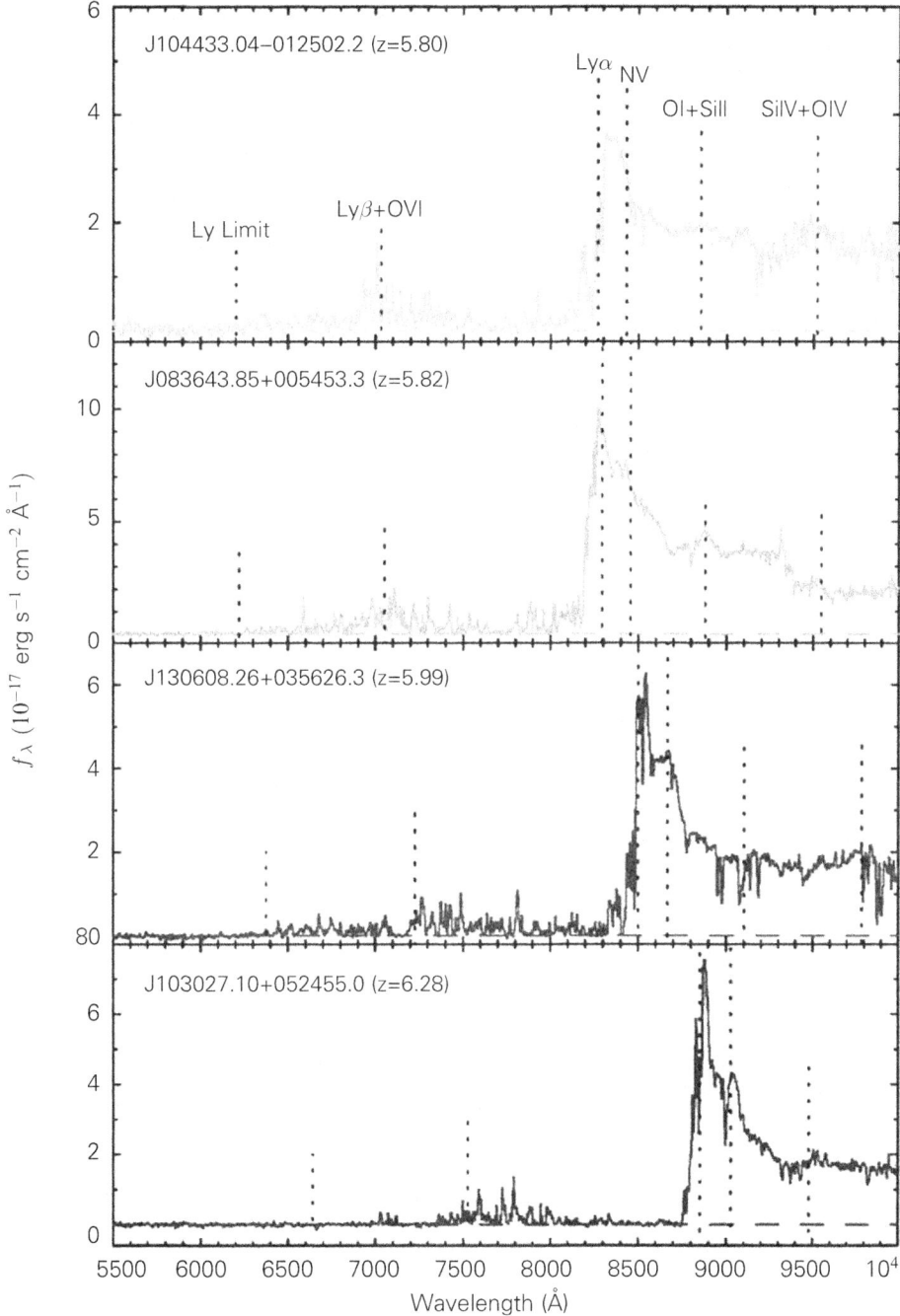

Figure 11.3. Spectra of $z \geq 5.8$ quasars. In each, expected wavelengths of prominent emission lines, as well as the Lyman limit, are indicated by the dashed lines. Note Gunn–Peterson Troughs in each compared to absence in Figure 11.2. (Becker et al. 2001).

Here pairs of quasars near to each other in the sky or single quasars lensed as double images are useful since the light from both lines of sight may have traversed the same cloud. The cloud sizes (at least at high redshift) turn out to be on the order of hundreds of kiloparsecs up to a Megaparsec. That makes the total baryon contents of the intergalactic medium very substantial. In fact, if the clouds were spherical then the clouds would contain more baryons than the universe as a whole. This paradox can be resolved by assuming that the gas clouds typically cannot be spherical but must be highly flattened, filamentary or sheet-like structures (Rauch and Haehnelt 1995). Numerical cosmological simulations predict that the whole universe is permeated by a "cosmic web" of gas in which the galaxies are embedded. The same cosmological simulations also predict the amount of baryons necessary to produce a Lyman alpha forest absorption pattern as strong as the one actually observed. From studies like this it is quite likely that most of the baryons were in hydrogen clouds in the early universe at times corresponding to when the redshift was $z \approx 2$ or higher.

While the main question may have been answered and we can now account for the baryons at high redshift, the problem has not gone away completely. We still don't know for sure what happened to the baryons in the present universe. Are they still in the form of gas, or in the form of structures that have since then been born in the universe? It is generally thought that the baryons are still in gaseous state, but now as tenous gas of temperature $T \approx 10^5$–10^7 K. Tenuous gas at such temperature is hard to detect "locally" by current instruments. Such gas is usually called warm-hot intergalactic matter (WHIM), and it is detected primarily by the absorption of lines of highly ionized oxygen. From such studies it has been concluded that WHIM may solve the missing baryon problem in the local universe where $z \approx 0$ (Prause et al. 2007). See Figure 11.3.

Chapter 12
Discovering dark matter

This chapter recounts the discovery of dark matter in different scales and of the principal detection methods. First we describe a "false alarm", Jaan Oort's (1932) claim that there is as much matter in the neighborhood of the Sun in dark form as in the luminous stars. Only one year later, Fritz Zwicky suggested that the Coma cluster of galaxies contains ten times more mass than the mass of its members. In that case, the cluster is actually bound together by dark matter. On a more local scale, Franz Kahn and Lodewijk Woltjer suggested that our Galaxy and the Andromeda Galaxy, the primary galaxies of the Local Group, contain ten times more mass than indicated by the combined visible light in these galaxies. In the 1970s these anomalies finally were finally recognized as the rule: there was dark matter in almost any galactic system that we could study, and its existence was revealed by many methods. Finally, it was realized that Oort's original claim was a "false alarm".

In discussing this process the **mass-to-light ratio** is a useful parameter. The Sun's mass is approximately $M_\odot = 2 \times 10^{33}$ g and its luminosity $L_\odot \cong 4 \times 10^{33}$ erg/sec. The basic unit of the mass-to-light ratio is $M_\odot/L_\odot$, the solar mass divided by its luminosity. For the Sun and for a system consisting of solar type stars the mass-to-light ratio is 1. Refining this concept, for old stars such as those in elliptical galaxies the mass-to-light ratio should typically be 6 units (Sargent et al. 1978). This ratio should be the global value of a galaxy or cluster of galaxies as a whole if there are no other significant mass components. Being more specific, the wavelength band where the light is measured must be specified. Hereafter, we refer to mass-to-light ratios which are based on measurements in the visual region of the spectrum.

12.1   Dark matter in the Milky Way disk near the Sun

We may determine the surface density of the local portion of our Galactic disk by observing the motions of stars (like the Sun) perpendicular to the disk, which was first done by Öpik (1915).

As an example, in Figure 12.1, we display the orbit of the Sun, up and down through the central plane of the Galaxy, and in radial direction inward and outward of the Galactic center. The speed perpendicular to the disk is about $Z_0 = 7$ km/s for the Sun. An oscillatory motion results, with the time from the plane to maximum equal to $t_{turn} = 16.5 \pm 1.5 \times 10^6$ yr, one quarter of the whole period. Simple derivation and detailed modeling of the solar motion has shown that the oscillation period

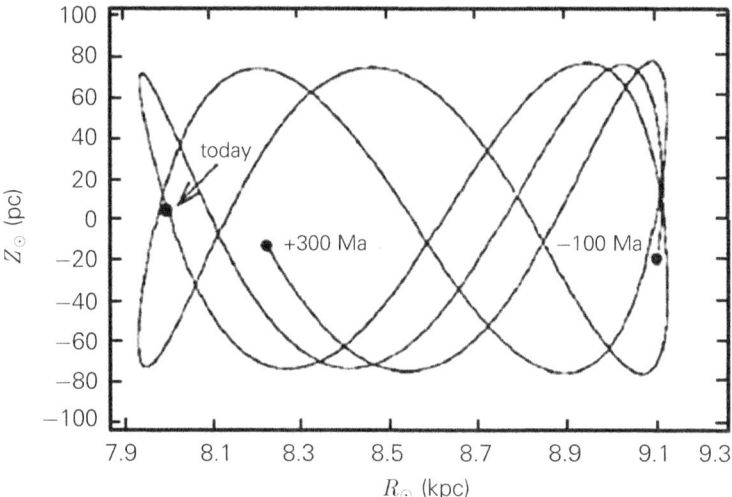

Figure 12.1. Motion of our Sun through the midplane of the Galaxy from 300 million years in the past to 100 million years in the future. Today's position is marked. Coordinates are the distance from the Galactic center along the horizontal axis and the distance from the Galactic midplane positive and negative along the vertical axis. Note that vertical motions are magnified.

and the disk gravitating matter are connected to each other (Matese et al. 1995) as shown in Figure 12.1.

To better understand how this motion arises, consider an idealized model, the motion along the axis, taken to be the z-coordinate of a homogeneous disk of outer radius R. Take the other two cylindrical coordinates to be r and ϕ, i.e., the radial distance from the axis and the angle around it.

A differential element of disk surface density Σ at $(r, \phi, 0)$ contributes a differential amount to the potential on the axis, at height z of

$$\Delta \Phi = -G\Sigma \frac{r\,d\phi\,dr}{\sqrt{z^2 + r^2}}.$$

Integrating over all radial distances from 0 to R and over ϕ, we get

$$\Phi(z) = -G\Sigma \int_0^R dr \int_0^{2\pi} \frac{r\,d\phi}{\sqrt{z^2 + r^2}} = -2\pi G\Sigma \int_0^R \frac{r\,dr}{\sqrt{z^2 + r^2}}$$

$$= -2\pi G\Sigma \sqrt{z^2 + r^2}\,\Big|_0^R = -2\pi G\Sigma \left(\sqrt{z^2 + R^2} - \sqrt{z^2} \right).$$

Let $z \ll R$. so we can specify the potential to the first order in z/R as

$$\Phi(z) = -2\pi G\Sigma(R - z)$$

Taking the derivative, the z force component is

$$F_z = -\frac{d\Phi}{dz} = -2\pi G\Sigma$$

It is interesting that this is independent of the height z. Applying this to the local neighborhood of the Sun and integrating the above gives

$$v_z = -2\pi G\Sigma t + Z_0.$$

where Z_0 is the z-component of the star's velocity at $z = 0$ at the time $t = 0$. The velocity is zero at time $t = t_{turn}$

$$t_{turn} = \frac{Z_0}{2\pi G\Sigma}$$

a turning point in the motion of the star. The second integration of the velocity over time gives

$$z = -\frac{1}{2}2\pi G\Sigma t^2 + Z_0 t.$$

After substituting $t = t_{turn}$, the maximum height of the orbit is

$$z_{max} = \frac{1}{2}\frac{Z_0^2}{2\pi G\Sigma}.$$

Solving for the surface density:

$$\Sigma = \frac{Z_0^2}{4\pi G z_{max}}$$

When Oort (1932, 1965) substituted the best observational values for the locations and speeds of stars into this formula, the result was a total surface density $\Sigma \approx 90\, M_\odot/\mathrm{pc}^2$. If this is correct, there is as much dark matter in the Galactic disk as there is visible matter whose contribution is about $\Sigma \approx 53 \pm 8\, M_\odot/\mathrm{pc}^2$ to the surface density, using modern data (Holmberg and Flynn 2004). Bahcall (1984a, b) used more modern extended data on local star locations and motions led to $\Sigma \approx 75\, M_\odot/\mathrm{pc}^2$, a bit closer to the census of local stars and gas.

This was a local dark matter problem. A better model connects the potential at each level z to the local matter density via the Poisson equation. Going through the proper theoretical modeling and substituting of the latest observational data, e.g., from the Hipparcos satellite, one obtains a result which agrees with the $\Sigma \approx 53\, M_\odot/\mathrm{pc}^2$ observations of the local luminous matter density (Holmberg and

Flynn 2004). As a result of improved data and models, the local dark matter problem has disappeared in recent years. It is interesting to note that there were estimates by Öpik (1915) and later by Kuzmin (1952, 1955) along the same lines which dispelled the dark matter problem, but they were not taken seriously by the research community. Local solar neighborhood dark matter proved to be "false alarm" which took a long time to resolve.

12.2   Dark matter discovery in clusters via the Virial Theorem

The first durable signature of dark matter was recognized by the Swiss-American astronomer Fritz Zwicky in 1933. He studied the Coma cluster of galaxies and used the peculiar (relative to the center of mass of the system) velocities of galaxies to estimate the mass of the cluster. This has been determined using the Virial Theorem in the form

$$2E_{KIN} + W = 0. \tag{12.1}$$

where E_{KIN} is the mean kinetic energy of the motions of galaxies (excluding internal stellar motions) and W is their mean potential energy. Because of its importance in the discovery of dark matter we will discuss the derivation of the Virial Theorem pointing out assumptions which are important in interpreting the results. The derivation of the Virial Theorem is found in Box 11.

The Virial Theorem is about averages, not about instantaneous values. Also for the derivation of the Virial Theorem, the assumption of boundedness is essential. If the system has nearby escapers, the theorem does not apply. Thus there is a problem with the Virial Theorem. Actually, many body systems do have escapers. One has to identify them and to leave them out of the calculation before applying the theorem. Due to the limits of observations, it is an impossible task and therefore the Virial Theorem usually gives only limiting values of the quantities calculated (Aarseth and Saslaw 1972). Finally, as we discuss in the latter part of this book, dark energy can modify the Virial Theorem. Although it does not have any strong effects in clusters of galaxies or within galaxies, it can have significant effects for groups of galaxies.

When we carry out an analysis of the mass in clusters of galaxies, or even in a sample of triple galaxies or binary galaxies, we have to compensate somehow for loss of information. One can only view the system from one direction so only two spatial coordinates and one velocity coordinate for a galaxy are known. Three coordinates are missing.

However, this lack of information can be overcome. Take the projected position vector on the sky of a galaxy be $\mathbf{R}_i$, projected from the true position vector $\mathbf{r}_i$.

Consider another galaxy, j, let the angle between $\mathbf{r}_i - \mathbf{r}_j$ and the line of sight be θ_{ij}. Then

$$|\mathbf{R}_i - \mathbf{R}_j| = |\mathbf{r}_i - \mathbf{r}_j| \sin \theta_{ij}. \tag{12.2}$$

When i and j move around, we may assume that the time average of $|\mathbf{R}_i - \mathbf{R}_j|$ is obtained by taking the average of $\sin \theta_{ij}$ over all directions. Assume that

$$\left\langle \frac{1}{|\mathbf{R}_i - \mathbf{R}_j|} \right\rangle = \frac{1}{|\mathbf{r}_i - \mathbf{r}_j|} \left\langle \frac{1}{\sin \theta_{ij}} \right\rangle$$

$$\left\langle \frac{1}{\sin \theta_{ij}} \right\rangle = \frac{\int_0^{2\pi} \int_0^\pi \frac{1}{\sin \theta} \sin \theta \, d\theta \, d\varphi}{\int_0^{2\pi} \int_0^\pi \sin \theta \, d\theta \, d\varphi} = \frac{\pi}{2} \tag{12.3}$$

and replace $1/r_{ij}$ in Equations (12.1) and (B11.8) by

$$\frac{2/\pi}{|\mathbf{R}_i - \mathbf{R}_j|}. \tag{12.4}$$

This latter quantity can be obtained from observations. Calculating the kinetic energy, we can take

$$\langle v_i^2 \rangle = 3 \langle (v_r)_i^2 \rangle \tag{12.5}$$

if $(v_r)_i$ is the radial (Doppler shift) velocity of galaxy i relative to the cluster center. Define the mean projected radius R of the cluster by

$$\frac{1}{3R} = \frac{2}{\pi M^2} \sum_{i=1}^n \sum_{j=i+1}^n \frac{m_i m_j}{|\mathbf{R}_i - \mathbf{R}_j|} \tag{12.6}$$

whereby

$$\langle W \rangle = -\frac{GM^2}{3R}. \tag{12.7}$$

Also define the radial velocity dispersion squared by

$$\sigma_r^2 = \frac{\sum m_i (v_r)_i^2}{M} \tag{12.8}$$

which gives

$$2 \langle E_{KIN} \rangle = 3 M \sigma_r^2. \tag{12.9}$$

Therefore Equation (12.1) leads to

$$M = \frac{9\sigma_r^2 R}{G}. \tag{12.10}$$

This equation is specifically valid for a system whose brightness distribution follows **de Vaucouleurs' Law**. Here R is the effective radius in which half of the light on the sky is projected inside this radius. Also σ_r is the dispersion of radial velocities at the central region. In terms of actual stellar systems, de Vaucouleurs' law is an observationally based relation between the surface brightness of a stellar system and the projected distance from the center of the system. It fits well elliptical galaxies and central spheroids of spiral galaxies.

It is common to use the **Plummer sphere** model with a spherically symmetric density distribution for clusters of galaxies.

$$\rho(r) = \frac{3a^2}{4\pi} \frac{M}{(r^2 + a^2)^{5/2}} \tag{12.11}$$

This distribution has the corresponding spherical gravitational potential

$$\Phi(r) = -\frac{GM}{\sqrt{r^2 + a^2}}. \tag{12.12}$$

For this potential, the potential energy is

$$W = -\frac{GM^2}{32a/3\pi}. \tag{12.13}$$

This potential agrees with Equation (12.7) if $R \cong 1.13a$. We see that Equation (12.10) applies to Plummer spheres with $R \cong a$.

Problem Show that the Plummer sphere of density distribution, Equation (12.11), has the potential and potential energy given by Equations (12.12) and (12.13).

Zwicky (1933) used the Virial Theorem in estimating the total mass M of the Coma cluster of galaxies. The Coma cluster is an approximately spherical system, which is regular in shape, so that it was reasonable to believe that the cluster is in virial equilibrium in its own gravity. With the modern measured velocity $\sigma_r \cong 1\,000$ km/s and the effective radius $R \cong 1$ Mpc, the virial (gravitational) mass of the cluster is $M_{vir} \approx 2 \times 10^{15} M_\odot$. This is approximately the same large mass Zwicky obtained but with up-to-date data.

Box 11 Derivation of the Virial Theorem

Take n galaxies each of mass m_i be in positions $\mathbf{r}_i$ and which move with velocity $\dot{\mathbf{r}}_i$. The virial is defined as

$$A = \sum_{i=1}^{n} m_i \mathbf{r}_i \cdot \dot{\mathbf{r}}_i \qquad \text{(B11.1)}$$

The virial's time derivative is

$$\dot{A} = \sum_{i=1}^{n} m_i \dot{\mathbf{r}}_i \cdot \dot{\mathbf{r}}_i + m_i \ddot{\mathbf{r}}_i \cdot \dot{\mathbf{r}}_i = 2T + \sum_{i=1}^{n} \mathbf{F}_i \cdot \mathbf{r}_i. \qquad \text{(B11.2)}$$

having made use of Newton's second law connecting the force $\mathbf{F}_i$ and acceleration $\ddot{\mathbf{r}}_i$ on galaxy i. Then take a time average of both sides over a long interval τ:

$$\langle \dot{A} \rangle = \frac{1}{\tau} \int_0^\tau \dot{A}\, dt = \langle 2E_{KIN} \rangle + \left\langle \sum_{i=1}^{n} \mathbf{F}_i \cdot \mathbf{r}_i \right\rangle. \qquad \text{(B11.3)}$$

For a system that remains bounded, $\langle \dot{A} \rangle \to 0$ when $\tau \to \infty$. And thus

$$2\langle E_{KIN} \rangle + \left\langle \sum_{i=1}^{n} \mathbf{F}_i \cdot \mathbf{r}_i \right\rangle = 0 \qquad \text{(B11.4)}$$

which is a general form of the Virial Theorem.

Assuming the Newtonian gravitational force:

$$\mathbf{F}_i = -G \sum_{j=1, j \neq i}^{n} m_i m_j \frac{\mathbf{r}_i - \mathbf{r}_j}{\left|\mathbf{r}_i - \mathbf{r}_j\right|^3} \qquad \text{(B11.5)}$$

Evaluating Equation (B11.4) for the Newtonian case,

$$\sum_{i=1}^{n} \mathbf{F}_i \cdot \mathbf{r}_i = -G \sum_{i=1}^{n} \sum_{j=1, j \neq i}^{n} m_i m_j \frac{\mathbf{r}_i - \mathbf{r}_j}{r_{ij}^3} \cdot \mathbf{r}_i \qquad \text{(B11.6)}$$

where $r_{ij} = |\mathbf{r}_i - \mathbf{r}_j|$. Writing the double summation twice and changing the role of summation indices, the two identical sums can be added together and divided by 2:

$$\sum_{i=1}^{n} \mathbf{F}_i \cdot \mathbf{r}_i = -\frac{1}{2} G \left[\sum_{i=1}^{n} \sum_{j=1, j \neq i}^{n} m_i m_j \frac{\mathbf{r}_i - \mathbf{r}_j}{r_{ij}^3} \cdot \mathbf{r}_i + m_j m_i \frac{\mathbf{r}_j - \mathbf{r}_i}{r_{ji}^3} \cdot \mathbf{r}_j \right]$$

$$\text{(B11.7)}$$

$$= -\frac{1}{2} G \left[\sum_{i=1}^{n} \sum_{j=1, j \neq i}^{n} m_i m_j \frac{\mathbf{r}_i - \mathbf{r}_j}{r_{ij}^3} \cdot (\mathbf{r}_i - \mathbf{r}_j) \right].$$

Due to force symmetry relative to i and j (Newton's third law), the double sum gives every term twice. To have every term once, the latter sum should start from $j = i + 1$. Doing this and multiplying the right hand side by two:

$$\sum_{i=1}^{n} \mathbf{F}_i \cdot \mathbf{r}_i = -G \left[\sum_{i=1}^{n} \sum_{j=i+1}^{n} m_i m_j \frac{\mathbf{r}_i - \mathbf{r}_j}{r_{ij}^3} \cdot (\mathbf{r}_i - \mathbf{r}_j) \right]$$

(B11.8)

$$= -G \sum_{i=1}^{n} \sum_{j=i+1}^{n} \frac{m_i m_j}{r_{ij}} = W.$$

Substituting into Equation (B11.4) gives the Virial Theorem

$$2 \langle E_{KIN} \rangle + \langle W \rangle = 0.$$ (B11.9)

Once the gravitating mass is estimated, the next step is the calculation of the mass-to-light ratio of the system whose total luminosity, L, has been measured. For a cluster of stars or a single galaxy, the L of all stars is measured. For a cluster of galaxies, the total L of the stars in all the galaxies is also measured. Then the mass-to-light ratio is calculated. **Mass-to-light ratio** is defined as follows: The Sun's mass is about $M_\odot = 2 \times 10^{33}$ g and its luminosity $L_\odot \cong 4 \times 10^{33}$ erg/sec. The unit is $M_\odot / L_\odot$. For the Sun or for any system of stars consisting of solar type stars the mass-to-light ratio is 1. For systems made only of old stars such as those in elliptical galaxies the mass-to-light ratio should be typically 6 units (Sargent et al. 1978). This should be the value for a galaxy or cluster of galaxies as a whole if no other signifi-cant mass components (such as dark matter or non-luminous baryonic matter) are present. Also to be exact, the wavelength band where light is measured must be specified. Typically, in this book mass-to-light ratios which are based on measure-ments in the visual region.

Using the Sun's $M_\odot / L_\odot$ as the unit, Zwicky (1933) applied the Virial Theorem to the Coma cluster of galaxies. Recall that the Coma cluster is approximately spherical, regular in shape so Zwicky (1933) used the Virial Theorem to estimate its gravitational mass, $M_{vir} \approx 2 \times 10^{15} M_\odot$. For a total luminosity of $L \approx 5 \times 10^{12} L_\odot$, the mass-to-light ratio is thus

$$M/L \sim 400.$$ (12.14)

If the galaxies were made mainly of ordinary stars, we would expect only

$$M/L \sim 5.$$ (12.15)

This huge difference was the first hint that much of the mass is in the form of dark matter.

12.3   Subclusters in rich clusters of galaxies

Recall our critical discussion of the assumption in the Virial Theorem that the cluster is bounded with no escapers. Actually, this assumption is not fully satisfied for the Coma Cluster since it has two dominant galaxies in its center. These two create a binary gravitational well which affects strongly the orbits of the smaller galaxies and even throws some of them out of the cluster. Superficially, clusters of galaxies appear to be relaxed systems similar to globular star clusters. However, in 1971 the American astronomer Herbert J. Rood and the Indian-American astronomer Gummuluru N. Sastry of Wesleyan University noted that there is a class of clusters, called "binary", dominated by two bright galaxies. The comparatively nearby Coma Cluster belongs to this class, which constitutes 10% of all rich clusters. In 1974 Rood noticed subclustering around the two bright supergiant galaxies of the Coma cluster, NGC4874 and NGC4889. Valtonen and Byrd (1979) concluded from this and other data that the two supergiants are independent dynamical units which have massive extensive dark matter halos. They studied the effect of a massive central binary to the cluster as a whole as shown in Figure 12.2.

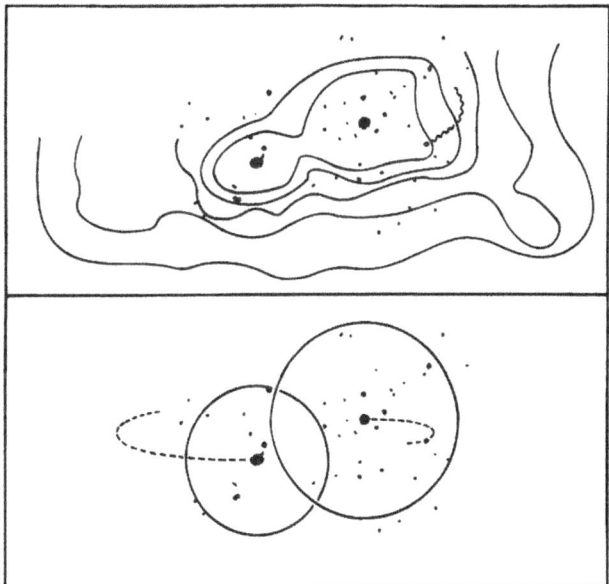

Figure 12.2. Central part of the Coma Cluster according to Valtonen and Byrd (1979). Indicated are two dominant galaxies NGC 4874 (right) and NGC 4889 (left) as well as their companions of known redshift out to 150 and 120 kpc, respectively. The upper frame shows contours of diffuse light and the approximate radio tail of NGC 4869. Circles in the lower diagram show the approximate extents of major galaxies. Dashed lines indicate possible orbits for the two.

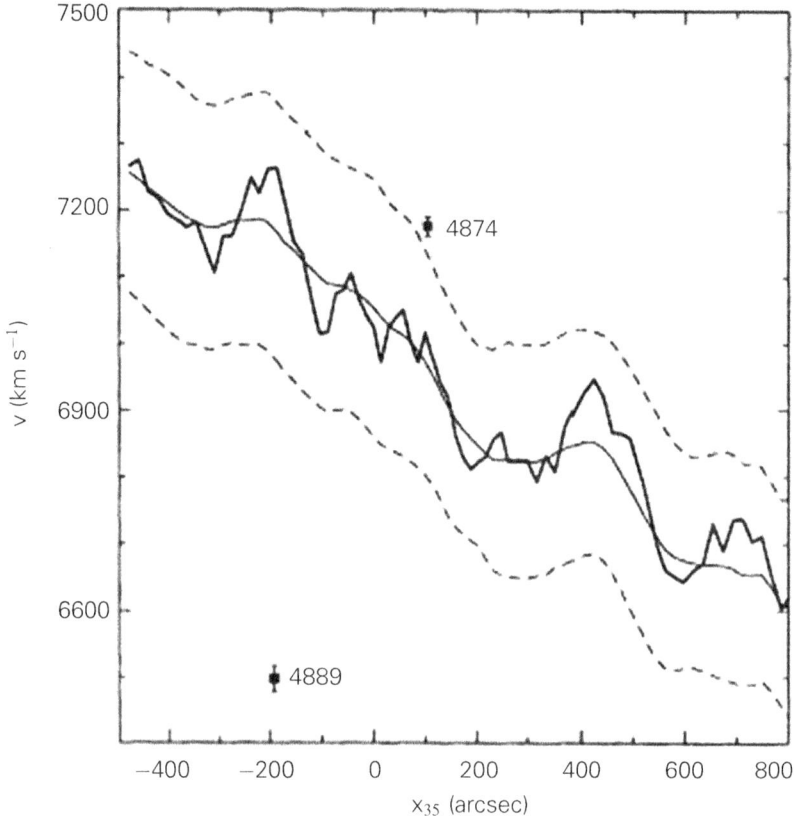

Figure 12.3. The average radial velocity of faint galaxies in the Coma cluster (solid line) displayed as a function of the position x along the cluster major axis. The slant of the curve is indicative of net rotation in the orbital motions. It is retrograde (in the opposite sense to the relative motion of the binary pair of supergiant galaxies NGC 4874 and NGC 4889). The binary members are also marked. Dashed lines indicate limits of uncertainty (Biviano et al. 1996).

Evidence of subclustering discussed by Rood became stronger by the mid-1990s when it was widely accepted. Not only are the galaxies in the central region of the Coma cluster well separated in two clumps, both in the sky and in redshift (Colless and Dunn, 1996), but in x-ray emission the two supergiant galaxies have their own, extensive gaseous halos. These gaseous halos are an order of magnitude cooler than the cluster gas as a whole (Vikhlinin et al. 2001). New galaxy velocity data also revealed an interesting consequence of the binary action on the cluster galaxies. The system of fainter galaxies appears to be in rotational orbital motion inside the 1 Mpc radius from the cluster center. The rotation is retrograde relative to the orbital motion of the two supergiant galaxies (Figure 12.3, Biviano et al. 1996).

This rotational orbital motion is a strong clue to its origin. Retrograde orbits are more stable near a binary than direct orbits (see e.g., Valtonen and Karttunen 2005). Galaxies from direct orbits would have been ejected further out from the central 1 Mpc volume. The remaining galaxies which are in stable orbits show net retrograde motion. The concept of Valtonen and Byrd (1979) is thus verified. More up-to-date modeling of the Coma cluster with a binary center has been carried out by Laine et al. (2004).

One important consequence of the binaries in clusters of galaxies is the fact that such binaries tend to throw out lesser galaxies by the **slingshot** mechanism. The speeds of the ejected galaxies are comparable to the orbital speed of the binary. If the orbital speed is greater than the escape velocity from the cluster, some of the cluster galaxies will be found in escape orbits but still near the cluster on the sky. It is thus difficult to know which ones are escaping. Thus, use of the Virial Theorem in determining the cluster mass is not valid unless the escapees are excluded in the statistical treatment. Using the Virial Theorem without regard to the existence of escapers, the calculated cluster mass may exceed the true mass by as much as a factor of two. There is evidence for this over-estimate in the Coma cluster where the x-ray mass determination gives a lower value than the Virial mass estimation (Cowie et al. 1987), but the error bars are too large for a definite conclusion (Balland and Blanchard, 1997).

More generally, at least a third of galaxy clusters contain substructure (Crone et al. 1996, Dahle et al. 2002). Cluster substructure arises readily in cosmological simulations (Moore et al. 1998). Structures in all scales develop during cosmological evolution. Substructures, being smaller in scale than the cluster structure, develop first, and some will be found inside the larger structures later. The subclusters' survival within the clusters depends on their densities. In the case of the Coma cluster, the binary sorts out the retrograde and direct orbits of the cluster member galaxies, preferentially ejecting the latter. Typically, during the evolution of the universe and the clusters, in it, the central subclusters form at redshifts $z = 1 - 3$ from the merging of a dozen halos, according to numerical simulations (Ghigna et al. 2000). Separate subclusters can survive for a long time (Moore et al. 1998).

Even though the existence of substructure leaves mass estimates for the whole cluster somewhat uncertain, there is strong observational evidence that the subclusters themselves are strongly dark matter dominated. The mass-to-light ratio of the substructures is as high as or even higher than in the cluster as a whole. Thus substructures do not negate the evidence for dark matter.

12.4   Dark matter discovery in clusters via the cluster gas

We already hinted above at another independent method of determining the cluster mass and its dark matter content, namely using the cluster gas as a probe of the

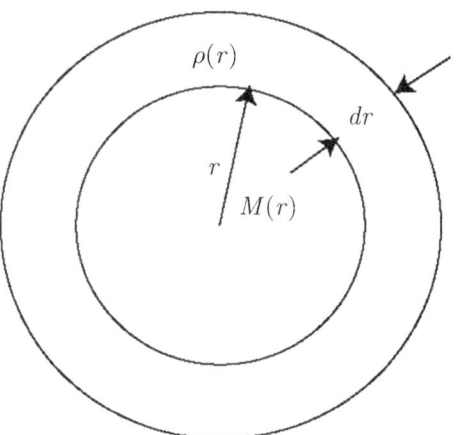

Figure 12.4. The hydrostatic equilibrium in a spherically symmetric potential.

gravitational potential of the cluster. We will discuss it in this section, and describe yet another method, gravitational lensing by the cluster mass in a later section. The results of different methods have been in remarkably good agreement with each other. In every method the dark matter has been found to be dominant. Therefore we will here consider the x-ray emitting gas in the potential well created by the dark matter.

Figure 12.4 shows a spherical shell of gas in a spherically symmetric cluster potential. Here the radius of the shell is r, thickness is dr, and gas density is $\rho(r)$.

Gravitational force towards the cluster center is

$$\frac{GM(r)}{r^2} 4\pi r^2 \rho(r)\, dr \tag{12.16}$$

where $M(r)$ is cluster mass within r. Gravitational force is balanced by that arising from the pressure difference dP across the thickness of the shell:

$$4\pi r^2\, dP. \tag{12.17}$$

Equating the above two gives the condition for hydrostatic equilibrium:

$$\frac{dP}{dr} = -\frac{GM(r)}{r^2}\rho. \tag{12.18}$$

Considering hydrogen gas, its pressure at temperature T_e is

$$P = \rho\,\frac{k_B T_e}{m_H}. \tag{12.19}$$

Thus

$$\frac{dP}{dr} = \frac{k_B}{m_H}\left(T_e\frac{d\rho}{dr} + \rho\frac{dT_e}{dr}\right)$$

$$= \frac{k_B\rho T_e}{m_H}\left(\frac{d\rho}{\rho dr} + \frac{dT_e}{T_e dr}\right) \qquad (12.20)$$

$$= \frac{k_B\rho T_e}{m_H r}\left(\frac{d(\ln\rho)}{d(\ln r)} + \frac{d(\ln T_e)}{d(\ln r)}\right).$$

Substituting Equation (12.20) into Equation (12.18) and solving for $M(r)$:

$$M(r) = -\frac{k_B T_e}{G m_H}r\left(\frac{d(\ln\rho)}{d(\ln r)}\frac{d(\ln T_e)}{d(\ln r)}\right). \qquad (12.21)$$

Assuming equilibrium, x-ray observations can give T_e. The two logarithmic derivatives enable calculation of the mass distribution.

Modern x-ray observations allow the separate determination of the density and the temperature profiles of clusters of galaxies. Earlier, it was common to assume a density profile of a particular kind and perhaps a constant temperature which are still convenient for discussion. The **beta model** (Cavaliere and Fusco-Femiano, 1976) was widely used in which the cluster mass as a function of r is given by

$$M(r) = 1.6\cdot10^{15}\beta\left(\frac{T_e}{10\text{ keV}}\right)\left(\frac{r}{\text{Mpc}}\right)\frac{(r/r_c)^2}{1 + (r/r_c)^2}M_\odot$$

where temperature T_e units are 10 keV and r is in Mpc. For a sample of 14 clusters Lewis et al. (1999) find $\beta \cong 0.72$, $T_e \cong 6.5$ keV, and $r_c \cong 145$ kpc (Figure 12.5).

The calculated cluster mass within 1 Mpc from the center is typically $7\times10^{14}M_\odot$. In these 14 clusters the Virial mass/x-ray mass ratio was calculated to be 1.04 ± 0.07. The fraction of mass of the cluster in gas was calculated to be about 8%; in some other work values like 10% have been found (Schmidt et al. 2001). Although this is greater than the total mass in all the stars of the member galaxies by a factor of 5 or so, only about 10% of the cluster is baryonic gas or stars, the great majority is dark matter (Girardi et al. 1998). As we have mentioned earlier, this dark matter superiority applies for the universe as a whole.

According to our discussions of Big Bang nucleosynthesis, baryonic matter can be only about 4% of the critical density. Approximately the same value is obtained by observing the interstellar gas at redshift $z \cong 2$ using the absorption lines of the Lyman α forest (discussed in Chapter 11, Cen and Ostriker 1999). Therefore, if 4% of the critical density equals 10% of the dark matter density (0.04 critical = 0.10 dark matter), dark matter can only contribute roughly 4/10 of the critical density

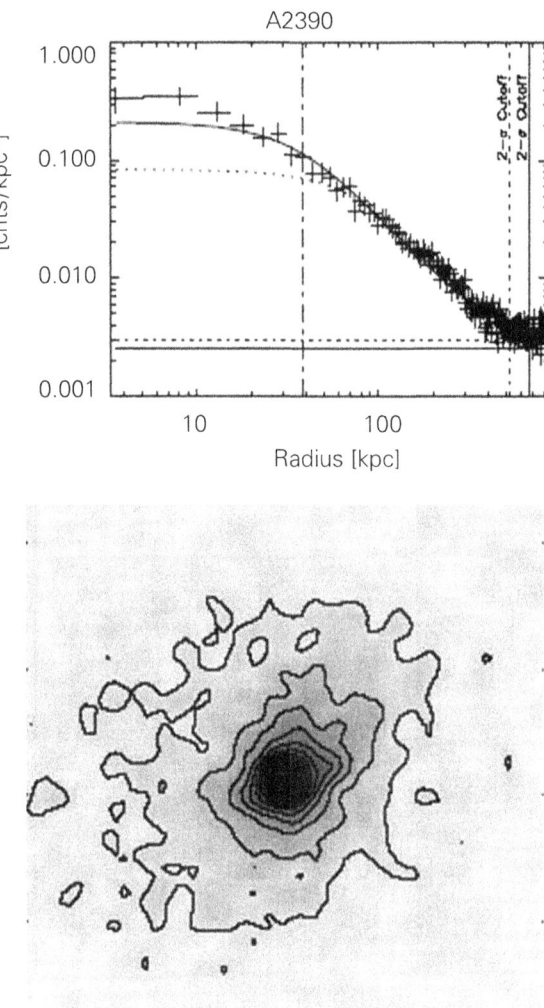

Figure 12.5. The beta model fit to the brightness profile of the cluster of galaxies Abell 2390 (upper). X-ray emission from the cluster of galaxies Abell 2390 (lower) representing a 4.5 × 4.5 degree field (Lewis et al. 1999). Model parameters are $\beta = 0.62$, $r_c = 145$ kpc, and $T_e = 8.9$ keV. X-ray mass out to $r = 740$ kpc is $M(r) = 4.5 \times 10^{14} M_\odot$; the ratio of dynamical/x-ray mass is 1.09 ± 0.13. Gas mass fraction is 15% inside this r. The jump of the x-ray brightness near the center of the cluster, above the beta model, indicates a flow of gas towards the cluster center of 500 $M_\odot$ per year (a "cooling" flow or a deviation from hydrostatic equilibrium).

of the universe. This assumes that a cluster of galaxies is a fair sample of the whole universe (White et al. 1993). In terms of the notation we have used, $\Omega_M \approx 0.4$ where M represents both baryonic and dark matter which are both gravitating matter.

It is interesting that these observations show that $\Omega_M = 1$ due to baryonic and dark matter is totally excluded. However, Ω_M could be lower than 0.4 since all the gaseous matter may not have been detected so far. Recall our discussion of the Lyman Alpha forest. Cooler gas at $T_e \approx 10^6$ K may be quite abundant, making a significant addition to the baryon budget (Valageas et al. 2002, Bonamente et al. 2003). Another sample of 45 clusters of galaxies shows a greater gas fraction placing a more stringent upper limit: $\Omega_M \leq 0.3$ (Mohr et al. 1999). Allen et al. (2004) find

$$\Omega_M = 0.24 \pm 0.04,$$

where a flat geometry of the universe was assumed along with a Hubble constant of $H_0 = 70$ km/s/Mpc.

12.5   Dark matter in the Milky Way disk and its halo

One of the first suspected locations of dark matter was in the disk of our Galaxy near the Sun. As we have discussed, after much work and final improvement in the data the suspected local disk dark matter is greatly reduced from what it was first thought to be. However, in the process of the work our Galaxy's dominant and extensive dark matter halo out of the disk plane has been discovered.

It is now known that the Sun circles the center of our Galaxy at the distance of about 8 kpc or 1.6×10^9 AU. In a summary of the results, Eisenhauer et al. (2003) reported a geometric measurement to the center with an uncertainty of about 5%. They combined astrometric and spectroscopic measurements of the star S2 orbiting the massive black hole candidate in the Galactic center. The solution for the best-fit Kepler orbit gave the needed parameters to calculate the distance $R_0 = 7.94 \pm 0.38$ kpc.

The Dutch astronomer Jan H. Oort of the University of Leiden and the Swedish astronomer Bertil Lindblad (1925) of Stockholm Observatory were the first to confirm that disk stars like the Sun have an orderly nearly circular motion about the Galactic center and also relative to the globular clusters and halo stars. Lindblad proposed an orbital velocity for the Sun of about 300 km/s; modern measurements have reduced this to $\sim$220 km/s. Our Galaxy is thus a disk of stars, plus gas and dust with a more spherical halo containing globular clusters and other objects.

Although dust blocks our view of the distant disk, in 1932, J. Oort showed that one could deduce information about the mass of the Galaxy from purely local disk observations. Oort used observations of both high velocity stars and disk stars close to the Sun. Because of the elongatioon of their orbits, high velocity stars show a "drift" relative to the solar motion. Oort pointed out that this drift direction was 90° from the direction of the center of the globular cluster system concluding that the Sun was orbiting about a center defined by the center of the globular cluster system.

Velocities of stars in the Galaxy are described by their three Cartesian components Θ, Π, and Z. The velocity along the Sun-Galactic center line Π is measured positive in the direction opposite to the Galactic center. The velocity $90°$ to the plane, Z is measured positive toward the north Galactic pole. This normal coincides with the hemisphere of the sky away from the Milky Way within which the Earth's north pole is located. Finally, the tangential component Θ is measured positive in the direction of the Sun's (circular) orbital motion around the center of the Galaxy, clockwise around the center looking down on the disk from the north.

Oort noted patterns in measured proper motions and radial velocities of nearby stars relative to the Sun at different Galactic longitudes that were predictable if the disk rotated differentially (with increasing orbital periods for increasing distance from the center). Radial velocities are measured using the Doppler effect while proper motions are angular motions across the sky in arcsec/yr. Oort noted the following patterns. Stars at longitude $0°$, $90°$, $180°$ and $270°$ show small radial velocities relative to the Sun. Galactic longitudes are the angle between the direction to the star in the disk plane from the direction of the Galactic center counter clockwise as as seen from far above the Sun in the norther half of the sky. Moreover, he found that stars at longitude $45°$, $135°$ and $225°$, and $315°$ showed **minima** (negative Doppler shifts) and **maxima** (positive Doppler shifts) respectively of radial velocities.

The above pattern can be derived. Consider stars (or gas clouds) traveling in circular orbits near the Sun in the Galactic plane at a distance R_0 from the center moving at orbital angular rate ω_0. The angular speed for a star at R is ω. The magnitude of the orbital velocity vector is $R\,\omega$ for the star, with a similar expression for the Sun. Indicate Galactic longitude with ℓ (Figure 12.6).

Take the star to be at the distance d from the Sun, with an angle γ between the Sun's circular orbit velocity vector and the Sun-star line. Extending d through the star, define another angle α between this line and the star's orbital velocity vector line. The size of the radial velocity difference along the line d is

$$v_r = (\omega R)\cos\alpha - \omega_0 R_0 \cos\gamma. \qquad (12.22)$$

In the Sun-star-Galactic center triangle, the angle at the Sun is ℓ and the interior angle at the star is β. Now $\cos\gamma = \sin\ell$, and by the law of sines,

$$\frac{\sin\ell}{R} = \frac{\sin\beta}{R_0}. \qquad (12.23)$$

Using

$$\beta = 90° + \alpha \rightarrow \sin\beta = \cos\alpha \qquad (12.24)$$

and Equation (12.23)

$$\cos\alpha = \sin\beta = \frac{\sin\ell}{R} R_0. \qquad (12.25)$$

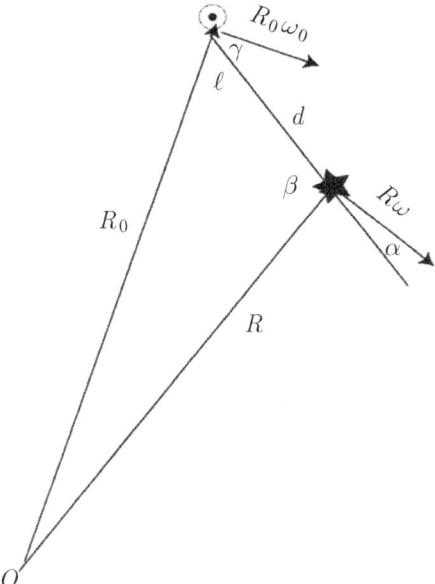

Figure 12.6. Schematic diagram showing the distance from the Galactic Center R_0, to the Sun and to a star R along with their respective velocity vectors in the galactic plane looking down from the North Galactic Pole.

The velocity difference along the line of sight or the radial velocity of the star relative to the Sun is

$$v_r = (\omega R_0)\sin\ell - \omega_0 R_0 \sin\ell = R_0(\omega - \omega_0)\sin\ell. \qquad (12.26)$$

There is a sinusoidal variation of radial Doppler shift velocity with ℓ corresponding to the observations by Oort described earlier. The two angular rates approach one another at the same distance from the center of the Galaxy. Then the variation with ℓ approaches zero. Oort only observed stars fairly close to the Sun so that

$$\omega - \omega_0 \cong (R - R_0)(d\omega/dR)_0. \qquad (12.27)$$

Here

$$\left(\frac{d\omega}{dR}\right)_0 = \frac{1}{R_0}\left(\frac{d\Theta}{dR}\right)_0 - \frac{\Theta_0}{R_0^2}, \qquad (12.28)$$

where $\omega = \Theta/R$. Evaluation at the solar orbital radius is indicated by the 0 subscript here and in Figure 12.6. Thus

$$v_r = \left[\left(\frac{d\Theta}{dR}\right)_0 - \frac{\Theta_0}{R_0}\right](R - R_0)\sin\ell. \qquad (12.29)$$

For objects at a small distance, d, close to the Sun and its orbital radius,

$$R_0 - R \approx d \cos \ell. \tag{12.30}$$

Substituting in the equation above and also using $\sin \ell \cos \ell = (1/2) \sin 2\ell$, one obtains as Oort observed a double sinusoidal variation with ℓ over $360°$,

$$v_r = A d \sin 2\ell \tag{12.31}$$

where

$$A = \frac{1}{2} \left[\frac{\Theta_0}{R_0} - \left(\frac{d\Theta}{dR} \right)_0 \right]. \tag{12.32}$$

This expression is **Oort's constant** A. Note that the radial velocity is zero toward the Galaxy center, at $90°$, and at $180°$.

As mentioned earlier, we can mathematically find a periodic pattern in proper motions with Galactic longitude which was also observed by Oort. The magnitude of the velocity difference transverse to the line of sight is

$$v_t = (\omega R) \sin \alpha - \omega_0 R_0 \sin \gamma = (\omega R) \sin \alpha - \omega_0 R_0 \cos \ell. \tag{12.33}$$

Now

$$R \sin \alpha = R_0 \cos \ell - d$$

which can be substituted in the tangential velocity,

$$v_t = \omega R_0 \cos \ell - \omega_0 R_0 \cos \ell - \omega d$$

$$= (\omega - \omega_0) R_0 \cos \ell - \omega d.$$

Substitute for

$$\omega - \omega_0 \cong \left(\frac{d\omega}{dR} \right)_0 (R - R_0)$$

$$= \left(\frac{1}{R_0} \left(\frac{d\Theta}{dR} \right)_0 - \frac{\Theta_0}{R_0^2} \right) (R - R_0)$$

Use Equation (12.29), and note that $\omega d \approx \omega_0 d$:

$$v_t = - \left(\left(\frac{d\Theta}{dR} \right)_0 - \frac{\Theta_0}{R_0} \right) d \cos^2 \ell - \left(\frac{\Theta_0}{R_0} \right) d$$

since $R - R_0 = -d \cos \ell$.

Use the identity, $\cos^2 \ell = (1/2)(1 + \cos 2\ell)$, to obtain

$$v_t = A d \cos 2\ell + B d = d(A \cos 2\ell + B).$$

Here we define **Oort's constant** B

$$B = -\frac{1}{2}\left(\frac{d\Theta}{dR}\right)_0 - \frac{1}{2}\frac{\Theta_0}{R_0}. \qquad (12.34)$$

The proper motion (corresponding to Oort's observations) is

$$\frac{v_t}{d} = A \cos 2\ell + B. \qquad (12.35)$$

Recent values using much better and more numerous radial velocity and proper motion observations from the Hipparcos satellite are $A = 11.7 \pm 0.8$ (km/s)/kpc and $B = -10.5 \pm 0.6$ (km/s)/kpc (Mignard 1998, Kovalevsky 1998).

The latest values of the Oort constants confirm our understanding of the rotation curve of the Galaxy

$$\left(\frac{d\Theta}{dR}\right)_0 = -(A + B) = -1.2 \pm 1.0 \text{ km/s/kpc}, \qquad (12.36)$$

a slow decline with radius compared to the ~ 220 km/s of the local circular speed, Θ_0. Actually, it may be flat. It should be noted that the determination of the rotation curve beyond the solar radius has been complicated by the lack of good distance indicators at large distances. Different kinds of methods have given complementary information (e.g., Blitz 1979, Petrovskaya and Teerikorpi 1986, Merrifield 1992) which support the view that the rotation curve is approximately flat up to about $2.5 \times R_0 = 20$ kpc.

If $(d\Theta/dR)_0 = 0$, then the disk may be modeled by a **Mestel disk** which has the property that the surface density Σ is

$$\Sigma(R) = \Sigma(R_0)R_0/R \qquad (12.37)$$

and a flat circular rotation speed Θ_0 where

$$\Theta_0^2 = 2\pi G\Sigma(R_0)R_0. \qquad (12.38)$$

If we put the current values of $R_0 = 8$ kpc and $\Theta_0 = 220$ km/s in the equation, one obtains:

$$\Sigma(R_0) \cong 220 \, M_\odot \, \text{pc}^{-2}.$$

In the calculation above, we assumed the orbital motion of the Sun is due entirely to the matter in the disk.

The observed value of the surface density 50 $M_\odot/\mathrm{pc}^2$ is far from the 220 $M_\odot/\mathrm{pc}^2$. Clearly, the local circular motion of the Sun must be primarily determined by something other than the disk matter. However, no other stellar or gaseous components observed via their light or radio emission in the Galaxy exist in amounts that can be responsible for the observed high value of the local rotation speed. The obvious answer is that there is much gravitating dark matter which is inside the solar orbit in our Galaxy. Moreover, this dark matter does not lie in the disk but is in the halo. As we have said, for the disk surface density itself, one can account for almost all the matter as being commonplace stars and gas clouds. So our Galaxy is indeed full of dark matter which dominates the orbital motion of disk material.

The methods described above do not tell us much about the matter outside the solar orbit about the center of the Galaxy. But there is other information that the rotation curve of our Galaxy is fairly flat (constant speed versus radius) much like other disk galaxies. Thus we conclude that the Galaxy has a relatively small baryonic disk and a central bulge which is confined inside a spheriodal dark matter halo of much greater extent and mass.

12.6   Dark matter discovery inside disk galaxies via rotation curves

Dark matter in the Andromeda Galaxy (M31) was detected unambiguously via optical rotation speed, v, measurements of the disk of M31 by Vera C. Rubin and W. Kent Ford, Jr. (Carnegie Institution of Washington, 1970) and via radio data by Morton S. Roberts (National Radio Astronomy Observatory) and Robert N. Whitehurst (University of Alabama). These were combined in 1975 to produce a rotation curve for M31 out to large distances. The rotation curve is a relation between the rotation velocity and the distance from M31 center as seen on the sky. In a disk galaxy like M31, similar to our own Galaxy, stars move along approximately circular orbits around the center of the system. For an approximately spherical distribution of matter, a star undergoes the gravity attraction by all the other stars and other gravitating material within its orbit producing an acceleration v^2/r, where v is its rotational velocity, and r is the orbital radius. The gravitational acceleration is GM/r^2, where M is the mass within the orbit. Equating, a relation for the interior mass as a function of radius of stellar system is:

$$M = v^2 r/G. \tag{12.39}$$

Since v is observed to be constant with respect to radius, the total interior mass of the Andromeda Galaxy increases more or less linearly with distance, out to the most distant point measured at 30 kpc from the center. The mass would probably increase even further if measurements could be continued. The mass cannot be due to stars since the emitted light is very weak at such distances. Thus the dark matter becomes more and more dominant over luminous matter the further we are from the center of the M31.

The dark matter story reached a culmination in the mid-1970s when two important studies appeared, one in Estonia (Einasto et al. 1974a), the other in the USA (Ostriker, Peebles and Yahil 1974). The structure and kinematics of individual spiral galaxies (like the Milky Way or Andromeda Galaxy) were studied using galaxy models and observational material on galactic stellar populations and motions. The conclusions of both studies were the same: there is an extended dark halo around a typical spiral galaxy whose mass is about ten times larger than the total mass of all the stars in the galaxy.

Jaan Einasto, Ants Kaasik, and Enn Saar of the Tartu observatory reported flat rotation curves for other galaxies, publishing it in *Nature* (1974). They used combined data for a half dozen galaxies with well measured rotation curves. In some spiral galaxies, rotation curves become flat at the periphery of the disk. Data on dwarf satellite galaxies rotating around the main galaxy enabled the extension of the rotation curve far beyond the visible stellar disk of the galaxy. Rotation curves can thus be traced to distances about ten times the visible disk radius of the main galaxy. The curves remain practically flat up to these distances. The rotational velocities turn out to be about the same, ~ 200 km/s, at the distance range 10–100 kpc, the same as within 10 kpc from the galaxy center, the typical radius of the stellar disk.

Detection and measurements of individual disk galaxy dark halo masses were made by V.C. Rubin, W.K. Ford, and N. Thonnard (1978) using a large collection of spiral galaxies for which the rotational velocities were measured beyond the visible stellar disks, out where hydrogen clouds and companion dwarf galaxies orbit around the center of the main galaxy.

The above discussion is observational combined with a bit of theory. American astronomers, Jeremiah P. Ostriker, P. James E. Peebles, and Amos Yahil of Princeton University argued in a 1974 *Astrophysical Journal* article that the stability of a galactic disk can be guaranteed by the gravity of a very massive halo in which the disk is embedded. If there is only the disk, the small observed random motions about the circular motion of stars results in a violently unstable disk. As a result of these two 1974 observational and theoretical studies, it became clear that a typical spiral galaxy has not only disk-like and central spherical sub-systems, but also an extended halo around it, so that the total size of each galaxy is about ten times greater than the sizes of its stellar sub-systems and the total mass is about ten times its total stellar mass. Later, similar halos were found around massive elliptical galaxies.

Thus roughly 90% of the galaxy mass is dark. Some of the dark matter is revealed by rotation curves, as far out from the center of the galaxy as it is possible to make measurements. It is now apparent that the observed rotation curve can be

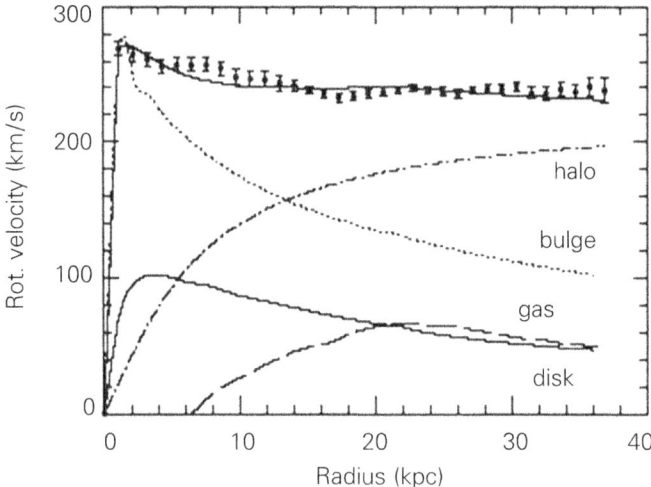

Figure 12.7. The observed rotation curve of NGC 7331. Estimates of the contributions from disk, bulge, and halo are fitted to the observed rotation curve given by the dots with the net model curve going through the dots. The rotation curve of the disk is fixed by the observed stellar velocity dispersions within the disk similar to Oort's procedure for our local Galaxy disk. The bulge and halo curve contributions dominate the mass distribution at all radii. (Bottema 1999).

explained by different mass components. At large radii, one may see that the typical total mass of a large spiral galaxy like M31 is near 10^{12} $M_\odot$. Thisis primarily due to dark matter around galaxies which contain extended massive halos (or coronas) as well as a disk and a nuclear bulge. The different components are plotted in Figure 12.7 in terms of the contribution of each to the observed rotation curve.

The dark matter distribution is isothermal in galactic halos. This implies that the velocity dispersion must be large so that the system is not unstable. Indeed, the flatness of the rotation curve at the distances where the dark matter dominates, i.e., $v = (GM(r)/r)^{1/2} = $ constant, means that the density of dark matter falls off a as the inverse square of the radius: $\rho_D \propto r^{-2}$. This applies to the dark halo of the Andromeda Galaxy as well as other disk galaxies.

12.7   Dark matter discovery in the Local Group

In 1959, new evidence for dark matter came from our Local Group of galaxies. The two major members of the group are our Galaxy and the Andromeda Galaxy; the rest are dwarf galaxies that contribute little to the total mass of the group. In the first approximation, the group may be considered a binary system of the two largest members. Franz D. Kahn of the University of Manchester and Lodewijk Woltjer of

Leiden University (1959) found a solution to the standard two-body problem in which the two galaxies were considered as point-like masses that move toward each other along the straight line connecting the two. The observed distance between the masses is $R_0 \cong 0.7$ Mpc, and their relative velocity is $\dot{R}_0 \cong 120$ km/s. This figure for the velocity is obtained when the directly measured velocity of about -300 km/s (found already by Slipher) is reduced to the center of mass of our Galaxy by taking into account the Sun's orbital motion.

One also needs to know the age of the system is necessarily less than the age of the universe). The total mass of the group is then calculated to be $M_{LG} \geq 2 \times 10^{12} \, M_\odot$. But the conventional luminous stellar mass of the Galaxy and the M31, i.e., the mass of all the stars in both galaxies, $M_{lum} \cong 2 \times 10^{11} M_\odot$ is an order of magnitude smaller. So, considerable non-stellar dark matter must be present in the system.

The Andromeda Galaxy is very similar to our Galaxy. It is often assumed that it is somewhat bigger and more massive than our Galaxy, but comparisons are difficult since we observe our Galaxy from inside and the Andromeda Galaxy from outside. All other galaxies in the local area within ≈ 2 Mpc from us are definitely smaller and less massive than these two galaxies. It is therefore a good approximation to view the Local Group as a two-body system with several dozen minor members, "test particles". In this view, the two galaxies formed as they moved apart from one another, reaching a maximum distance and now are approaching each another. The infall of one galaxy radially towards the other is a dynamical problem whose solution can be obtained in parametric form.

With the Newtonian inverse square law, the equation of motion is

$$\ddot{R}(t, M) = -GM/R^2, \tag{12.40}$$

where R is the distance between the two and M is the mass of the pair. Double dots indicate the second derivative with respect to time.

The equation of motion is a nonlinear second-order differential equation with respect to time which does not enter the equation explicitly. The first integration can be done easily. Introduce two new variables: one of them is the radial velocity, $\dot{R} = v$. The other is R, which is used instead of time t. With the new variables, the left-hand side of the equation of motion can be written as $\ddot{R} = \dot{v}$. Using the derivative v' with respect to R, one has

$$\dot{v} = \frac{\partial v}{\partial t} = \frac{\partial v}{\partial R} \frac{\partial R}{\partial t} = v'\dot{R} = vv'. \tag{12.41}$$

Now Equation (12.40) takes a form:

$$vv' = -GM/R^2. \tag{12.42}$$

The first integral of the equations is:

$$(1/2)v^2 = GM/R + E(M). \qquad (12.43)$$

which gives kinetic versus potential energy. Here $E(M)$ is an arbitrary constant of integration, the total energy per unit mass. It is the sum of the kinetic energy of the pair,

$$E_{KIN}(t, M) = (1/2)v^2,$$

and its potential energy,

$$W(t, M) = -GM/R,$$

per unit mass. As usual, potential energy is defined so that it is zero at infinitely large distance: $W(\infty) = 0$.

In analogy with the Kepler problem for motions in the gravity field of a central mass, the energy $E(M)$ may be positive, negative, or zero. We are interested in the bound motion ($E < 0$).

Jumping ahead to dark energy, note that classifying the types of orbits according to the sign of the total energy is generally invalid in the presence of dark energy with its anti-gravity. For this case the orbits may be infinite for all the three types of motion as we will discuss later in this book. For M31 and our Milky Way, we can approximate the motion well with no dark energy but we will quantify its effect in our later discussion.

Analytical solutions for bound (also called elliptic $E < 0$) motions can be found in terms of elementary functions in a parametric form.

The elliptic solution is

$$R = \frac{GM}{2|E|}(1 - \cos\eta), \qquad (12.44)$$

$$t - t_i = \frac{GM}{(2|E|)^{3/2}}(\eta - \sin\eta). \qquad (12.45)$$

Here η is the eccentric anomaly familiar from celestial mechanics and t_i is the initial moment of time corresponding to $R = 0$ at the origin of the universe when the material of the two was close together.

Take the derivative of each of the above equations with respect to η. Then divide the first equation by the second equation, which leads to the radial velocity in the orbit

$$\dot{R} = \frac{(2|E|)^{1/2}\sin\eta}{(1 - \cos\eta)}. \qquad (12.46)$$

M is the mass of the system and

$$GM/|E| = R_{max}$$

where R_{max} is the maximum separation of the two galaxies.

At present the observed values are

$$R_0 = 0.73 \text{ Mpc},$$

$$\dot{R}_0 = -120 \text{ km/s}, \quad \text{and}$$

$$t_0 = 13.2 \times 10^9 \text{ yr},$$

where t_0 is the time since the two galaxies started to separate from each other (the age of the oldest stars). We solve Equations (12.44)–(12.46) with these parameters for three unknowns η, R_{max} and M. The solution is not simple. It requires iteration giving $\eta = 243°$ as a solution to the equations. The mass derived from this solution, the Local Group mass, is $M_{KW} = 4.45 \times 10^{12} M_\odot$ (Kahn and Woltjer 1959, Einasto and Lynden-Bell 1982).

One of the major uncertainties in this calculation comes from $\dot{R}_0$ which depends strongly on the adopted circular rotation velocity of the Sun in the disk. The Andromeda Galaxy happens to lie more or less in the direction of motion of the Sun in its Galactic orbit. The observed approach speed of the M31 is -300 km/s. Taking the Sun speed to be 220 km/s in its orbit, 180 km/s of this is towards the Andromeda Galaxy, and then the relative speed of the two galaxy centers is -120 km/s. A greater rotation speed Θ would reduce the relative speed and the group mass.

Since its publication in 1959, because it uses the Local Group where distances and velocities are so well known, the Kahn Woltjer results have been accepted as strong evidence for large amounts of dark matter even in galaxies like our own. Taking the luminosity of our Galaxy as $\times 10^{10} L_\odot$ and the Andromeda Galaxy as $5 \times 10^{10} L_\odot$, the Local Group $M/L \sim 56$ in solar units.

Moreover, there is a possiblility that our Local Group and other groups may have had a more chaotic past than is envisioned in the simple-minded radial recession and then approach of the Kahn-Woltjer model (e.g., Zheng et al. 1991, Valtonen et al. 1993, Byrd et al. 1994, Chernin et al. 2004). The studies which consider a more violent past do not greatly change the mass estimates.

12.8   Dark matter in binary galaxy systems

In the case of stars, the most reliable method of mass determination is for binary stars. For the visual binary both members of the system are seen individually, the orbits of the stars are followed in the sky over time, sometimes requiring observations over decades or even longer. In this way a segment of the relative orbit is found, or even the whole orbit. The observational points are fitted to an ellipse, as Kepler's laws require.

Kepler's laws indicate the primary should be at the focus of the observed orbital ellipse of one relative to the other member. The projection of an ellipse to the plane of the sky is also an ellipse, but the focus of the orbital ellipse (marked by the primary star) does not project to the focus of the projected ellipse. The displacement between the original and new foci gives the orientation in space of the original orbit, after some rather non-trivial calculations. When the 3-dimensional shape of the binary orbit has thus been found, with the radial velocity between the two components, the total mass of the system is easily calculated using Newtonian mechanics.

The binary stars can be generalized to binary galaxies. In this case, the apparent orbital motion is so slow the galaxies do not cover a significant fraction of the orbital ellipse track during the modern astronomy era. For tailed radio galaxies, orbital tracks can be observed, or at least is it possible to reconstruct them. Besides our own Local Group, the binary galaxy 3C129, provides a unique case for estimating dark matter because, in this binary, we actually see the orbit track of one large Galaxy around another. In contrast to the assumed motion for M31 and the Milky Way, in this binary, the transverse motion appears to be large. Here we consider the tail of 3C129 (Miley et al. 1972). Figure 12.8 shows the radio tail contours overlaid on an optical photograph. The radio galaxy 3C129 is seen at the head of the tail, and the supergiant primary galaxy in the general area of the focus of the elliptical arc.

The orbital ellipse can be reconstructed under two alternative assumptions: that the density inside the radio tail is small compared with the density of the intergalactic gas, and that the densities inside and outside the tail are the same. In the latter case the tail tracks the orbit, in the former, buoyancy of the tail shifts the tail away from the primary galaxy. The resulting original orbits are shown in Figure 12.9 (Byrd and Valtonen 1978).

The two solutions represent the extreme cases. From other evidence, the buoyancy model is more likely to be correct. The shift of the primary from the focus of the projected orbital ellipse gives the 3-dimensional original orbit, and together with the measured radial velocity differences between the galaxies (Nilsson et al. 2000), the total mass of the galaxy system obtained is $4 \times 10^{13} M_\odot$ for the buoyancy model, with twice as much for non-buoyancy. The amount of light for these galaxies is not known exactly. On the basis of the general galaxy types it should be $\sim 2 \times 10^{11} L_\odot$ indicating the mass-to-light ratio is about 200. The primary galaxy is a giant elliptical galaxy, quite different from the Andromeda Galaxy and our own Milky Way Galaxy. This leaves little doubt about existence of dark matter in the 3C129 system. The dark matter halo in this case is more typical of subcluster halos than galaxy halos.

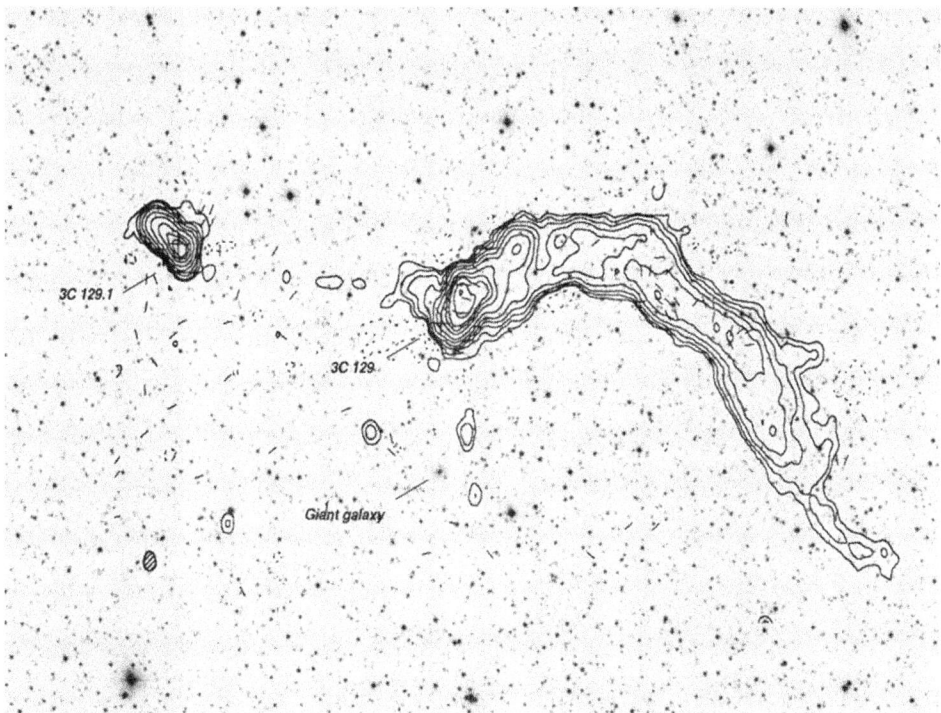

Figure 12.8. The radio tail of 3C129 from Jägers (1987), shown by contours, overlaid on an optical photograph from POSS Digital Sky Survey (Nilsson et al. 2000).

Figure 12.8 also displays another radio galaxy 3C129.1 to the left of 3C129 which is thought to be at the center of an x-ray emitting cluster of galaxies about 18 Mpc behind the radio galaxy 3C129 so it has no effect on the orbital track of the pair.

However, for other systems of binary galaxies one has available nothing else than the "snapshot" observation of the projections of the mutual distance and the mutual velocity of the two galaxies. If one uses these raw quantities together with Kepler's III Law, one derives an underestimated total mass. For a sample of binary galaxies one has to use statistical approaches which attempt to take into account the projection factors and various selection effects, and do averaging as in the case of the Virial Theorem (Box 11). With this approach, in samples of binary and triple galaxies (including the halos)

$$M/L \sim 50 \qquad (12.47)$$

(Peterson 1979, Zheng et al. 1993, Wiren et al. 1996, Teerikorpi 2001). Thus very roughly 90% of the mass is dark matter even in small systems. Clusters of galaxies contain even more dark matter ($M/L \sim 400$).

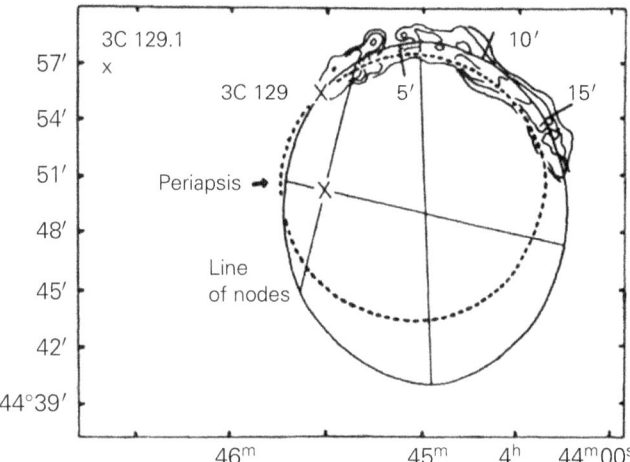

Figure 12.9. Two orbital solutions for 3C129 binary system. The solid line is without buoyancy model, the dashed line is for maximal buoyancy (Byrd and Valtonen (1978)).

As we have seen, some dark matter is revealed by rotation curves of individual galaxies, as far out from the center of the galaxy as it is possible to make measurements. But it is now apparent that dark halos of galaxies go beyond the outer edge of the stellar and gaseous disk.

12.9   Dark matter discovery via gravitational lensing

Dark matter halos are generally much more extensive than baryonic components of galaxies by an amount difficult to know exactly. A promising method of probing the dark matter uses gravitational lensing. Lensing responds to all gravitating matter, baryonic or dark. We thus continue the discussion of Section 3.3 and extend it beyond deflection of light near mass points. An extensive discussion of gravitational lensing is found in Mollerach and Roulet's (2002) book.

For galaxies or clusters of galaxies rather than points, the distribution of mass between the source and the observer is extended in the sky and in the line of sight. Often one can identify the lens with a system of bodies, stars or galaxies, which are clustered at a definite distance from us. Then we can use another approximation, a **thin lens** approximation rather than a point mass. At the average distance of the lensing objects We may define an **image plane** in which the background objects are seen. The background objects themselves are thought to be in the **source plane**. Then the problem is to find a source plane mapping to the image plane. Then after comparing with observations, one can deduce the true appearance of the source plane is like. The source plane is what would be observed in the sky with no lenses.

A model of the mass distribution in the lens must be found to create a correct transformation between the source plane and the image plane. Often this model is the most important finding, when a map of the extent of dark matter is required. According to Equation (3.61), the deflection angle $\Delta\phi$ due to the point mass M is

$$\Delta\phi = \frac{4GM}{c^2 b} \tag{12.48}$$

where b is the impact distance of the light ray in the image plane.

Observations of the amount of deflection permit calculation of the lens mass M. Even for a non-point lens, it is possible to model the surface mass distribution e.g., by assuming it to be circularly symmetric. A galaxy or a cluster of galaxies may generate such a symmetric distribution to the first approximation. For such a symmetry, only the mass inside the circle of radius b contributes to the deflection. This is like the Newtonian mechanics result that outside any spherically symmetric object the gravitational force is the same as if all its mass were concentrated at the center. The lensing result may be derived using similar reasoning.

Let the source have a uniform surface brightness over its subtended solid angle. Surface brightness is conserved during travel of light through space. However, deflections may change the shape and the solid angle of the object so that the total brightness (the product of the surface brightness and the solid angle) also changes. The quantity

$$A = \frac{d\Omega}{d\Omega_0} \tag{12.49}$$

is called the **amplification**, where $d\Omega$ is the observed solid angle and $d\Omega_0$ is the solid angle in the absence of lenses. The lens acts somewhat like a telescope.

Consider a point-like lens first as shown in Figure 3.2. In the source plane, the solid angle is a slice of width $\beta d\phi$ of the ring of thickness $d\beta$. In the image plane, it has width $\theta d\phi$ and thickness $d\theta$ as shown in Figure 12.10.

Here $d\phi$ is unchanged. Now

$$A_\pm = \frac{d\Omega_\pm}{d\Omega_0} = \frac{\theta_\pm d\theta_\pm}{\beta d\beta} \tag{12.50}$$

Here $\pm$ refers to the two images shown in Figure 3.2. Using Equation (3.72)

$$\frac{\theta_\pm}{\beta} = \frac{1}{2} \pm \frac{1}{2\beta}\sqrt{\beta^2 + 4\theta_E^2}$$

$$\frac{d\theta_\pm}{d\beta} = \frac{1}{2} \pm \frac{\beta}{2\sqrt{\beta^2 + 4\theta_E^2}} \tag{12.51}$$

$$A_\pm = \frac{\theta_\pm}{\beta} \cdot \frac{d\theta_\pm}{d\beta} = \frac{1}{2} \pm \frac{\beta^2 + 2\theta_E^2}{2\beta\sqrt{\beta^2 + 4\theta_E^2}}.$$

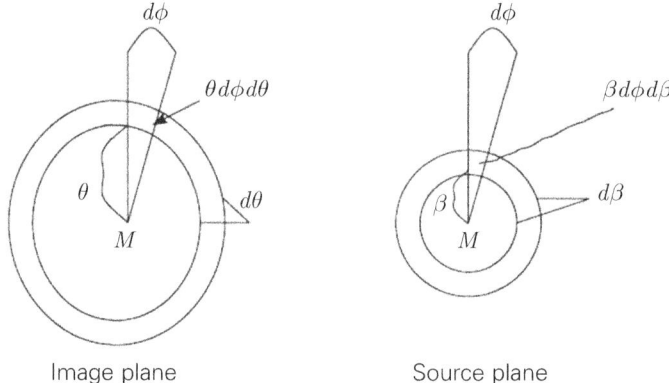

Figure 12.10. Source and the image plane around a point mass M.

If amplification is negative, the image is inverted. This happens for A_-. If the two cannot be resolved (that is seen as separate images), the overall amplification

$$A = A_+ + |A_-| = \frac{\beta^2 + 2\theta_E^2}{\beta\sqrt{\beta^2 + 4\theta_E^2}}. \qquad (12.52)$$

When $\beta \to 0, A \to \infty$, the image, an Einstein ring, has a theoretically infinite amplification. In reality, the amplification has a finite large value because actual astronomical objects are not point-like and an average must be taken over the surface of the object. The Einstein ring is an example of what is called a **critical line**.

For a realistic extended lens, critical lines have the form of closed curves of different shapes. For the source plane the corresponding directions are called **caustics**, a concept for long known also in ordinary optics as bright patterns produced by light rays reflected or refracted by some object (Sir William Rowan Hamilton studied this topic in the 1820s). To have an Einstein ring, the caustic must be a single point behind the lensing mass M.

For points far away on the sky from the lensing mass distribution, a single image of the source is seen. However, if the source is brought closer in the sky to the lensing object, and crosses the caustic, then two new additional images will appear besides the single image. Thus the total number of images is therefore always odd (**Burke's Theorem**, Burke 1981). These two additional images are inverted relative to one another. These are strongly magnified when the source is close to the caustic. For a source further from the caustic, images are further apart and away from the critical line. One of the images is usually strongly "de-magnified" lying close to the lensing mass. It is thus hard to see. In practice, only two images may be observed as seen in Figures 12.11 and 12.12.

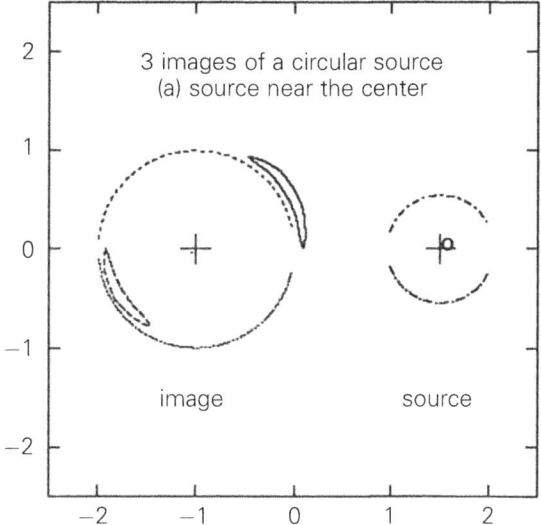

Figure 12.11. The process of how a circular source almost directly behind an extended iso-thermal sphere lens is imaged. The center of the lens is marked by a cross. The two elongated images lie close to the Einstein ring indicated by two semi-circles. The third is weak and close to the center. In the source plane, parts of the caustics are shown as two smaller dotted semi-circles). The circular source is close to the center making a point-like caustic. Figure adapted from Mollerach and Roulet (2002).

For a circularly symmetric lens Equations (3.65)–(3.67) become

$$\beta = \theta \left(1 - \frac{\langle \Sigma \rangle}{\Sigma_{crit}} \right) \tag{12.53}$$

where

$$\langle \Sigma \rangle = \frac{M(<b)}{\pi b^2} \tag{12.54}$$

$$\Sigma_{crit} = \frac{c^2}{4\pi G} \frac{D_{OS}}{D_{OL} D_{LS}}. \tag{12.55}$$

Σ_{crit} is the **critical density**, while $\langle \Sigma \rangle$ is the average density of the lens system within the impact radius b, and $M\,(<b)$ is mass within a radius equal to b.

Problem. Verify Equation (12.53).

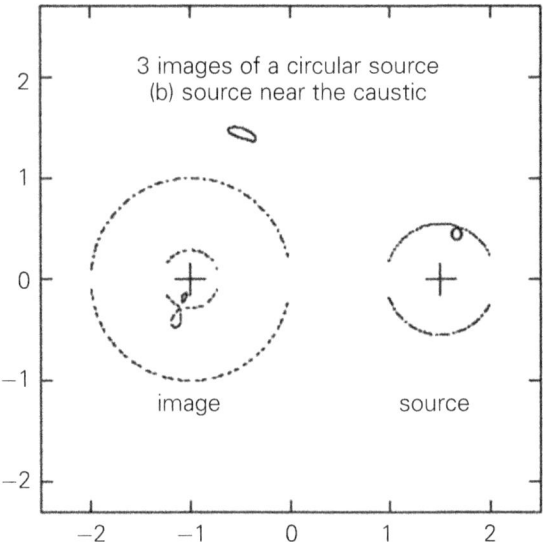

Figure 12.12. Like Figure 12.11, but the circular source is close to the caustic. The radial critical line of the caustic curvec has been added to the image plane as two smaller semicircles. Images within the Einstein ring are radially elongated while the third image is now elongated tangentially. Figure adapted from Mollerach and Roulet (2002).

Typically, the matter density of a lens increases towards its center. Thus $\langle \Sigma \rangle$ is a function of b, and it reaches its maximum value Σ_{center} at $b = 0$. If within the lens $\Sigma_{center} > \Sigma_{crit}$, then at some $b = b_E$ the density equals the critical density:

$$\langle \Sigma \rangle (b = b_E) = \Sigma_{crit}. \qquad (12.56)$$

This condition gives one possible solution of the lens equation: $\beta = 0$ and

$$\theta_E = \frac{b_E}{D_{OL}}. \qquad (12.57)$$

By the symmetry condition, the image is an Einstein ring which is at angular distance θ_E. Here the condition for obtaining the Einstein ring is, in addition to $\beta = 0$, that $\Sigma_{center} > \Sigma_{crit}$.

What is the value of Σ_{crit} and how does it compare with Σ_{center} for galaxies. Multiple images are possible only when the Einstein ring forms, thus $\Sigma_{center} > \Sigma_{crit}$, applies for multiple images also.

Suppose that the object is twice as far away from us as the lens:

$$D_{OS} = 2D_{LS} = 2D_{OL}. \qquad (12.58)$$

Then

$$\Sigma_{crit} = \frac{c^2}{4\pi G} \frac{2}{D_{OL}}$$

$$= \frac{(3 \cdot 10^5 \text{ km/s})^2}{4\pi \cdot 4.3 \cdot 10^{-3} \left(\frac{\text{km}}{\text{s}}\right)^2 \frac{\text{pc}}{M_\odot} \frac{D_{OL}}{1\,000\,\text{Mpc}} \cdot 10^9 \text{ pc}} \tag{12.59}$$

$$= 3.3 \cdot 10^3 \left(\frac{D_{OL}}{1\,000\,\text{Mpc}}\right)^{-1} \frac{M_\odot}{\text{pc}^2}.$$

Now observationally we know that densities of $10^4\, M_\odot/\text{pc}^2$ are typical for galaxy centers. Using the equation above, it is thus possible that Σ_{center} exceeds the critical density at large lens distances, $D_{OL} \geq 1\,000$ Mpc or $z \geq 0.2$.

Such cases have been found. The famous "Einstein Cross" is a quasar at $z = 1.695$ that is seen as five images due to a lensing galaxy whose $z = 0.039$. Four images lie in a circle whose radius is $0.9''$. This angle corresponds to 700 pc at the galaxy lens distance. Therefore, take $b_E = 700$ pc. Since $D_{OS} \approx D_{LS} \gg D_{OL}$, the value of Σ_{crit} is half of the value in the above formula while $D_{OL} \cong 165$ Mpc. Taking the typical value, $\Sigma_{crit} \cong 10^4\, M_\odot/\text{pc}^2$, we get

$$M(<b_E) \cong 1.5 \times 10^{10} M_\odot \tag{12.60}$$

which is reasonable for the center of a giant galaxy.

A simple model for the surface mass density that may be appropriate for galaxies is the projection of an isothermal sphere on a plane. This model for the predominant dark matter halo creates a flat rotation curve. The projected surface density on the sky is

$$\Sigma(\theta) = \frac{\sigma^2}{2GD_{OL}\theta} \tag{12.61}$$

where σ is the one dimensional velocity dispersion of the sphere along the line of sight. The mass inside the radius θ is

$$M(\theta) = \frac{\pi\sigma^2 D_{OL}\theta}{G} \tag{12.62}$$

It turns out that the reduced deflection angle is independent of θ:

$$\alpha = \frac{D_{LS}}{D_{OS}} \frac{4\pi\sigma^2}{c^2}. \tag{12.63}$$

The deflection angle can be calculated as follows.

$$\Delta\varphi = \frac{4\pi\sigma^2}{c^2} = 4\pi \frac{(\sigma/250\,\text{km/s})^2 (250)^2 (\text{km/s})^2}{(300\,000)^2 (\text{km/s})^2} 206265'' = 1.8''(\sigma/250\,\text{km/s})^2.$$

$$(12.64)$$

The lens equation is

$$\beta = \theta - \alpha, \tag{12.65}$$

The Einstein angle, when $\beta = 0$, is

$$\theta_E = \alpha. \tag{12.66}$$

For the "Einstein Cross", we have observationally $\theta_E \cong 0.''9$ and $D_{LS}/D_{OS} \cong 1$. Equation (12.64) implies $\sigma = 177$ km/s, which is not unreasonable in a central region of a major galaxy.

Recall that if the source is closer on the sky to the center of the mass distribution than θ_E, there are two images at

$$\theta_\pm = \beta \pm \theta_E, \tag{12.67}$$

otherwise there is only one at θ_+.

Let us illustrate the situation with the following illustrations. Take the source to be much further than the lens so rays of light from it arrive parallel to each other at the lens. In the first illustration, Figure 12.13, $\beta = 0$ and the Einstein ring forms.

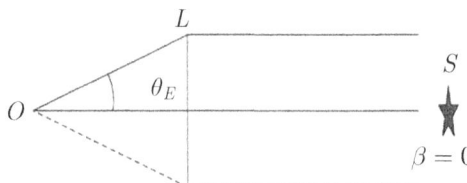

Figure 12.13. Two parallel light rays from a distant source S arrive at the observer at O through a comparatively nearby gravitational lens L. Here $\beta = 0$.

The second example has $\beta = \theta_E$. Two images just barely form in Figure 12.14.

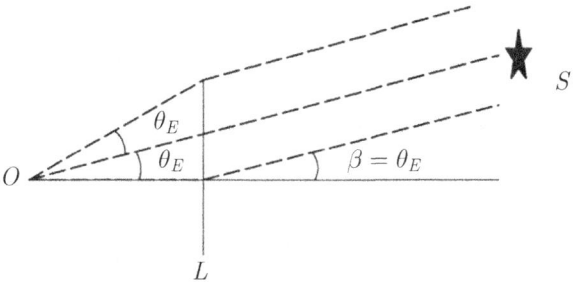

Figure 12.14. As Figure 12.13, except that $\beta = \theta_E$.

The third example, in Figure 12.15, has $\beta > \theta_E$. Here the ray of light coming closest to the center of the lens does not reach the observer. Thus only one image is seen.

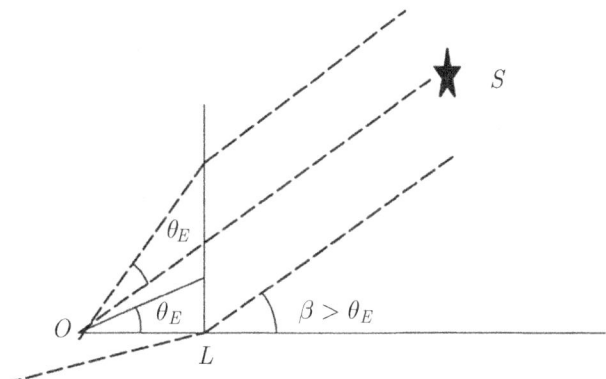

Figure 12.15. Like Figure 12.13, except that $\beta > \theta_E$.

The fourth category, in Figure 12.16, is for $\beta < \theta_E$.

In Figure 12.16, two images form at angles $\theta_E + \beta$ and $\theta_E - \beta$ relative to the center of the lens.

For this example, since $d\theta_\pm / d\beta = 1$, the amplification

$$A_\pm = \frac{\theta_\pm}{\beta} = 1 \pm \frac{\theta_E}{\beta}. \tag{12.68}$$

Now since A_- is negative, the secondary image must be inverted.

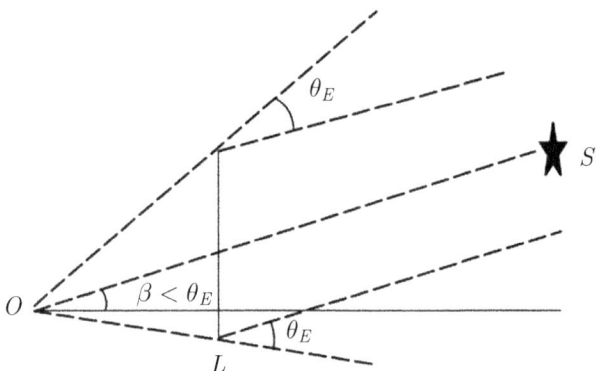

Figure 12.16. As Figure 12.13, except that β is non-zero and $\beta < \theta_E$.

The constant deflection angle, which is independent of θ, means that the image is not deformed in the radial direction. For this case, amplification is entirely due to tangential deformation.

The lens equation (Equation (12.65)) can be solved for any other radially symmetric mass distribution. One may represent the solution as

$$\beta = \beta(\theta), \tag{12.69}$$

In other words, this is a curve in the $\beta - \theta$ plane. Take the Plummer sphere as an example since it is useful for actual galaxies. A sphere of mass M is defined so that its gravitational potential is "softened" from the point Newtonian potential; i.e., the potential approaches a finite value (rather than infinity) when the distance from the center $r \rightarrow 0$. See Section 12.2, Equations (12.11)–(12.12). The surface mass density in the xy-plane is

$$\Sigma(R) = \int_{-\infty}^{+\infty} \rho\left(\sqrt{R^2 + z^2}\right) dz = \frac{M}{\pi} \frac{a^2}{\left(a^2 + R^2\right)^2}. \tag{12.70}$$

Now the radial distance R in the xy-plane is: $R = \sqrt{x^2 + y^2}$.

Problem. Verify Equations (12.11) and (12.70).

In Figure 12.17, the solution curve $\beta = \beta(\theta)$ for the Plummer sphere is shown.

Now, at large $\pm\theta$ the curve approaches the $\beta = \theta$ line. We can see that the zero point $\beta = 0$ corresponds to the solutions in the Einstein ring crossing the horizontal axis at $\theta - \theta_E$ and $\theta + \theta_E$ as well as to a third image at $\theta = 0$. The magnitude of θ_E in this model is

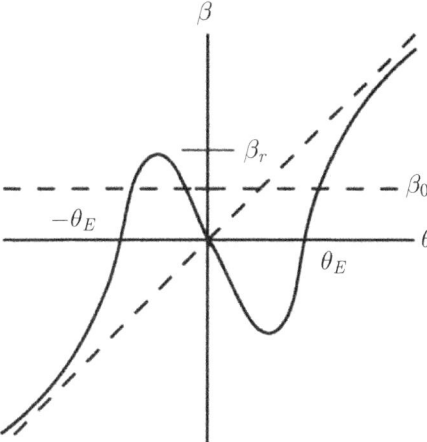

Figure 12.17. Functional relation (solid line) between the position of the lensed galaxy (β) and its image (θ) assuming a Plummer sphere model. The curve approaches the tilted dashed, no lensing $\beta = \theta$ line at large distances from the lens. Details of this approach depend on the mass model assumed for the lens. At the fixed value β_0 (horizontal dashed line) we get three images, at the three crossing points with the solid curve. Adapted from Sparke and Gallagher (2000).

$$\theta_E = \sqrt{5}\,\frac{a}{D_{OL}}. \qquad (12.71)$$

We see that source close to the center, say at β_0, forms three images. There are two on the opposite side of the center of the lens and one on the near side. The closest image has $d\beta/d\theta < 0$ which means that the image is inverted. As β increases, the two images on the opposite side from the center approach each other until they coincide at $\beta = \beta_r$. For large values of β, only one image remains on the near side of the center of the lens.

At small angles β the amplification A_+ is large as in Equation (12.51). The images close to $\theta = \theta_E$ are extended arcs. The curve $\theta = \theta_E$ is the tangential critical curve. The image close to the center, $\theta \cong 0$, is faint from Equation (12.50). See Figure 12.11.

For a lens which is not exactly round, the arcs may split up into multiple arcs. A single source creates an odd number of images although the image near the center may be too weak to be observed.

When β is just below β_r, where $d\beta/d\theta \to 0$, the amplification by Equation (12.50) again becomes a large value. Now, the image is stretched in a radial direction. The curve $\beta = \beta_r$ is designated as the radial caustic. Corresponding values of θ, called θ_r, outline a circle $\theta = \theta_r$. This is called the **radial critical curve** which always lies inside the Einstein ring (Figure 12.12).

The shape of a lensed galaxy is modified. A circular galaxy appears elliptic. If the image is close to the Einstein ring radius, the elongation is tangential. Close to the radial critical curve the major axis elongation is in the radial direction. To measure the difference in compression in the two directions we define

$$\gamma = \frac{1}{2}\left[\frac{d\beta}{d\theta} - \frac{\beta}{\theta}\right] \tag{12.72}$$

Here $d\beta/d\theta$ represents the radial compression (see Figure 12.10) and β/θ is the compression in the tangential direction.

Now we calculate the shear using Equation (12.53) for a circularly symmetric lens. We have for the radial and tangential compressions

$$\frac{d\beta}{d\theta} = 1 - \frac{\langle\Sigma\rangle}{\Sigma_{crit}} - \frac{\theta}{\Sigma_{crit}}\frac{d\langle\Sigma\rangle}{d\theta} \tag{12.73}$$

$$\frac{\beta}{\theta} = 1 - \frac{\langle\Sigma\rangle}{\Sigma_{crit}}. \tag{12.74}$$

Therefore,

$$\gamma = \frac{1}{2}\left[-\frac{\theta}{\Sigma_{crit}}\frac{d\langle\Sigma\rangle}{d\theta}\right]. \tag{12.75}$$

Since

$$\langle\Sigma\rangle = \frac{M(<b)}{\pi b^2} = \frac{\int_0^b 2\pi r \Sigma(r)\,dr}{\pi b^2} \tag{12.76}$$

$$\frac{d\langle\Sigma\rangle}{db} = -\frac{2M(<b)}{\pi b^3} + \frac{2\pi b \Sigma(b)}{\pi b^2} \tag{12.77}$$

Using $\theta/d\theta = b/db$,

$$\gamma = \frac{1}{\Sigma_{crit}}\left[\frac{M(<b)}{\pi b^2} - \Sigma(b)\right] = \frac{\langle\Sigma\rangle(<b) - \Sigma(b)}{\Sigma_{crit}}. \tag{12.78}$$

For a given lens, a large number of background galaxies at different b's suffices to estimate the γ. Thus we recover the mass distribution as a function of b. Typically the distortions are small. This is called *weak lensing*.

Tangential arcs are often seen on the sky around rich clusters of galaxies. Figure 12.18 shows the cluster $0024 + 1654$ in which eight images of the same background galaxy are seen! The lensed images are of the single background galaxy

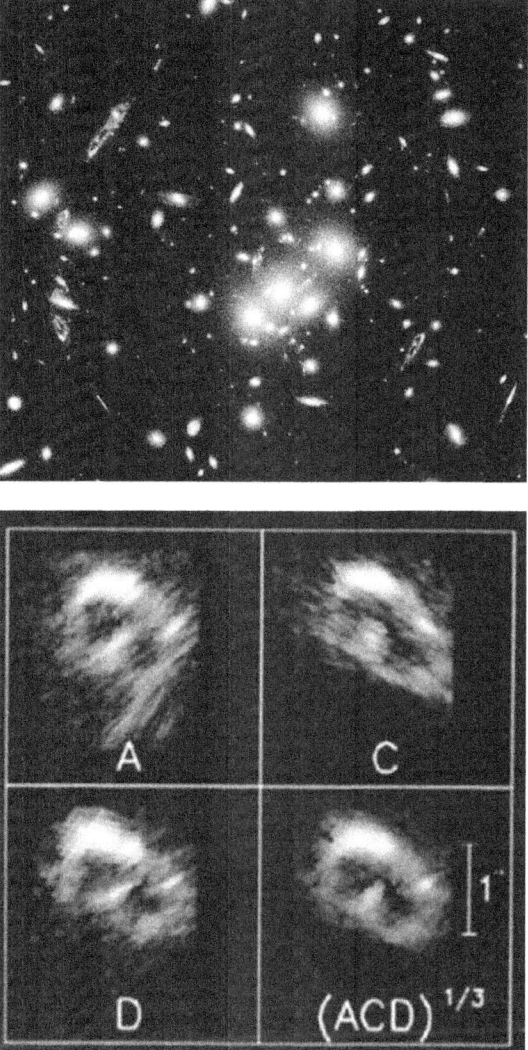

Figure 12.18. Top frame (a): The strong lensing of a background blue galaxy by a foreground cluster of galaxies 0024 + 1654. Eight images of the blue galaxy are seen in this Hubble Space Telescope picture. The tangentially elongated blue galaxy images mark the approximate location of the Einstein ring. Bottom frame (b) Frames of images with distortion removed and a combined image with 1″ angular scale. From NASA and STScI. For details see Colley et al. (1996). http://hubblesite.org/newscenter/archive/releases/1996/10/image/a/.

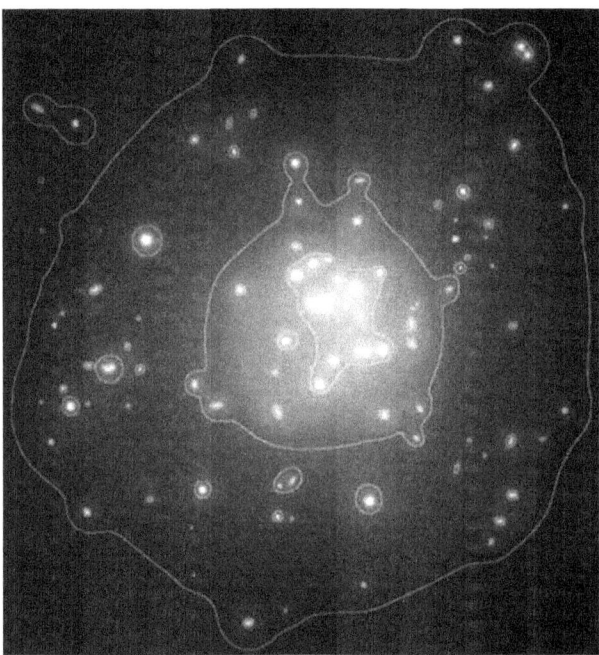

Figure 12.19. Color-coded map of the mass reconstruction of CL 0024 + 1654. Galaxy images are shown. The contour lines are a detailed reconstruction of the mass profile of the cluster. (Tyson et al. 1998).

which is at redshift $z = 1.675$. The lensing model shows that about 83% of the mass is distributed smoothly; the remainder is in galaxies. Within about 150 kpc from the center (within the circle defined by the arcs) there is about $2 \times 10^{14}\ M_\odot$ of matter (Tyson, et al. 1998), a very nice mass determination separate from the Virial Theorem via galaxy motions or X-ray emission.

In this cluster's model, the mass-to-light ratio of the whole cluster is about 230 in solar units. The value has been corrected to the redshift $z = 0$ since the cluster is at a relatively high redshift of $z = 0.395$, and the mass-to-light ratio is expected to evolve with cosmic time. The mass-to-light ratio for the individual galaxies is about 50 solar units. Because we are discussing here the total mass of galaxies, out to their outermost edge, results like this are the best determinations of galaxy mass-to-light ratios.

Outside the giant arcs there are only single images of background galaxies but these are also slightly deformed in the tangential direction. This is weak lensing, which was first detected by Tyson et al. 1990. The weak lens images are in contrast to the giant arcs and multiple images (*strong lensing*, discovered by Lynds and Petrosian 1989). Weak lensing facilitates studies of cluster mass distribution out to very large distances ($\sim$4.5 Mpc) from the center (Figure 12.19).

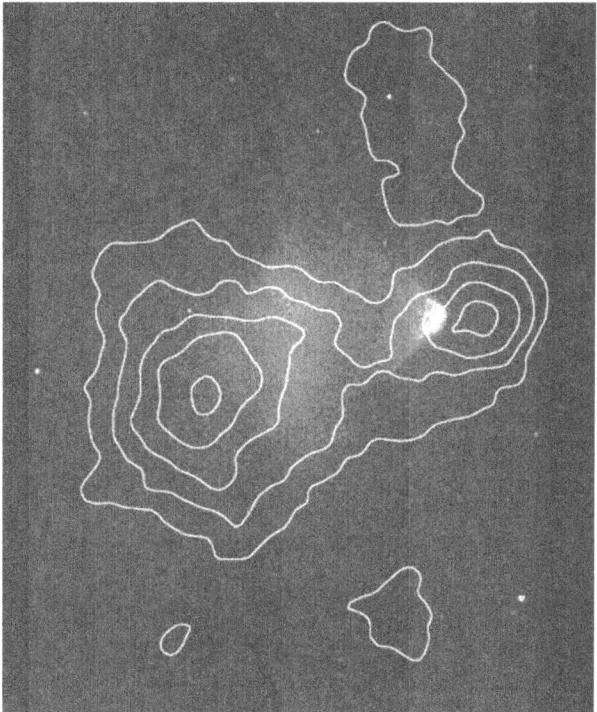

Figure 12.20. X-ray photo (white shading) by Chandra X-ray Observatory of the Bullet Cluster (1E0657-56). Gravitating matter (mainly dark matter and also stars) is shown as contours. NASA image. For details see Clowe et al. 2006).

Aside from the previously described strong lensing, the cluster $0024 + 1654$ is also where the weak lensing study was extended further than in any other cluster previously because of the Hubble Space Telescope (Kneib et al. 2003). At ~ 4.5 Mpc from the center of the cluster a significant 1% shear signal was found. The weak lensing all the way from 0.5 Mpc to 4.5 Mpc was used to construct a mass model for the whole cluster. The innermost 0.15 Mpc had already been studied by strong lensing, as mentioned. Weak and strong lensing covers more this cluster in a unique way. Figure 12.19 shows density contours of the dark and luminous matter, overlaid with an optical image. Dark matter is also associated with individual halos of galaxies. In this sense we are actually "seeing" dark matter just like one can "see" a quartz crystal by its refraction of light passing through it.

From the above studies, the cluster has two subclusters, with a mass ratio about 2:1 and separation of more than 1 Mpc. The bigger of the two clusters shows strong gravitational lensing, the smaller one is more diffuse. The mass-to-light ratio was studied using the near infrared K-waveband, and was found to be surprisingly constant throughout the cluster. Any tendency of galaxies forming preferentially either in denser or less dense environments, does not show up in $0024 + 1654$. These results are very similar to the Coma cluster (Geller et al. 1999, Rines et al. 2001). In scales greater than individual galaxies, baryonic matter and dark matter are very closely coupled. However, in galaxies there is relatively more baryonic matter near the center, less in the outskirts (Figure 12.7).

Also, the value of the mass-to-light ratio turns out to be similar to the Coma cluster, after correction to $z = 0$. Extrapolated to visual light, both clusters have mass-to-light ratios of about 200 solar units. If this value is true for the universe, the

$$\Omega_M \cong 0.2. \tag{12.79}$$

The weak lensing signal is not strong enough to be observed from individual galaxies. However, the effects of galaxy halos can be studied statistically. From many studies of groups and clusters of galaxies; both individual and statistical average mass-to-light ratios have been derived. The results are in agreement with the findings of the cluster $0024 + 1654$ described above (Squires et al. 1996, Mellier 1999, Wilson et al. 2001a and 2001b, Hoekstra et al. 2001, 2002, 2003 and 2004). Nevertheless, the question of the masses of individual galaxies is still open.

A remarkable example of the separation of baryons from dark matter is found in the "bullet" cluster of galaxies (Clowe et al. 2006). There two galaxy clusters composed of galaxies of stars and dark matter plus hot gas have collided. As shown by the contours in Figure 12.20, the galaxies and dark matter in each cluster passed through one another and continued past the impact point. Two contour peaks in dark matter and matter density are seen separated from one another as revealed by gravitational lensing. However, displayed as a "glow", the extremely hot x-ray emitting gas in each has collided and via the short mean free path stayed at the collision area between the two clusters today. In contrast, the dark matter, the widely separated galaxies (and even the stars within them) did not collide in an inelastic manner.

12.10   Dark matter in different scales

To summarize the dark matter determinations, we display the determinations M/L in different scales in Figure 12.21. It appears that the mass-to-light ratio increases with the increasing scale from small galaxies to large ellipticals and spirals. But then the increase levels off at about $M/L \cong 200$ in the scale of groups and clusters of galaxies. A sample of 105 clusters gives a typical value of $M/L \cong 170$ solar units (Girardi et al. 2000). The "standard" view until the late 1990s was that the extrapolation from galaxies continues on a straight line which cuts the $\Omega_M = 1$ level at $R \cong 1$ Mpc. Many attempts were made in trying to prove this, but failed. The need to find enough dark matter to make $\Omega_M = 1$ vanished with the discovery of dark energy in 1998.

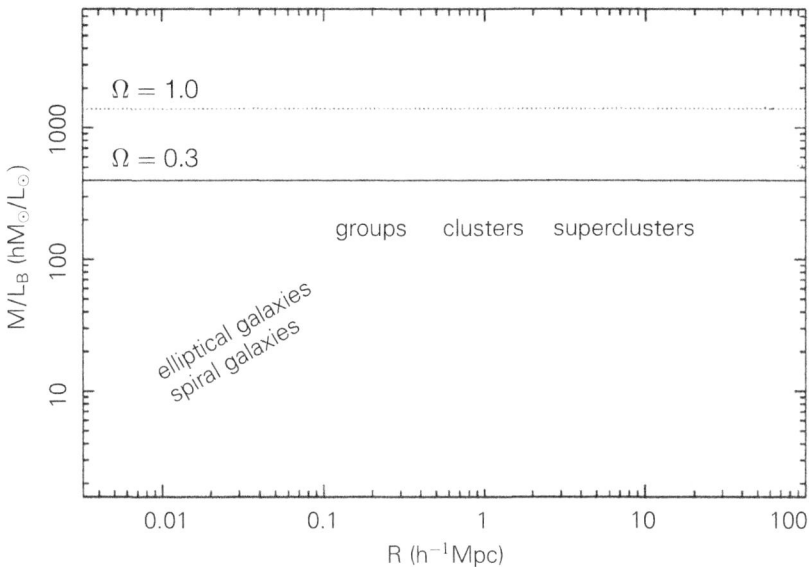

Figure 12.21. Median values for the mass-to-light ratio M/L for systems in different scales R. The Hubble constant of 100 km/s/Mpc is assumed; for $H_0 = 71$ km/s/Mpc the two horizontal lines giving the values of Ω_M should move downwards by the scale factor of 0.71.

Note that $M/L = 200$ over the universe corresponds to $\Omega_M \cong 0.2$ if clusters of galaxies provide a fair sample of the universe for Ω_M. This is supported by measurements of mass-to-light ratios in the fields surrounding the rich Coma cluster, using the velocities of infall towards the cluster (Rines et al. 2001). There is no evidence that M/L (and hence the inferred Ω_M) would change when going from the virialized cluster to its non-virialized outskirts. Interestingly, for a less rich nearby cluster, the Virgo cluster (the center of the Local Supercluster), a result from such infall studies (e.g., Tully and Shaya 1984, Teerikorpi et al. 1992, Mohayaee and Tully 2005) has been that the mass-to-light ratio of the spiral rich field population around the Virgo cluster has to be significantly less (around 1/5) than the $M/L \approx 500$ for the Virgo cluster. In the Local Supercluster within the radius of about 40 Mpc about 1/5 of the light comes from the Virgo cluster and 4/5 from the surroundings. Therefore, one may estimate that $M/L \approx 180$ and we again recover the density parameter $\Omega_M \approx 0.2$ for a scale of 10 Mpc. The WMAP data of CMB are best fitted with $\Omega_M = 0.28$ (Bennett et al. 2003).

The value of Ω_M is also constrained by extrapolating the cluster evolution. For rich clusters of galaxies like the Coma, their co-moving number density should increase very strongly with time if $\Omega_M = 1$ but much more slowly if we have a smaller $\Omega_M = 0.3$ (Eke et al. 1996). Fitting calculations with observations of rich clusters, the value $\Omega_M = 0.34 \pm 0.13$ has been determined (Bahcall et al. 1997,

Bahcall and Fan 1998, Donahue et al. 1998, Tran et al. 1999). This is a powerful method of discriminating between different cosmological models when more samples of high redshift clusters became available.

12.11   The importance and nature of dark matter versus baryonic matter

Dark matter has thus been recognized as an ingredient of structures at all levels— from galaxies to galaxy groups and clusters. The dark matter mass is deduced to be approximately an order of magnitude larger than the mass of ordinary matter. Also dark matter seems to follow the isothermal law of distribution everywhere in galaxies and in clusters.

Dark matter appears to dominate by mass over ordinary matter not only in galaxy systems, but also in the universe. Today, its overall mean cosmological density is about 5 times larger than the total density of ordinary matter and the CMB. As discussed previously, the dark matter contribution to the total cosmic density is usually quoted in terms of the fraction of the critical density: $\Omega_D = 0.233 \pm 0.013$ versus baryonic matter $\Omega_B = 0.0462 \pm 0.0015$ according to the WMAP data. The same data indicate that the total cosmic density is about 1, in the units of the critical density: $\Omega \cong 1$, so that the dark matter contributes about 23% to the total energy of the universe, while the "ordinary" matter contributes not more than about 5%. As we shall see, the rest is due to dark energy dominating the overall behavior of the universe today.

What could dark matter be? Dark matter is not thought to be made of ordinary baryonic matter. With sufficient precision, ordinary matter may be considered a mixture of ionized or neutral atoms made of protons and neutrons, as well as the same number of electrons as there are protons. Dark matter emits neither visible light nor other electromagnetic waves, and in practice it does not interact at all with electromagnetic radiation. Recall our discussion earlier in this chapter of the Bullet Cluster of galaxies in which the collisionless nature of dark matter is shown. Ordinary baryonic gas in each of the two clusters is dissipative and winds up concentrated near the collision site via electromagnetic interaction. In contrast, each cluster's dark matter passes through the other's dark matter freely. We should point out that each cluster's galaxies and their component stars (occupying very little space) also pass through the other cluster.

Could dark matter be baryonic matter concentrated in small bodies? Dark matter is evidently non-baryonic, as was first realized by the American theorist Robert V. Wagoner of Stanford University (Wagoner, Fowler and Hoyle 1967). Evidence comes from the Big Bang nucleosynthesis. As discussed earlier in this book, the deuterium calculated to emerge from the cosmological thermonuclear reactions is far less than that expected on the basis of the observed gravitating matter density ρ_M, if it was baryonic matter (Figure 11.1). The fractional density of baryons constrained by Big Bang nucleosynthesis is $\Omega_B \cong 0.05$. Thus baryons cannot account for dark matter whose fractional density is almost an order of magnitude larger: $\Omega_D \cong 0.23$. So we cannot propose the dark matter is composed of particles, say, black holes, that were once made of baryonic matter.

Instead dark matter probably is composed of new, yet unknown elementary particles. The direct detection of dark matter particles and progress in theoretical particle physics will hopefully clarify the problem. A popular assumption treats dark matter as being made of weakly interacting massive long-lived particles. Such particles are presumably like neutrinos, but with a considerably larger mass m_D (say, $m_D c^2 \approx 1\,\text{TeV}$). In comparison, the mass of the hydrogen atom m_H, or the proton, is $m_H c^2 \approx 10^{-3}\,\text{TeV}$. Here TeV symbolizes tera-electronvolts, i.e., 10^{12} electronvolts ($1\,\text{TeV} \cong 1.6\,\text{erg}$). Why dark matter particles, whatever they are, make up the largest non-dark energy contribution to the mass and energy of the universe is one of the crucial open questions of modern cosmology and also of fundamental physics (see the review by Feng (2010)).

Chapter 13

Dark matter and baryonic structures

Galaxy formation, the origin of galaxy clusters and superclusters are all related to physical processes that developed in the expanding cosmic matter at early stages of the cosmic evolution. **Gravitational instability** plays a major role in these processes. Rather than creating them, instabilities of various types usually destroy structures. However, initial seed proto-galactic perturbations in the homogeneous and uniform cosmic medium are amplified by gravitational instability. Via this amplification, perturbations can become stronger and eventually give rise to observed cosmic structures.

13.1   Newton's concept of gravitational instability

As mentioned earlier in this text, Isaac Newton of Cambridge University wrote in 1692 that a homogeneous and uniform distribution of gravitating matter would be unstable. Pieces of different sizes and masses would fragment under their own gravity. These would then collapse. Newton proposed in a few lines of a letter to a colleague that this may account for the origin of the Sun and stars.

Much later, in 1902, James Jeans, also at Cambridge University, clarified the conditions under which this collapse could develop. Jeans realized that thermal pressure of the cosmic medium could resist gravitational collapse of over-dense fragments. Gravitational instability is thus only possible if the gravity is stronger than the pressure that emerges due to gradients in the regions.

In a quantitative way, consider a region of density ρ_0 which is slightly enhanced with the size R surrounded by a uniform medium. The pressure force per unit mass is about

$$-\nabla P/\rho_0 \cong (P - P_1)/(\rho_0 R) \approx P/(\rho_0 R).$$

where the enhanced pressure inside the region is P, and the pressure at its boundary is P_1. Since P_1 is less than P, one can neglect it in comparison to P for a rough approximation. The pressure force is directed outwards, so that thermal pressure acts against further contraction.

Oppositely, the gravity force per unit mass within the region is

$$-GM/R^2 \cong -G\rho_0 R$$

where the total mass of the region is M. The gravitation is directed inwards to the center. Gravity is > the pressure if $R > R_J \cong (P/(G\rho_0^2))^{1/2}$. Now $P \cong \rho_0 c_s^2$, taking c_s is the speed of sound, the critical length is

$$R_J \cong c_s/(G\rho_0)^{1/2}.$$

Now R_J is the **Jeans critical length** outside which gravitational instability occurs. Instability can develop, if the size of the over-dense region is larger than R_J. This is designated as the **Jeans Criterion** for gravitational instability. If the size of the over-dense region is less than R_J, then thermal pressure will support it against self-gravity. In this case the perturbation will oscillate like sound waves.

The Jeans Criterion is valid for the expanding universe, as first pointed out in 1939 by George A. Gamow and the Hungarian-American physicist Edward Teller, both at George Washington University. Eugeny M. Lifshitz developed a complete relativistic theory of gravitational instability based on the Friedmann model in 1946 by in terms of General Relativity.

Gravitational instability as the key to large-scale structure formation has been observationally confirmed. Cosmic microwave background (CMB) studies have enabled us to see the state of the universe at an early epoch of less than one million years. No galaxies or clusters of galaxies could have formed by then, and the cosmic mass distribution was almost perfectly uniform. Only very weak irregularities, or perturbations, in the matter density existed. In 1992 seed cosmic perturbations were discovered observationally in the CMB by the Russian orbiting observatory RELICT and the American COBE.

Gravitational instability amplifies the initial perturbations so they are strong enough to form the observed cosmic structures. We will give the physics of gravitational instability within the standard cosmological model.

13.2 Basic hydrodynamics

In the early universe, the gas that filling it may be treated as a perfect fluid which is expanding uniformly in which the basic laws of hydrodynamics apply. Giving a brief derivation, let us focus on a fluid element whose size is $\Delta V = \Delta x \Delta y \Delta z$ at the position x. The density at this position is $\rho(\mathbf{x}, t)$ and velocity is $v(\mathbf{x}, t)$ both a function of time t.

Most commonly used is the **Lagrangian** frame which differs from the usual Eulerian frame of reference. Lagrangian coordinate values are fixed to the element of fluid, i.e., the coordinates are co-moving with the fluid. In the universe's expansion, the Lagrangian system follows the mean motion of the matter. This most natural system looks at the evolution from the point of view of a particle that follows the general expansion.

In the **Eulerian** system the fluid flows through points in a fixed reference frame. For specific points in this frame and to the properties of the fluid at these points, we may study a property called $Q(\mathbf{x}, t)$ which has a time rate of change $\partial Q/\partial t$ at

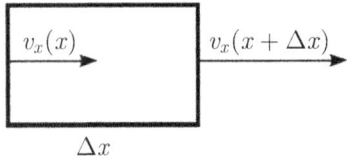

Figure 13.1. Time rate of change of a rectangular volume element. The opposite faces move with different velocities.

a fixed point **x**. In the Lagrangian coordinates, the point **x** follows the flow. Therefore, in this system, the time change of Q is

$$\frac{dQ}{dt} = \frac{\partial Q}{\partial t} + \frac{\partial Q}{\partial x^i}\frac{dx^i}{dt} = \frac{\partial Q}{\partial t} + \mathbf{v} \cdot \nabla Q \qquad (13.1)$$

Here i is a compactly stated summation index, $i = 1, \ldots, 3$ convention used earlier. First on the right hand side is the Eulerian time derivative. Second term is the change due to the specific movement of the fluid from one point to another nearby point.

Now what happens to the fluid element ΔV? Considering the change along the x-axis (Figure 13.1), the flow velocity at the left edge of the volume is $\mathbf{v}(\mathbf{x})$ and at the right edge $\mathbf{v}(\mathbf{x} + \Delta \mathbf{x})$. The length of the volume changes by

$$d(\Delta x) = \frac{d(\Delta x)}{dt}dt = \Delta\frac{dx}{dt}dt = (v_x(\mathbf{x} + \Delta\mathbf{x}) - v_x(\mathbf{x}))dt$$

$$= \left(v_x(\mathbf{x}) + \frac{\partial v_x(\mathbf{x})}{\partial x}\Delta x + \cdots - v_x(\mathbf{x})\right)dt \qquad (13.2)$$

$$\cong \frac{\partial v_x(\mathbf{x})}{\partial x}\Delta x dt$$

using the Taylor-series expansion of $v_x(\mathbf{x} + \Delta\mathbf{x})$ for small $\Delta\mathbf{x}$. The volume changes at the rate

$$\frac{d}{dt}(\Delta V) = \frac{\partial v_x}{\partial x}\Delta x\Delta y\Delta z = \frac{\partial v_x}{\partial x}\Delta V.$$

Corresponding changes occur in y and z directions, all of which are added up. Thus

$$\frac{d}{dt}(\Delta V) = \left[\frac{\partial v_x}{\partial x} + \frac{\partial v_y}{\partial y} + \frac{\partial v_z}{\partial z}\right]\Delta V = (\nabla \cdot \mathbf{v})\Delta V. \qquad (13.3)$$

Because we are in Lagrangian coordinates, during the flow the mass inside a fluid element is conserved, i.e., unchanging, so

$$\frac{d}{dt}(\rho \Delta V) = 0. \tag{13.4}$$

It follows that

$$\frac{d\rho}{dt}\Delta V + \rho\frac{d}{dt}(\Delta V) = 0. \tag{13.5}$$

The last term on the left hand side is from Equation (13.3) giving

$$\frac{d\rho}{dt}\Delta V + \rho(\nabla \cdot \mathbf{v})\Delta V = 0$$

which is independent of the volume ΔV. This is the **mass continuity equation** in Lagrangian form:

$$\frac{d\rho}{dt} + \rho\nabla \cdot \mathbf{v} = 0. \tag{13.6}$$

The Eulerian form can be obtained using Equation (13.1)

$$\mathbf{v} \cdot \nabla\rho + \rho\nabla \cdot \mathbf{v} = 0. \tag{13.7}$$

Applying the vector identity

$$\nabla \cdot (\rho\mathbf{v}) = \mathbf{v} \cdot \nabla\rho + \rho\nabla \cdot \mathbf{v} \tag{13.8}$$

gives

$$\frac{\partial\rho}{\partial t} + \nabla \cdot (\rho\mathbf{v}) = 0. \tag{13.9}$$

The motion of the fluid is controlled by forces which may be divided in two categories: pressure P and all other forces which may be written as $\boldsymbol{f}\,\Delta V$ where $\boldsymbol{f}$ is the **body-force density** and ΔV is the fluid element volume of the fluid. The pressure's effect is obtained by integrating the component P_n normal to the surface element dA over the whole surface as shown in Figure 13.2.

From the generalized form of the Gauss Theorem,

$$-\int P_n dA = -\int_{\Delta V} \nabla P dV. \tag{13.10}$$

If the volume ΔV is so small that ∇P may be considered constant, the total force from pressure is

$$-\nabla P \Delta V$$

On this the force per unit volume is $-\nabla P$.

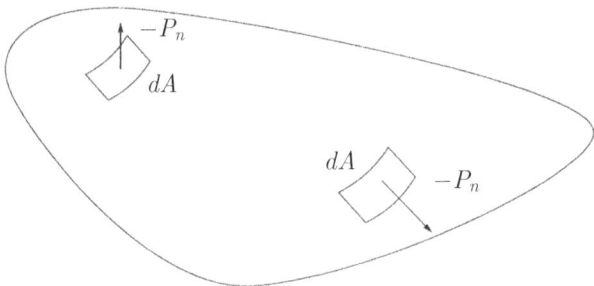

Figure 13.2. Volume element ΔV, its surface element dA and the negative of the normal component of pressure. Note P_n is directed inward.

Applying Newton's second law to the unit volume ΔV,

$$\rho \frac{dv}{dt} = \boldsymbol{f} - \nabla P. \tag{13.11}$$

Here the mass inside the volume is replaced by mass/ΔV which is the mass density ρ. This is consistent with using the force/ΔV on the right hand side. We designate Equation (13.11) as the **equation of motion** in Lagrangian form. The corresponding Eulerian equation can be obtained with the help of Equation (13.1):

$$\rho \frac{\partial \mathbf{v}}{\partial t} + \rho (\mathbf{v} \cdot \nabla) \mathbf{v} = \boldsymbol{f} - \nabla P. \tag{13.12}$$

If the gravitational force is the only body-force to be considered, it may be obtained from the gravitational potential Φ:

$$\boldsymbol{f} = -\rho \nabla \Phi. \tag{13.13}$$

More generally, there may be contributions from magnetic and electric fields which will not be studied here. Substituting Equation (13.13) into Equation (13.12) and dividing by ρ:

$$\frac{\partial \mathbf{v}}{\partial t} + (\mathbf{v} \cdot \nabla) \mathbf{v} = -\nabla \Phi - \frac{1}{\rho} \nabla P. \tag{13.14}$$

Now this is the usual expression of the Eulerian equation of motion.

13.3   Jeans Criterion

A stationary fluid is defined to have zero flow speed, $\nu_0 = 0$, with zero gradients of density and pressure: $\nabla \rho_0 = 0$ and $\nabla P_0 = 0$. Also neglecting the gravitational force:

$\nabla \Phi = 0$. Let us introduce a small perturbation ρ_1 to the constant density ρ_0 corresponding to a small perturbation P_1 in the constant pressure P_0. Equation (13.12) becomes:

$$(\rho_0 + \rho_1)\frac{\partial \mathbf{v}}{\partial t} + (\rho_0 + \rho_1)\mathbf{v}_1 \cdot \nabla \mathbf{v}_1 = -\nabla P_1 \tag{13.15}$$

where $\mathbf{v}_1$ is the small fluid velocity which results. Under the same assumptions, the continuity Equation (13.7) gives under the same assumptions

$$\frac{\partial \rho_1}{\partial t} + \mathbf{v}_1 \cdot \nabla \rho_1 + (\rho_0 + \rho_1)\nabla \cdot \mathbf{v}_1 = 0. \tag{13.16}$$

Assuming that ρ_1, $\nabla \rho_1$, P_1, $\mathbf{v}_1$, $\partial \mathbf{v}_1/\partial t$ and $\nabla \mathbf{v}_1$ are all small, their products are even smaller and can be neglected. Thus, the lowest order terms in Equations (13.15) and (13.16) give

$$\rho_0 \frac{\partial \mathbf{v}_1}{\partial t} = -\nabla P_1 \tag{13.17}$$

$$\frac{\partial \rho_1}{\partial t} + \rho_0 \nabla \cdot \mathbf{v}_1 = 0. \tag{13.18}$$

The pressure and density are connected in adiabatic processes by

$$P = K\rho^{\Gamma} \tag{13.19}$$

where K and Γ are constants. Thus

$$\nabla P = \frac{\partial P}{\partial \rho}\nabla \rho \tag{13.20}$$

and because $\nabla P_0 = \nabla \rho_0 = 0$,

$$\nabla P_1 = \frac{\partial P}{\partial \rho}\nabla \rho_1. \tag{13.21}$$

Substituting this into Equation (13.17):

$$\rho_0 \frac{\partial \mathbf{v}_1}{\partial t} = -\frac{\partial P}{\partial \rho}\nabla \rho_1 \tag{13.22}$$

we take divergence of both sides:

$$\rho_0 \frac{\partial}{\partial t}\nabla \cdot \mathbf{v}_1 = -\frac{\partial P}{\partial \rho}\nabla^2 \rho_1. \tag{13.23}$$

The quantity on the left hand side of Equation (13.23) can be obtained in another way by taking the time derivative of Equation (13.18):

$$\frac{\partial^2 \rho_1}{\partial t^2} + \rho_0 \frac{\partial}{\partial t} \nabla \cdot \mathbf{v}_1 = 0. \tag{13.24}$$

Equations (13.23) and (13.24) combined leads to

$$\frac{\partial^2 \rho_1}{\partial t^2} = c_s^2 \nabla^2 \rho_1. \tag{13.25}$$

which is a wave equation for ρ_1. The wave speed in this equation is

$$c_s = \left(\frac{\partial P}{\partial \rho} \right)^{1/2}. \tag{13.26}$$

This depends on the constant density ρ_0 for given values of K and Γ (see Equation (13.19)) which is called the **sound speed**. The oscillations of density and pressure propagate with this speed from the origin.

Now consider a perturbation under a gravitational potential Φ and a velocity field $\mathbf{v}_0$. Again, the small density perturbation, ρ_1, induces small perturbations, P_1 and $\mathbf{v}_1$. In addition, the original gravitational potential Φ_0 is perturbed by the amount Φ_1, i.e., $\Phi = \Phi_0 + \Phi_1$. The gravitational potential Φ can be derived from the density distribution ρ by using the **Poisson equation** (Box 3)

$$\nabla^2 \Phi = 4\pi G \rho. \tag{13.27}$$

The unperturbed density distribution is thus

$$\nabla^2 \Phi_0 = 4\pi G \rho_0, \tag{13.28}$$

and the perturbed density distribution is

$$\nabla^2 \Phi_0 + \nabla^2 \Phi_1 = 4\pi G \rho_0 + 4\pi G \rho_1. \tag{13.29}$$

We subtract Equation (13.28) from Equation (13.29) to give

$$\nabla^2 \Phi_1 = 4\pi G \rho_1. \tag{13.30}$$

The unperturbed continuity equation (Equation (13.9)) is

$$\frac{\partial \rho_0}{\partial t} + \nabla \cdot (\rho_0 \mathbf{v}_0) = 0 \tag{13.31}$$

Now the corresponding equation for the perturbed case is

$$\frac{\partial \rho_0}{\partial t} + \frac{\partial \rho_1}{\partial t} + \nabla \cdot (\rho_0 \mathbf{v}_0) + \nabla \cdot (\rho_0 \mathbf{v}_1 + \rho_1 \mathbf{v}_0) = 0 \tag{13.32}$$

We neglect the product of the two small quantities ρ_1 and $\mathbf{v}_1$. Subtraction of Equation (13.31) from Equation (13.32) gives

$$\frac{\partial \rho_1}{\partial t} + \nabla \cdot (\rho_0 \mathbf{v}_1 + \rho_1 \mathbf{v}_0) = 0. \tag{13.33}$$

Similarly, we write the equation of motion for the unperturbed system (Equation (13.14))

$$\frac{\partial \mathbf{v}_0}{\partial t} + (\mathbf{v}_0 \cdot \nabla)\mathbf{v}_0 = -\nabla \Phi_0 - \frac{1}{\rho_0}\nabla P_0 \tag{3.34}$$

and then the equation of motion for the perturbed system:

$$\frac{\partial}{\partial t}(\mathbf{v}_0 + \mathbf{v}_1) + (\mathbf{v}_0 + \mathbf{v}_1) \cdot \nabla(\mathbf{v}_0 + \mathbf{v}_1) = -\nabla(\Phi_0 + \Phi_1) - \frac{1}{\rho_0 + \rho_1}\nabla(P_0 + P_1). \tag{13.35}$$

We can subtract Equation (13.34) from Equation (13.35) and put $\rho_0 + \rho_1 \cong \rho_0$, $\mathbf{v}_1 \cdot \nabla \mathbf{v}_1 \cong 0$:

$$\frac{\partial}{\partial t}\mathbf{v}_1 + \mathbf{v}_0 \cdot \nabla \mathbf{v}_1 + \mathbf{v}_1 \cdot \nabla \mathbf{v}_0 = -\nabla \Phi_1 - \frac{1}{\rho_0}\nabla P_1. \tag{13.36}$$

To progress, we take the initially uniform medium at rest, i.e., $\rho_0 = $ constant, $\mathbf{v}_0 = 0$, so that from Equation (13.33) we get

$$\frac{\partial \rho_1}{\partial t} + \rho_0 \nabla \cdot \mathbf{v}_1 = 0 \tag{13.37}$$

and from Equation (13.36)

$$\frac{\partial \mathbf{v}_1}{\partial t} + \nabla \Phi_1 + \frac{1}{\rho_0}\nabla P_1 = 0. \tag{13.38}$$

For adiabatic perturbations (Equation (13.26))

$$P_1 = c_s^2 \rho_1 \tag{13.39}$$

This is substituted into Equation (13.38):

$$\frac{\partial \mathbf{v}_1}{\partial t} = -\nabla \Phi_1 - \frac{c_s^2}{\rho_0}\nabla \rho_1. \tag{13.40}$$

Taking the divergence:

$$\nabla \cdot \frac{\partial \mathbf{v}_1}{\partial t} = \frac{\partial}{\partial t}(\nabla \cdot \mathbf{v}_1) = -\nabla^2 \Phi_1 - \frac{c_s^2}{\rho_0}\nabla^2 \rho_1. \tag{13.41}$$

The term in the middle of the above equation may be obtained in another way by taking the time derivative of Equation (13.37):

$$\frac{\partial}{\partial t}(\nabla \cdot \mathbf{v}_1) = -\frac{1}{\rho_0}\frac{\partial^2 \rho_1}{\partial t^2}. \tag{13.42}$$

Equations (13.41) and (13.42) may now be combined together a third Equation (13.30):

$$\frac{1}{\rho_0}\frac{\partial^2 \rho_1}{\partial t^2} = 4\pi G\rho_1 + \frac{c_s^2}{\rho_0}\nabla^2 \rho_1 \tag{13.43}$$

or

$$\left(\nabla^2 - \frac{1}{c_s^2}\frac{\partial^2}{\partial t^2} + \frac{4\pi G\rho_0}{c_s^2}\right)\rho_1 = 0. \tag{13.44}$$

This is a wave equation like Equation (13.25), but with an additional term $4\pi G\rho_0/c_s^2$. We will now try to solve it, attempting a solution whose form is

$$\rho_1 = Ae^{i(\mathbf{k}\cdot\mathbf{r}-\omega t)} \tag{13.45}$$

Here ω is the frequency and $\mathbf{k}$ the wave vector. Substituting this solution into Equation (13.44) and carrying out the calculations:

$$(\nabla\rho_1)_x = \frac{\partial}{\partial x}Ae^{i(k_x x+k_y y+k_z z-\omega t)} = ik_x\rho_1,$$

$$(\nabla\rho_1)_y = ik_y\rho_1,$$

$$(\nabla\rho_1)_z = ik_z\rho_1$$

From these,

$$\nabla\rho_1 = (\nabla\rho_1)_x\hat{\mathbf{i}} + (\nabla\rho_1)_y\hat{\mathbf{j}} + (\nabla\rho_1)_z\hat{\mathbf{k}} = i(\hat{\mathbf{i}}k_x + \hat{\mathbf{j}}k_y + \hat{\mathbf{k}}k_z)\rho_1 = i\mathbf{k}\rho_1. \tag{13.46}$$

Here $\hat{\mathbf{i}}$, $\hat{\mathbf{j}}$, and $\hat{\mathbf{k}}$ are Cartesian unit vectors along the x, y, and z axes. Repeating the "delta" operation, we get

$$\nabla^2\rho_1 = -k^2\rho_1. \tag{13.47}$$

Similarly we calculate

$$\frac{\partial}{\partial t}\rho_1 = -i\omega\rho_1,$$

$$\frac{\partial^2}{\partial t^2}\rho_1 = -\omega^2\rho_1. \tag{13.48}$$

The wave equation becomes

$$\left(-k^2 + \frac{\omega^2}{c_s^2} + \frac{4\pi G\rho_0}{c_s^2}\right)\rho_1 = 0$$

or

$$\omega^2 = k^2 c_s^2 - 4\pi G\rho_0. \tag{13.49}$$

If $\omega^2 < 0$, ω must be imaginary. Setting $\omega = i\eta$ (where η is a real number), then

$$\rho_1 = A e^{i\mathbf{k}\cdot\mathbf{r} + \eta t} \tag{13.50}$$

In this second case, in the equation above, ρ_1 increases exponentially with t.

In another case, $\omega^2 > 0$, ω is a real number and ρ_1 is a pure oscillating sine wave. At the limit of $\omega^2 = 0$ the limiting value, $k = k_J$, is

$$k_J^2 = \frac{4\pi G\rho_0}{c_s^2}. \tag{13.51}$$

We have the first case if $k < k_J$, $\omega^2 < 0$ causing the density perturbation to grow larger exponentially. On the other hand, if $k > k_J$, $\omega^2 > 0$, the second case. Thus, the perturbation survives as a wave of constant amplitude.

The limiting value k_J is interesting. It is related to the minimum size of an unstable region λ_J through the equation

$$\lambda_J = \frac{2\pi}{k_J} = \sqrt{\pi}\frac{c_s}{(G\rho_0)^{1/2}}. \tag{13.52}$$

Here λ_J is the wavelength and k_J the wave number at the limit of stability. Instability occurs for regions that exceed λ_J in size. Stability occurs for smaller regions which are stable against collapse. The critical value λ_J is known as the **Jeans length**. We can calculate a corresponding mass

$$M_J = \frac{4\pi}{3}\rho_0\left(\frac{\lambda_J}{2}\right)^3 \tag{13.53}$$

which is called the **Jeans mass**. Except for the $\sqrt{\pi}$ factor, this λ_J agrees with the Jeans critical length R_J derived more simply in Section 13.1.

13.4   Jeans Criterion for collisionless dark matter gas

We shall see that dark matter plays a central part in gravitational instability when it dominates over radiation in the universe. Ordinary baryonic matter is a minor constituent. The physical nature and content of dark matter are unknown. It is most probably a gas of particles that interact with each other and with other matter and energy only via gravity and weak force. This gas is thus practically collisionless i.e., particle-particle collisions are very rare. The weak-force collision cross-section is very small and the particle density is low soon after the dark matter annihilation freeze-out. At later times, the mean free path of dark matter particles is large; it is much larger than, say, a current observable horizon radius.

The major difference of dark matter from "ordinary" baryonic gas is that the latter has frequent collisions and very short mean free path of particles. A mass element of ordinary gas moves with a bulk velocity that is common for all the particles. The requirement is that the size of the element is considerably larger than the particle mean free path. In a collisionless gas, like dark matter, there is no common velocity. In other words, particles move freely with their individual velocities. A bulk motion is still possible as a sum of free individual motions of particles. Thus, a description in terms of individual particles, rather than in terms of gas elements, is appropriate for such a collisionless gas. Instead of a hydrodynamic description, dark matter needs a kinetic description. The basic quantity in the kinetic description is the **distribution function**, which is the particle density $f(\mathbf{r}, \mathbf{v}, t)$ in phase space.

Now $f(\mathbf{r}, \mathbf{v}, t)$ tells the number of particles in a unit volume $d\mathbf{r} = dx\,dy\,dz$ of space and in a unit range of velocities $d\mathbf{v} = dv_x\,dv_y\,dv_z$, centered on position $\mathbf{r}$ and around velocity $\mathbf{v}$:

$$dN = f(\mathbf{r}, \mathbf{v}, t)\,d\mathbf{r}\,d\mathbf{v}. \tag{13.54}$$

The $f(\mathbf{r}, \mathbf{v}, t)$ can also depend on time t. If it does not, the system is stationary. The individual particles can move around in such a system, but with no change in the distribution function.

Thus, we define the particle density of collisionless gas by means of an integral over all the particle velocities:

$$n(\mathbf{r}, t) = \int f(\mathbf{r}, \mathbf{v}, t)\,d\mathbf{v}. \tag{13.55}$$

For simplicity, assume that the particle masses m are equal, so the mass density

$$\rho(\mathbf{r}, t) = mn(\mathbf{r}, t). \tag{13.56}$$

The gravitational potential Φ can be obtained from the Poisson equation:

$$\nabla^2 \Phi(\mathbf{r}, t) = 4\pi G \rho(\mathbf{r}, t)$$

The system is described by points in 6-dimensional phase space, with three space coordinates $\mathbf{r}$ and three velocity coordinates $\mathbf{v}$. The position of a particle in this phase space can be represented by a 6-dimensional vector $\mathbf{w}$:

$$\mathbf{w} = (\mathbf{r}, \mathbf{v}). \tag{13.57}$$

whose corresponding velocity vector is

$$\dot{\mathbf{w}} = (\dot{\mathbf{r}}, \dot{\mathbf{v}}) = (\mathbf{v}, -\nabla\Phi). \tag{13.58}$$

Here we assume that the acceleration $\dot{\mathbf{v}}$ arises from the gravitational potential Φ.

Now we extend the continuity equation (Equation (13.9)) to the 6-dimensional phase space:

$$\frac{\partial f}{\partial t} + \frac{\partial}{\partial x_k}(f\dot{w}_k) = 0. \tag{13.59}$$

Again, the repeated index k implies summation, $k = 1, \ldots, 6$. Dividing the terms in this sum into two parts:

$$\frac{\partial}{\partial x_k}(f\dot{w}_k) = f\frac{\partial}{\partial x_k}\dot{w}_k + \dot{w}_k\frac{\partial f}{\partial x_k}. \tag{13.60}$$

The first right hand term may be written using the ordinary space and velocity coordinates:

$$\frac{\partial}{\partial x_k}\dot{w}_k = \frac{\partial v_i}{\partial x_i} + \frac{\partial \dot{v}_i}{\partial v_i} \tag{13.61}$$

where the summation index $i = 1, \ldots, 3$. In phase space x_i and v_i are independent of one another. Thus $\partial v_i/\partial x_i = 0$. Now the acceleration $\dot{v}_i = -\partial\Phi/\partial x_i$. Now since Φ does not depend upon the velocity v_i, the derivative $\partial\dot{v}_i/\partial v_i = 0$. As a result, the first term on the right side of Equation (13.60) is zero.

Equation (13.59) then is

$$\frac{\partial f}{\partial t} + \dot{w}_k\frac{\partial f}{\partial w_k} = 0 \tag{13.62}$$

And using 3-dimensional components,

$$\frac{\partial f}{\partial t} + v_i\frac{\partial f}{\partial x_i} + \dot{v}_i\frac{\partial f}{\partial v_i} = 0. \tag{13.63}$$

In vector notation, and using $\dot{v}_i = -\partial\Phi/\partial x_i$,

$$\frac{\partial f}{\partial t} + \mathbf{v}\cdot\nabla f - \nabla\Phi\cdot\frac{\partial f}{\partial \mathbf{v}} = 0. \tag{13.64}$$

The above is the **Vlasov** or **collisionless Boltzmann equation**. Equation (13.62) is, in fact, the Lagrangian derivative (Equation (13.1)) in six dimensions. Thus

$$\frac{df}{dt} = 0. \tag{13.65}$$

This equality is called the **Liouville Theorem**. It tells us that particles move in the 6-dimensional phase space in such a way that the density of particles (in 6 dimensions) is a constant.

The above is the kinetic equation for collisionless gas. The velocity's derivative with the gravity force affecting the particle is connected by the equation of motion:

$$\dot{\mathbf{v}} = -\nabla \Phi. \tag{13.66}$$

With this set of equations, consider the Jeans problem for a uniform self-gravitating collisionless non-relativistic gas of dark matter particles. We assume that the unperturbed matter distribution is time-independent.

We can represent a weak linear perturbation (as in the section above, Equation (13.45)) by a wave propagating along, say, the axis x. The wave is characterized by δf, $\delta \rho$, $\delta \Phi$. The dependence of these quantities on time and spatial coordinates has the form of a wave. Thus these quantities are proportional to

$$F(t, x) = \exp i(kx - \omega t). \tag{13.67}$$

One has, putting $\delta f = A(\mathbf{v})F$, $\delta \rho = BF$ and $\delta \Phi = CF$,

$$f = f_0 + A(\mathbf{v})F,$$
$$\rho = \rho_0 + BF, \tag{13.68}$$
$$\Phi = \Phi_0 + CF.$$

Here B and C are constant while A varies as a function of velocity $\mathbf{v}$. Also f_0 depends on the velocity. Since f_0 does not depend on the direction of motion it should be a function of v^2 : $f_0 = f_0(v^2)$. The constants A and B are connected by (see Equations (13.55) and (13.56))

$$B = m \int A d\mathbf{v}. \tag{13.69}$$

Equation (13.36) is the kinetic equation in the x-direction,

$$\frac{\partial f}{\partial t} + v_x \frac{\partial f}{\partial x} + \dot{v}_x \frac{\partial f}{\partial v_x} = 0.$$

Derivatives are calculated as in the previous section (Equations (13.46) and (13.48)):

$$\frac{\partial f}{\partial t} = -i\omega \delta f = -i\omega A F,$$

$$\frac{\partial f}{\partial x} = ik\delta f = ikAF.$$

Now, the derivative $\frac{\partial f}{\partial v_x}$ may be written

$$\frac{\partial f}{\partial v_x} = \frac{2v_x \partial f}{\partial (v_x^2)} \simeq 2v_x \frac{df_0(v^2)}{d(v^2)}$$

where the term proportional to $\frac{\partial \delta f}{\partial (v^2)}$ has been neglected as a small quantity. The value of $\dot{v}_x$ can be obtained using Equation (13.66) in a linear form

$$\dot{v}_x = -\frac{\partial \Phi}{\partial x} = -ik\delta\Phi. \tag{13.70}$$

Poisson's equation (Equation (13.37)) gives in a linearized form

$$-k^2 \delta\Phi = 4\pi G\delta\rho.$$

Substituting $\delta\Phi$ into Equation (13.70):

$$\dot{v}_x = \frac{i4\pi G\delta\rho}{k} = \frac{i4\pi GBF}{k}.$$

The kinetic equation can be reduced to:

$$(\omega - kv_x)A(\mathbf{v}) - 8\pi GBk^{-1}v_x \frac{df_0(v^2)}{d(v^2)} = 0. \tag{13.71}$$

The last equation integrated over the velocities gives:

$$(\omega - kv_x) - 8\pi Gmk^{-1}v_x \int \frac{df_0(v^2)}{d(v^2)} d\mathbf{v} = 0. \tag{13.72}$$

The Jeans critical wave number, $k = k_J$, comes from this equation when $\omega = 0$:

$$k_J^2 = -8\pi Gm \int \frac{df_0(v^2)}{d(v^2)} d\mathbf{v}. \tag{13.73}$$

Carrying out the integration in parts, using $d\mathbf{v} = 4\pi v^2 dv$:

$$\int_0^\infty \frac{df_0}{d(v^2)} 4\pi v^2 dv = \frac{4\pi}{2} \int_0^\infty \frac{v df_0}{d(v^2)} d(v^2) = \frac{4\pi}{2} vf_0 \Big|_0^\infty - \frac{4\pi}{2} \int_0^\infty f_0 dv = -\frac{1}{2} \int \frac{f_0}{v^2} d\mathbf{v}.$$

At large v, f_0 goes to zero faster than v^{-1}, generally true for distributions like the Maxwellian distribution. We introduce the mean inverse square velocity

$$\langle v^{-2} \rangle \equiv \frac{\int (f_0/v^2) d\mathbf{v}}{\int f_0 d\mathbf{v}}, \tag{13.74}$$

so that finally

$$k_J = (4\pi G\rho_0)^{1/2}\sqrt{\langle v^{-2}\rangle}. \tag{13.75}$$

This is similar to that obtained in the section above for ordinary gas with hydro-dynamics. The only difference is that instead of the simple inverse sound velocity, the square root of the mean inverse square velocity appears. This defined typical velocity plays the same role in collisionless gas as the sound speed in ordinary gas which represents the elasticity of this gas. The relation to the gas temperature is simple: $m\big(\langle v^{-2}\rangle\big)^{-1} \cong k_B T$. The general concept of gravitational instability can be applied to collisionless gas in practically the same way as instability is applied to ordinary collisional gas (Gennadiy S. Bisnovatyi-Kogan and Zeldovich 1970). In the collisionless gas of dark matter, its behavior in gravity fields can be for most purposes described in hydrodynamic terms (using density, bulk velocity, tempera-ture, etc.)—in a close similarity to ordinary gas.

13.5   Jeans Criterion in the expanding universe

For an expanding universe the above derivation needs to be modified. The free fall collapse of a region of radius $\lambda_J/2$ occurs in a time

$$t_{coll} = \frac{\lambda_J/2}{c_s} = \frac{\sqrt{\pi}/2}{(G\rho_0)^{1/2}} \approx (G\rho_0)^{-1/2} \tag{13.76}$$

On the other hand, universal expansion time span is (Equations (8.20), (8.24) and (8.26))

$$t_{exp} = \frac{2}{3}H_0^{-1} = \frac{2}{3}\sqrt{\frac{3}{8\pi G\rho_0}} \approx (G\rho_0)^{-1/2}. \tag{13.77}$$

We see that the expansion of the region that is trying to collapse may prevent the collapse or at least slow the collapse will be slowed. The calculations of the previous section for the expanding universe must be redone.

Consider a local density fluctuation in a smooth medium. The position of a mass element nearby is described by

$$\mathbf{r} = R(t)\mathbf{r}_0 \tag{13.78}$$

which is Equation (8.46) with $\mathbf{r}(t_0) = \mathbf{r}_0$. Its velocity

$$\mathbf{v} = \frac{\dot{R}(t)}{R(t)}\mathbf{r} + \mathbf{u}(\mathbf{r}, t). \tag{13.79}$$

On the right side the first term represents universal expansion (Equations (8.44) and (8.48)). The second term on the right is due to motions induced by the density perturbation. The density perturbation may be represented by

$$\rho(\mathbf{r}, t) = \rho(t)[1 + \delta(\mathbf{r}, t)] \tag{13.80}$$

Representing the mean mass density of the expanding universe by $\rho(t)$ while the fractional perturbation $\delta(\mathbf{r}, t)$ is assumed to be much less than unity. The density perturbation in Equation (13.30) is $\rho_1 = \rho\delta$, and Poisson's equation is

$$\nabla^2\Phi_1 = 4\pi G\rho\delta. \tag{13.81}$$

Substitute $\rho_1 = \rho_0(t)\delta(\mathbf{r}, t)$, $\mathbf{v}_0 = Hr$, $H \equiv \dot{R}(t)/R(t)$ and $\mathbf{v}_1 = \mathbf{u}$ in Equations (13.33) and (13.36). After some algebra, the equation of continuity becomes

$$\dot{\delta} + \nabla \cdot \mathbf{u} + \frac{\dot{R}(t)}{R(t)}\mathbf{r} \cdot \nabla\delta = 0 \tag{13.82}$$

and putting $R = R(t)$ for short the equation of motion is

$$\frac{\partial\mathbf{u}}{\partial t} + \frac{\dot{R}}{R}\mathbf{u} + \frac{\dot{R}}{R}\mathbf{r} \cdot \nabla\mathbf{u} = -\frac{1}{\rho}\nabla P_1 - \nabla\Phi_1. \tag{13.83}$$

The right side of Equation (13.83) contains the perturbed gravitational potential Φ_1 and the perturbation of pressure P_1 over the constant values Φ_0 and P_0.

We will now transform into Lagrangian coordinates. The Lagrangian radial coordinate corresponding to $\mathbf{r}$ is $\mathbf{r}_0$ as seen in Equation (13.78). In component form, the Eulerian x^i, $i = 1, \ldots, 3$, components are related to the Lagrangian x_0^i, $i = 1, \ldots, 3$, by

$$x^i = R(t)x_0^i, \quad i = 1, \ldots, 3. \tag{13.84}$$

The divergence $\nabla \cdot \mathbf{u} = \partial u^i/\partial x^i$ transforms as

$$\frac{\partial u^i}{\partial x^i} = \frac{\partial u^i}{\partial x_0^k}\frac{\partial x_0^k}{\partial x^i} = \frac{\partial u^i}{\partial x_0^k}\frac{1}{R(t)}\delta_k^i = \frac{1}{R(t)}\frac{\partial u^i}{\partial x_0^i} \tag{13.85}$$

and similarly for the gradient. Writing symbolically,

$$\nabla_r = \frac{1}{R}\nabla_{r_0} \tag{13.86}$$

where ∇_r in the Eulerian and ∇_{r_0} the Lagrangian derivatives.

From now on we assume that the Lagrangian system is used, and we put simply $\nabla_{r_0} = \nabla$. After considerable amount of algebra (see e.g., Byrd, Chernin and Valtonen 2007), we arrive at

$$\frac{d^2\delta}{dt^2} + 2\frac{\dot{R}}{R}\frac{d\delta}{dt} - \frac{c_s^2}{R^2}\nabla^2\delta - 4\pi G\rho\delta = 0. \tag{13.87}$$

Again, we have again something like the wave equation in Equation (13.25) with two additional terms. Accordingly, we try a wave-like solution

$$\delta = \delta_{\mathbf{k}}(t)\,e^{i\mathbf{k}\cdot\mathbf{r}} \tag{13.88}$$

where $\mathbf{k}$ is a co-moving wave vector; the corresponding wavelength λ stretches with the expansion of the Universe. Therefore, the time dependence is in the amplitude $\delta_{\mathbf{k}}(t)$ and the spatial variation is in the exponential factor of Equation (13.88). Equation (13.47) (with the perturbation $\rho_1 \to \delta$) is still valid considering that we have ∇ as a Lagrangian operator and $\mathbf{k}$ as a Lagrangian wave vector:

$$\nabla^2\delta = -k^2\delta. \tag{13.89}$$

Substitute the above together with the time derivatives from Equation (13.88) into Equation (13.87). Canceling the common factor $e^{i\mathbf{k}\cdot\mathbf{r}}$ in both sides, we are left with

$$\ddot{\delta}_{\mathbf{k}} + 2\frac{\dot{R}}{R}\dot{\delta}_{\mathbf{k}} = \left(4\pi G\rho - \frac{c_s^2 k^2}{R^2}\right)\delta_{\mathbf{k}}. \tag{13.90}$$

On the right side, the two terms in the parentheses show the competition between gravity $(4\pi G\rho)$ and the pressure gradient $(c_s^2 k^2/R^2)$. These two are equal if

$$\lambda_J \equiv \lambda R = 2\pi\frac{R}{k} = \sqrt{\pi}\,\frac{c_s}{(G\rho)^{1/2}}. \tag{13.91}$$

Here λ_J is in the Eulerian system and λ in the Lagrangian system. Consider a perturbation which is initially growing, i.e., $\dot{\delta}_{\mathbf{k}} > 0$. In order that the growth would stop, at some point in time we must have $\dot{\delta}_{\mathbf{k}} = 0$ and $\ddot{\delta}_{\mathbf{k}} < 0$. At that time the right hand side of Equation (13.90) must be negative. This situation never occurs if

$$4\pi G\rho > \frac{c_s^2 k^2}{R^2}, \tag{13.92}$$

i.e., for wavelengths greater than λ_J. Then the growth continues. We have arrived again at the Jeans length of Equation (13.52) after transformation back to the Eulerian system or the proper wavelength $\lambda R(t)$.

Analysis of stability similar to Section 13.2 shows that again λ_J separates the unstable region where $\lambda > \lambda_J$, from the region of stable oscillations where $\lambda < \lambda_J$. The conclusions of the previous section also apply to the present case.

We may use the same expressions for the Jeans length and Jeans mass both in a static and in an expanding universe, but the exact forms of the solutions are different. The "sound waves" have slowly decreasing amplitudes in the expanding case, contrary to constant amplitude in a static universe. Growth of perturbations in an expanding medium is not exponential but instead is slower. As we shall see, the characteristics of growth depend on the details of the cosmological model. In the Einstein–de Sitter model $k = 0$, $\frac{\dot{R}}{R} = \frac{2}{3}t^{-1}$ from Equation (8.17), along with

$$4\pi G\rho = \frac{3}{2}H_0^2 = \frac{3}{2}\left(\frac{\dot{R}}{R}\right)^2 \tag{13.93}$$

(Equations (8.20), (8.24) and (8.26)). Thus Equation (13.90) becomes

$$\ddot{\delta}_{\mathrm{k}} + \frac{4}{3}t^{-1}\dot{\delta}_{\mathrm{k}} - \frac{2}{3}t^2\delta_{\mathrm{k}} = 0. \tag{13.94}$$

The solution can easily be verified as

$$\delta_{\mathrm{k}} = C_1 t^{2/3} + C_2 t^{-1} \tag{13.95}$$

where C_1 and C_2 are arbitrary constants. The first term on the right side represents a growing mode. The second is a damping mode that will become insignificant at large t. Although the growing mode wins over time, its growth is much slower than the exponential growth in a static universe (Equation (13.50)).

Since we skipped over the detailed calculations in the derivation of Equation (13.95), we may demonstrate this result in a simple model suggested by Yakov B. Zeldovich (1965).

Consider non-relativistic gas expansion, the second gravitating matter stage of the cosmic expansion. We will assume that the expanding medium is a non-relativistic gas, a good approximation for the second epoch of the cosmic expansion. This is because the density of non-relativistic matter (cold dark matter and baryons) is larger than the radiation and vacuum densities. We can neglect radiation and the vacuum, so we can consider this non-relativistic gas moving in its own self-gravity field.

We also assume that the unperturbed expansion is parabolic. This is described by the cosmological model in which the scale factor $R = R(t)$ and density $\rho = \rho(t)$ follow

$$R = R_F \propto t^{2/3}, \quad \rho = \rho_F = (6\pi G t^2)^{-1}. \tag{13.96}$$

This is also a good approximation to what we think is the real cosmological situation at the same epoch, as we discussed in Chapter 8's Equations (8.17) and (8.18).

Assume the perturbation region is a finite spherical part of a Friedmann model in which expansion is elliptic since the region will collapse. The radius of the region R grows according to the equation (Equation (8.14))

$$(1/2)\dot{R}^2 = GM/R - |E| \tag{13.97}$$

where M is the mass of the region. This equation is the energy conservation relation (E = energy per unit mass) in Newtonian mechanics.

We compare the expansion rate of the perturbation region with the parabolic expansion of an "unperturbed" region of the same mass M in which $(1/2)\dot{R}_F^2 = GM/R_F$. Clearly, the difference between these two is small if $(1/2)c^2/(GM/R) \ll 1$. Defining $\delta R = R - R_F \ll R_F$, in the linear approximation with the ratio $\delta R/R \ll 1$ we have

$$\frac{1}{R} = \frac{1}{R_F} - \frac{1}{R_F^2}\delta R, \quad \dot{R}^2 = \dot{R}_F^2 + 2\dot{R}_F\frac{\partial \delta R}{\partial t} \tag{13.98}$$

$$\frac{1}{2}\left(\dot{R}^2 - \dot{R}_F^2\right) - \left(\frac{GM}{R} - \frac{GM}{R_F}\right) = \dot{R}_F\frac{\partial}{\partial t}\delta R + \frac{GM}{R_F}\frac{\delta R}{R_F} = -|E|. \tag{13.99}$$

To first order put $R \cong R_F$ and $\dot{R} \cong \dot{R}_F$. Then

$$\frac{\partial \delta R}{\partial t}\dot{R} + (GM/R)\delta R/R = -|E|. \tag{13.100}$$

Because the above differential equation does not contain the independent variable t explicitly, R can be used as a new independent variable, giving

$$\frac{\partial \delta R}{\partial t} = \frac{\partial \delta R}{\partial R}\frac{\partial R}{\partial t}.$$

Then

$$\frac{\partial \delta R}{\partial R}\dot{R}^2 + \left(\frac{GM}{R}\right)\frac{\delta R}{R} = -|E|. \tag{13.101}$$

Since $\dot{R}^2 = 2GM/R$, the final differential is

$$\frac{\partial \delta R}{\partial R} + \frac{1}{2}\frac{\delta R}{R} = -|E|R/2GM. \tag{13.102}$$

Checking shows that this equation has a power-law solution

$$\delta R = -\frac{|E|}{5GM}R^2 \tag{13.103}$$

With this solution, and since

$$\rho = M\left[(4\pi/3)R^3\right]^{-1}, \tag{13.104}$$

the amplitude of density perturbation is

$$\delta = \delta\rho/\rho = -3\delta R/R = \frac{3|E|}{5GM}R. \tag{13.105}$$

The solution may be rewritten in the form:

$$\delta = \delta\rho/\rho \propto R \propto t^{2/3}. \tag{13.106}$$

The initially small density amplitude can grow. In the linear analysis, it increases as a positive power of time. In General Relativity the linear analysis was first performed by Eugeny M. Lifshitz (1946). The result for the growing perturbations in non-relativistic gas is exactly the same as above.

Importantly, the growth rate of linear perturbations does not depend on the shape of the perturbation region. This is due to the symmetry of the unperturbed state, where the medium is uniform. This together with the far reaching analogy between General Relativity linear analysis and the linear analysis in terms of the Newtonian mechanics, makes the result of Equation (13.106) rather general.

Next, consider radiation dominated plasma expansion, the earlier stage of the cosmic expansion. For radiation dominated plasma at the earlier stage of the cosmological expansion a similar linear analysis is also possible. At this earlier stage, the baryonic plasma is coupled with the radiation photons. Together, they move as a single fluid. Perturbations in which plasma and radiation move together are known as adiabatic perturbations. Instability analysis for adiabatic perturbations during radiation domination can be developed, similarly to Newtonian mechanics. The major difference between these two is the fact that the effective gravitating density in the Friedmann equations include radiation pressure $P_R = \frac{1}{3}\rho_R c^2$:

$$\rho_{eff} = \rho + 3P_R/c^2 \cong 2\rho_R. \tag{13.107}$$

The basic equations for the perturbation growth have now the form:

$$R_F \propto t^{1/2}, \quad \rho_F = 3/(32\pi G)t^{-2}, \tag{13.108}$$

$$(1/2)\dot{R}^2 = C/R^2 - |E|; \quad C = \text{Const} > 0. \tag{13.109}$$

In a relativistic photon dominated gas, considering linear perturbation, the solution for the perturbation growth is

$$\delta\rho/\rho \propto R^2 \propto t. \tag{13.110}$$

Thus, adiabatic perturbations at gravitating dark matter domination grow somewhat slower than the above. Recall that the Equation (13.106), gravitating solution has the form:

$$\delta = \delta\rho/\rho \propto R \propto t^{2/3}.$$

With a smaller time exponential. In both cases, however, this is a power-law growth, in contrast to the exponential growth in the Jeans original analysis for an ordinary, non-dark matter gas. The exponential time law is possible only if the unperturbed state of the fluid is static.

13.6   Evolution of density perturbations

Suppose that we have a density perturbation of wavelength λ in an expanding universe. First we ask how big λ is in relation to the Hubble radius R_H which is defined as

$$R_H(t) = c\left(\frac{\dot{R}(t)}{R(t)}\right)^{-1}. \tag{13.111}$$

This is about the size of the region $R_0(t_0)$ crossed at the speed of light within the current age t of the universe (Equation (3.6)). At the present t_0 using Equation (3.4),

$$R_H(t_0) = cH_0^{-1}. \tag{13.112}$$

Physical processes may operate coherently only over this size or smaller. Larger regions have not yet had time to share their physical properties. However, the first moments of the universe may be an exception. At that very early time inflationary expansion may have unified regions which are well beyond the current Hubble radius. To do this, we will here consider growth of perturbations where communication between different parts of the perturbation is essential. Then $R_H(t)$ is the practical sphere of influence.

Before, we have seen that the universal expansion

$$R(t) \propto t^n \tag{13.113}$$

where $n < 1$ ($n = 2/3$ in Equation (8.17) and $n = 1/2$ in Equation (10.20)). Therefore

$$R_H(t) = c\frac{R(t)}{\dot{R}(t)} \propto ct. \tag{13.114}$$

Now the Hubble radius grows linearly with time. On the other hand, as we have seen, the proper wavelength of the perturbation grows

$$\lambda \propto R(t) \propto t^n, \quad n < 1 \tag{13.115}$$

i.e., more slowly than the Hubble radius. Thus λ may be initially greater than R_H. However, at a certain moment of time t_{enter}, R_H catches up:

$$\lambda = R_H(t_{enter}). \tag{13.116}$$

The region of radius λ enters inside the Hubble radius at that moment. At later times, the region may be dealt with as a physical unit, possibly using Newtonian concepts. As an example, different parts of this region may adjust their pressure by emitting sound waves creating a coherent pressure structure.

As an example, consider the early expansion up to $t = t_{rec}$ where we follow Equation (10.20).

$$R_H(t_{rec}) = 2ct_{rec} = 2 \times \left(3 \times 10^5 \, \text{km/s}\right) 3.8 \times 10^5 \, \text{yr} = 230 \, \text{kpc}. \tag{13.117}$$

Relative to the above size, that same size region will be expanded by today by a factor

$$1 + z_{rec} \cong 1.1 \times 10^3. \tag{13.118}$$

The current size is ~ 250 Mpc, much larger. This size or smaller could have been coherent structures at recombination. They may have evolved to something distinctive that is observable today.

Currently, the Hubble radius is

$$R_H(t_0) = c/H_0 \cong 4225 \, \text{Mpc} \tag{13.119}$$

much greater than the region mentioned above.

The density excess over the mean density may or may not grow with time. Now, what happens to a region which is greater than the Hubble radius, $\lambda > R_H$? The mean density background universe can be represented by a $k = 0$ Friedmann model while the region of excess density must be of the type $k = +1$. This latter region expands less rapidly than the background model. If given enough time, it will ultimately collapse on itself (Figure 8.3). The early stages of galaxy formation follow this process.

Besides this **rising mode** of gravitational instability described above, there is also a **falling mode** which scales with time as

$$\delta R/R \propto 1/t. \tag{13.120}$$

For instance, if one compares two parabolic solutions, the first the Friedmann unperturbed expansion, $R_F \propto t^{2/3}$, and the second for a perturbation region which starts to expand earlier or later than the unperturbed medium. Then perturbation size $R \propto (t - t_i)^{2/3}$, where $t_i = \text{constant}$. If a perturbation region starts to expand later $(t_i > 0)$ than the background does, the difference in the expansion rate can be described by

$$\delta R = -R_F(2t_i/3t),$$

from which the above Equation (13.120) follows directly. Because the falling mode is related to a shift in time, its behavior is the same for both non-relativistic and ultra-relativistic fluids. The behavior of an arbitrary perturbation generally is described by a linear combination of both growing and falling modes (Equation (13.95)).

As we will see later, the falling mode may also be interesting in the theory of perturbation evolution.

Problem. Find the result of Equation (13.110) from Equations (13.108) and (13.109). What is the time behavior of velocity perturbations, $\delta v = \frac{\partial \delta R}{\partial t}$, in the process of gravitational instability in both non-relativistic and ultra-relativistic expanding fluids.

Perturbations which have $\lambda > R_H$ always grows. Growth is faster in the radiation dominated era and slower in the matter dominated era.

We see matter dominated era perturbations with $\lambda_J < \lambda < R_H$ should always grow stronger and finally collapse. The radiation dominated phase behavior is not so obvious. The time for expansion t_{exp} is very short and may be shorter than the collapse time t_{coll} of the density perturbation. The two time scales are given by Equation (10.21).

$$t_{exp} = \sqrt{\frac{3}{32\pi}}(G\rho_R)^{-1/2} \tag{13.121}$$

and

$$t_{coll} = \frac{\lambda}{c_s} = \sqrt{\pi}(G\rho)^{-1/2}. \tag{13.122}$$

Here ρ_R is radiation density and ρ matter density.

Collapse where $t_{coll} < t_{exp}$ leads to

$$\sqrt{\pi}\rho^{-1/2} < \sqrt{\frac{3}{32\pi}}\rho_R^{-1/2} \tag{13.123}$$

or

$$\frac{\rho}{\rho_R} > \frac{32}{3}\pi^2 \approx 10^2. \tag{13.124}$$

By definition of the radiation dominated era, $\rho_R > \rho$. Thus the above statement cannot be true and the major growth of sub-horizon perturbations is not possible during the radiation era.

13.7   Jeans mass in the early universe

The λ_J depends on sound speed c_s and density ρ. Equation (6.11), with small R_α, appropriate for non-relativistic particles, gives a relation for the general decrease of the velocity $v = c_s$,

$$c_s \propto R^{-1}. \tag{13.125}$$

where $R = R(t)$ is the scale factor. This relation applies to the dark matter particles, except in the case of very early times when the particles are relativistic ($t \leq t_D$). It applies also to baryons after recombination. Now, prior to recombination, the baryons are coupled to radiation photons. The sound speed of the relativistic photon-baryon fluid is $c_s = c/\sqrt{3}$.

Since $\rho \propto R^{-3}$ during the **matter domination**,

$$\lambda_J \propto R^{-1}/R^{-3/2} = R^{1/2}. \tag{13.126}$$

referring to dark matter at $t > t_*$, or to baryons at $t > t_{rec}$. When $t_* < t < t_{rec}$, for baryons $\lambda_J \propto \text{const}/R^{-3/2} \propto R^{3/2}$.

We see that the corresponding Jeans mass is

$$M_J \propto \rho\lambda_J^3 \propto R^{-3}R^{3/2} = R^{-3/2}. \tag{13.127}$$

Again this applies to dark matter at $t > t_*$ and baryons at $t > t_{rec}$. For baryons at $t_* < t < t_{rec}$ becomes $M_J \propto R^{-3}(R^{3/2})^3 = R^{3/2}$.

Also relevant is the mass within the Hubble radius (matter era; $\rho = \rho_M$)

$$M_H \propto \rho R_H^3 \propto \rho t^3 \propto R^{-3}(R^{3/2})^3 = R^{3/2} \tag{13.128}$$

(Equations (8.16) and (13.114)). Now, the corresponding results for the **radiation era** are

$\rho = \rho_R \propto R^{-4}$; $\lambda_J \propto R^{-1}/R^{-2} = R$ (dark matter),

$\lambda_J \propto \text{const}/R^{-2} \propto R^2$ (baryons);

$M_J \propto \rho_D\lambda_J^3 \propto R^{-3}R^3 = \text{constant}$ (dark matter),

$M_J \propto \rho_B\lambda_J^3 \propto R^{-3}R^6 = R^3$ (baryons), and

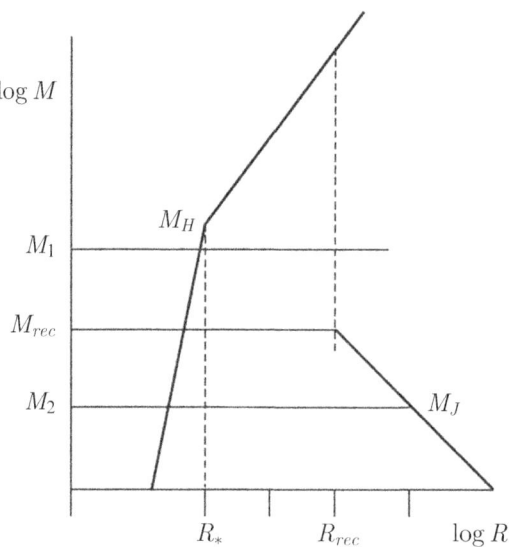

Figure 13.3. Baryonic mass within the Hubble radius, M_H, and the Jeans mass, M_J after re-combination, versus the scale factor R. Values of the scale factor at the end of the radiation domination R_* and at the recombination R_{rec}, are dashed lines. Three different values for the masses within the co-moving volume (M_1, M_2, and M_{rec}) are shown by thin horizontal lines. Evolution of the density perturbations following these constant mass lines is described in the text.

$$M_H \propto \rho t^3 \propto R^{-3} R^6 \propto R^3. \tag{13.129}$$

Examining Figure 13.3, we show a diagram of the above limits in the mass vs. scale factor plane. This graph can be used to study the evolution of a perturbation of mass M. Initially it is to the left of the M_H-border. Subsequently, the perturbation grows either as R^2 or R, depending on whether $R < R_*$ or $R > R_*$, respectively. Now, R_* separates the radiation dominated era from the matter dominated era. Right of the M_H-boundary the growth of perturbations in baryonic matter stops and may even be reversed. But when recombination comes, at $R = R_{rec}$, the growth of baryonic perturbations resumes, assuming that $M > M_{rec}$. This is true, e.g., for the mass value M_1 in Figure 13.3. For the opposing case, the resumption of growth is delayed for a while until the system is to the right of the M_J-border. The mass M_2 in Figure 13.3 refers to this case.

In a contrasting behavior, dark matter perturbations keep on growing through radiation and dark matter domination. Eventually these dark matter concentrations drag baryonic matter with them as will be discussed in detail below.

Now calculate the value of M_{rec}, the Jeans mass at recombination. The speed of sound at the time is

$$c_s^2 = \frac{5}{3}\frac{k_B}{m_H}T_{rec} = \frac{5}{3}\frac{1.38 \cdot 10^{-16}\,\mathrm{erg/K}}{1.67 \cdot 10^{-24}\,\mathrm{g}}\,3\,000\,\mathrm{K} \tag{13.130}$$

where m_H is the mass of the hydrogen atom and k_B is the Boltzmann constant. We get in km/s,

$$c_s \cong 6.4. \tag{13.131}$$

Scaling backward, from its present value, the density at the time is:

$$\rho = 0.26 \times 10^{-29}\,\mathrm{g\,cm^{-3}}\left(1.1 \times 10^3\right)^3 \cong 0.34 \times 10^{-20}\,\mathrm{g\,cm^{-3}}. \tag{13.132}$$

For this density, the Jeans length is

$$\lambda_J = \sqrt{\pi}\frac{6.4\,\mathrm{km/s}}{\sqrt{6.67 \cdot 10^{-8}\,\mathrm{cm^3\,g^{-1}\,s^{-2}} \cdot 0.34 \cdot 10^{-20}\mathrm{g\,cm^{-3}}}} \cong 0.75 \times 10^{20}\,\mathrm{cm}. \tag{13.133}$$

The corresponding Jeans mass is

$$M_{rec} = \frac{\frac{4}{3}\pi(0.38 \cdot 10^{20})^3 \cdot 0.34 \cdot 10^{-20}}{2 \cdot 10^{33}}M_\odot \cong 4 \times 10^5 M_\odot. \tag{13.134}$$

As soon as recombination is over, perturbations of larger mass than $4 \times 10^5\,M_\odot$ are expected to start evolution towards structures. Today's galaxies have the mass range of these preferred scales of structure. Structures less massive than M_{rec} will start development later and may lose out in the "collapse competition."

By way of contrast, just prior to the recombination, the Jeans length is greater than λ_J by a factor of $c/\sqrt{3}/c_s \cong 2.7 \times 10^4$ since the sound speed in the relativistic photon-baryon fluid is $c/\sqrt{3}$. As a result, the Jeans mass just prior to the recombination is much greater than the mass of any bound structure in the universe.

During the matter domination era, $G\rho = (6\pi t^2)^{-1}$ (Equation (8.18)) and $\lambda_J = \sqrt{\pi}(c/\sqrt{3})(6\pi t^2)^{1/2} \cong ct \cong R_H$ (Equation (13.91)).

Thus we may estimate that at $t_* < t < t_{rec}$,

$$M_J \approx M_H \approx 1.6 \cdot 10^{16}\,M_\odot(t/t_{rec}) \tag{13.135}$$

where $t_{rec} \approx 10^{13}$ s or roughly $300\,000$ years (Equations (13.131), Section 10.5). Similarly $M_J \cong M_H$ in the radiation era. Thus, between the dashed lines in Figure 13.3, M_H can be taken as the Jeans mass for baryons. For the cold dark matter, since it is so much more abundant, the Jeans mass is likely to be very much smaller than for baryons. This would be so small as to have no practical influence on the perturbation growth of the visible structures.

There are many possible reasons why perturbations cannot grow effectively between the M_H border and the $R = R_{rec}$ limit in Figure 13.3 which are related to the damping of oscillations during the limited growth period. One, relevant to dark matter, is **free streaming**. Dark matter particles appear to interact with each other and other material primarily via gravity. As a result, each dark matter particle moves along a geodesic in the space-time which is what is meant by free streaming.

Due to free streaming, dark matter particles tend to move out of over-dense regions and the density distribution is smoothed out. The free streaming scale λ_{FS} is therefore the minimum size of structures that may develop. Compare this with λ, the perturbation scale. Returning to this topic below we will find out that free streaming is a significant damping mechanism for hot dark matter but of no importance to cold dark matter.

Several other mechanisms specifically affect **baryons** during the radiation era. Radiation photons and baryons are interlocked through electron scattering. This causes **radiation damping** in baryonic density perturbations.

13.8   Free streaming in dark matter

Recapitulating, non-baryonic dark matter is a relic of the early universe. Dark matter particles are thought to be stable or long lived with a lifetime exceeding the present age of the universe. They are, most probably, weakly interacting massive particles (WIMPs).

Dark matter gas appears to be collisionless. In practice, the particles do not scatter on each other and do not interact with other particles or fields either. In particular, they do not interact with radiation or baryonic matter. Gravity is the only force that dark matter particles respond to, at least at the epoch when they are non-relativistic. This corresponds to when the cosmic temperature T satisfies $k_B T < m_D c^2$. If this inequality is met at the epoch when dark matter decouples from other cosmic particles and fields, it is called **cold dark matter** (CDM).

Recall that perturbations in dark matter do not undergo radiation damping. Radiative dissipation that can smooth out non-uniformity in baryonic plasma and radiation leaves weakly interacting dark matter unaffected. Free streaming is the only dissipation process that can affect dark matter perturbations. It is effective when the particles are still relativistic and smooths out any irregularities in dark matter distribution and motion at spatial scales $R \leq ct$ within the particle horizon.

Consider free streaming in cold dark matter and assume that the dark matter particle masses are about ≈ 1 TeV $\approx 10^{-21}$ g. The particles become non-relativistic at the early cosmic age $t \approx 10^{-12}$ sec and redshift $z \approx 10^{15}$. At that age, the baryonic density is about $\approx 10^{15}$ g/cm^3, and the baryonic mass within the horizon must be about 10^{22} g. This is the maximal baryonic mass scale that can be affected by free streaming. But this mass is negligibly small compared even to the mass of the

Earth. Thus free streaming has actually no cosmological significance for the most probable type of cold dark matter (WIMPs).

Free streaming is a relativistic process in a collisionless gas which may be approximated in terms of the general kinetic considerations of radiative diffusion whose damping scale can be introduced in analogy with the radiation diffusion scale:

$$\lambda_{FS} = (D_{FS}t)^{1/2} \tag{13.136}$$

where $D_{FS} = \frac{1}{3}cL_{FS}$ is the diffusion coefficient, and L_{FS} is the effective free path of the relativistic collisionless particles. Since WIMPs are collisionless, the effective free path is not limited by other particles. Only gravity can "curve" the trajectory of the ultra-relativistic particle. Such a change of the trajectory is significant only over distances comparable to the horizon so the effective free path $L_{FS} \approx ct$, and $D_{FS} \approx c^2 t$. This effective free path is the characteristic damping scale of the process.

We have seen that damping by free streaming is insignificant for WIMP cold dark matter. However, it might be much more effective for **hot dark matter**. Recall that hot dark matter particles have small masses so they can be ultra-relativistic long after decoupling from the baryon-photon medium. If the particle mass is, say, 30 eV $\cong 3 \times 10^{-32}$ g, the particles become non-relativistic at the redshift $z \approx 10^6$. At that time, the baryonic density is $\approx 10^{-15}$ g/cm^3 and the cosmic age $\approx 10^{10}$ sec. The baryonic mass within the horizon is $\approx 10^{13}$ $M_\odot$ at that time. We see that any perturbations in the hot collisionless particles are smoothed out in all mass scales that are smaller than the masses of galaxy groups and clusters.

The difference between hot and cold dark matter is therefore fairly significant for structure formation. If dark matter is hot, the process of structure formation must start with the largest systems like clusters or superclusters. Structure formation is "top down". On the contrary, if dark matter is cold, the process can start with smaller masses (like the masses of globular star clusters $\approx 10^5$ $M_\odot$). From these small structures larger structures must form in a "bottom up" process. Correspondingly, both "top-down" and "bottom-up" scenarios of structure formation are possible. Observational evidence is strongly in favor of cold dark matter (Blumenthal et al. 1982, 1984, Bond et al. 1982, Peebles 1982, Melott et al. 1983, White et al. 1983, Bond and Efstathiou 1984). Henceforth, we will consider dark matter as a cold non-relativistic collisionless gas.

13.9   Dark matter perturbations

Initial perturbations of collisionless dark matter follow the gravitational pull of radiation during radiation domination. From our discussion above, they grow— together with radiation and baryons—at super-Jeans scales, $R > R \approx ct$, in accordance with the general law: $\delta\rho_D/\rho_D \propto R^2$.

Dark matter sub-Jeans scale perturbations behave differently. Oppositely to plasma-radiation perturbations that are sound waves which do not grow and only dissipate due to the radiation damping, dark matter perturbations can grow at sub-Jeans scales, $R < R_J \approx ct$. Dark matter on sub-Jeans scales can be studied with a two-fluid model which assumes that there are two fluids that freely interpenetrate one other and interact only via gravity. One fluid is a radiation-plasma fluid, which is unperturbed and expands in the Friedmann parabolic regime $R_1(t) \propto R(t) \propto t^{1/2}$. The other is perturbed non-relativistic dark matter. Take the perturbation regime to be spherical in shape and uniform in density. The perturbation in dark matter is weak in amplitude and characterized by the scale factor $R_2(t)$.

Now we write down the equation of motion for each of the two fluids (put $\rho_V = 0$, divide Equation (8.2) by 3 and subtract from Equation (8.3))

$$\ddot{R}_i(t) = -\frac{4\pi}{3} G\rho_{eff} R_i(t), \quad i = 1, 2, \tag{13.137}$$

where $\rho_{eff} = \rho_R + 3P_R/c^2 + \rho_D = 2\rho_R + \rho_D$, $\rho_R = C_1/R_1(t)^4$, $\rho_D = C_2/R_2(t)^3$, C_i are positive constants. These two fluids are taken to be in motion relative to each other everywhere except at the center of the spherical perturbation region.

Total mass inside a spherical layer co-moving with a given fluid is not constant with time. Following the equations of motion, each fluid moves in a variable gravity field. Thus the mechanical energy of each fluid is not conserved separately. Because of the changes, the first integral of the equation of motion does not express energy conservation. Instead, it has a special symmetrical form:

$$\dot{R}_1 R_2 - \dot{R}_2 R_1 = C = \text{Const.} \tag{13.138}$$

This is the exact result of the integration of the above equations of motion. We see this readily by differentiating Equation (13.138): $\ddot{R}_1 R_2 = \ddot{R}_2 R_1$. Substituting $\ddot{R}_1$ and $\ddot{R}_2$ from Equation (13.137) proves the identity.

Now use the assumption above that the dominant fluid (which is radiation plus baryons) is unperturbed and does not feel the existence of the other fluid. In such an assumption, $R_1(t) = At^{1/2}$ for radiation, and the dark matter acts like test particles. This is obviously a quite reasonable approximation for the radiation domination epoch. Substituting R_1 into Equation (13.138), the equation for the second dark matter fluid takes the form:

$$\dot{R}_2 - \frac{1}{2t} R_2 = -C/(At^{1/2}). \tag{13.139}$$

It is easy to show a solution of the equation is

$$R_2 = At^{1/2}\left[1 - \frac{C}{A^2 t_i}\left(1 + \ln\frac{t}{t_i}\right)\right]$$

Thus the second integration can be presented in the form:

$$R_2(t) = At^{1/2}\left[1 - \frac{1}{3}\delta(t_i)(1 + \ln(t/t_i))\right]. \qquad (13.140)$$

Examining the above, the first term on the right side scales as in the unperturbed expansion, $R(t) \propto t^{1/2}$. The second term represents the relative change in the dark matter region $\delta R_2/R_2$ showing that the perturbation in the dark matter grows logarithmically as a function of time. Thus, one has for the density perturbation amplitude:

$$\delta = -3\delta R_2/R_2 = \delta(t_i)[1 + \ln(t/t_i)], \qquad (13.141)$$

where $\delta(t_i)$ is the amplitude at what we shall call the "initial" moment t_i which is actually when the scale of the perturbation becomes equal to the Jeans scale. The constant C of the first integration is, as a function of $\delta(t_i)$,

$$C = \frac{1}{3}A^2\delta(t_i)t_i. \qquad (13.142)$$

Thus, the sub-Jeans perturbations in cold dark matter survive radiation domination and even undergo logarithmic growth. Now we estimate the effect of growth for a perturbation with, for example, the baryonic mass $\approx 10^6 \, M_\odot$. For this value from the relations above, one may find that this mass becomes equal to the Jeans mass at a cosmic age of $t \cong 10^5$ s (Equation (13.128)). Identifying this equality age with t_i in the solution above, the amplitude δ increases about 10 times from t_i to the end of the epoch of radiation domination.

The logarithmic growth is not due to gravity of dark matter; this is a purely kinematical effect. Generally, if $R_1 \propto t^n$ for the dominant fluid, the perturbation in the test fluid can grow, if only $n \leq 1/2$.

Problem. Taking $R_1 = At^n$, $n \neq 1/2$, show

$$R_2 = At^n\left[1 - \frac{C}{(-2n+1)A^2t_i^{2n}}\left(1 + \left(\frac{t}{t_i}\right)^{(-2n+1)}\right)\right]$$

is a solution of Equation (13.138) and that the perturbation in the test fluid (2) can grow only if $n < 1/2$.

Since the density perturbation grows logarithmically, the perturbation in velocity can only fall:

$$\delta v_2 \propto t^{-1/2}[1 + \ln(t/t_i)]. \qquad (13.143)$$

Recalling that the time evolution of weak perturbations does not depend on the shape of the perturbation regions in linear analysis, the result is valid for not only spherically symmetrical perturbations, but also perturbations of any arbitrary shape.

Since the start of matter domination, dark matter perturbations grow at all the mass scales from the largest down to the Jeans mass at $t = t_*$. The critical mass is $M_{DJ}(t_*) \cong \rho_* \lambda_{DJ}(t_*)^3$ where the Jeans length is $\lambda_{DJ}(t_*) = \sqrt{\pi} c_s (G\rho_*)^{-1/2} \cong c_s(t_*)t_*$ (see Equation (13.121) with $t_* = t_{exp}$ and $\rho_* = \rho_R$). In this case, the equivalent of the sound speed c_s is actually the velocity dispersion in the collisionless dark matter medium. This speed can be estimated with the use of the standard relation $c_s(t_*) \cong \langle v^{-2} \rangle^{-1/2} \cong (k_B T_D / m_D)^{1/2}$, where m_D is the mass of the weakly interacting dark matter particles. If, for example, $m_D \approx 1\,\mathrm{TeV}$, the velocity dispersion was about the speed of light at $t \approx 10^{-12}$ sec when the WIMPs decoupled from the other particles and became collisionless and non-relativistic. During the time since then, the dark matter has undergone adiabatic cooling, and its temperature has dropped as $T_D \propto R^{-1} \propto t^{-1/2}$. The velocity dispersion has dropped as $t^{-1/4}$. As a result, the velocity $c_s(t_*) \approx 10^4$ cm/sec at $t = t_*$. This is a very modest velocity resulting a Jeans length $\lambda_J(t_*) \approx 10^{16}$ cm, and a Jeans mass of $M_J(t_*) \approx 10^{30}$ g. The low values mean that the dark matter perturbations have grown since $t = t_*$ at practically every mass scale in the universe.

13.10   Dark matter drag

After recombination, gravitational instability comes to its final stage at which perturbations in cold dark matter play the central part. Dark matter dominates completely at this stage. Thus dark matter perturbations grow freely and determine the evolution of perturbations in the baryonic matter. The most important of the new processes at this stage is the generation and amplification of perturbations in baryonic matter by the gravity drag of dark matter perturbations.

Up to the recombination era, initial perturbations in baryons decay in all the scales less than the Silk mass scale $M_S \approx 10^{13} M_\odot$. However the initial perturbations in the cold dark matter survive and even grow all the time in all the scales of interest, as we discussed above. Now, when baryons become free from radiation control, dark matter perturbations are able to excite perturbations in baryonic matter at the scales which are larger than the Jeans scale in baryons, $M > M_J(t_{rec}) \cong 4 \times 10^5 M_\odot$ (Equation (13.134)). The result is that the gravity of dark matter makes baryons follow spatial irregularities in the dark matter distribution.

A detailed calculation (Chernin 1981) shows that perturbations in baryons grow with time from zero amplitude at $t = t_{rec}$ to a level comparable to the dark matter perturbations. Increasing time by an order of magnitude, from $t = t_{rec}$ to $t = 10\,t_{rec}$, the density amplitude in baryons rises from zero to about half the density amplitude

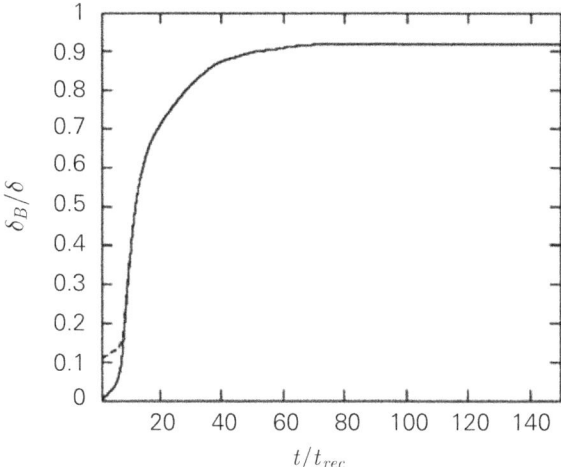

Figure 13.4. Dark matter drag: generation is the solid line and amplification is the dashed line which is above the generation line at small t ratio values. The drag is on baryonic density perturbations by dark matter perturbations.

of the dark matter perturbations. The perturbations in both fluids at $t = 200\, t_{rec}$, the density amplitudes of perturbations differ by no more than 10%. Well before the dark matter perturbations become nonlinear, the perturbation amplitude in baryons reaches practically the level of the dark matter amplitude, as is seen from Figure 13.4.

Again, recall that the time evolution of weak perturbations does not depend on shape. Therefore, the result is valid for not only spherically symmetrical perturbations, but also linear perturbations of any arbitrary shape.

Dark matter drag perturbations produce gravitational potential wells into which baryonic fluid fall in a rather short time. Baryonic perturbations reach the dark matter amplitude at a time a factor $<10^3$ over recombination. This makes possible formation of baryonic systems $\sim$ galaxies. Dark matter drag amplifies baryonic perturbations in the range from the Silk mass to the horizon-scale mass at recombination (where perturbations survive as sound waves). Theory without drag predicts an unacceptably high CMB temperature variation (from baryonic perturbations at recombination). Drag from dark matter predicts the observed amplitude (Sec. 14.6).

13.11   Termination of gravitational instability

Discovery of the dark energy cosmic vacuum introduces a new feature to the physics of gravitational instability in the universe. Now we will show that dark energy with its antigravity terminates the development of gravitational instability when it dominates in the expanding universe. The dark energy vacuum has perfectly constant density that does not change in space-time. Perturbations in the vacuum density as framed in General Relativity are not possible. Gravitational instability does not affect it and cannot develop in it. The vacuum dominates in the universe from the moment $t = t_V \cong 7.1$ Gyr, $z = z_V \cong 0.75$ onwards when the effective gravitating density,

$$\rho_{eff} = \rho_M - 2\rho_V = C/R_V^3 - 2\rho_V, \qquad (13.144)$$

(Equation (9.24)) becomes zero. At the same moment, $\ddot{R}(t_V) = 0$ (Equation (9.22)), and the vacuum domination begins.

The growth of weak linear perturbations terminates at the end of the matter domination. The dark energy antigravity puts a limit to the development of gravitational instability in the universe. A general solution for perturbation evolution on the cosmic vacuum background confirms this conclusion.

The first galaxies are seen at a redshift of about 10. Perturbations must become strong and nonlinear in the redshift interval $1 \leq z_N \leq 10$. To create a system, the value of the "initial" amplitude $\delta(t)$ must provide $\delta \approx 1$ at the redshifts z_N. For instance, at the recombination era of redshift $\sim 1\,000$ one has:

$$\delta(t_{rec}) \cong (1 + z_N)/(1 + z_{rec}) \cong 0.01 - 0.002. \qquad (13.145)$$

which results in a value of one at redshift 10. We shall see that the perturbations must be extremely finely tuned at $t = t_{rec}$ to reach the unity level in amplitude at the appropriate time interval. Why was there such a remarkably fine connection between local perturbations of various spatial scales and the universe as a whole? This is one of the most challenging problems in modern cosmology.

13.12   Dark matter and baryonic structures

We will now explore what kinds of structures are expected to arise from dark and baryonic matter. We will outline the theory of equilibrium distribution functions and their evolution. They serve as models of globular clusters, elliptical galaxies as well as clusters of galaxies. Via these models the total amount of gravitating (baryonic and dark) matter can be found with the goal of determining its cosmological importance i.e., the parameter Ω_M.

The basic units are considered to be point-like particles which are small in size compared to their mutual separations, regardless whether they are in fact dark matter particles, stars, or galaxies. We can also assume that they are **collisionless**. The assumption means more than the statement that direct hits are excluded. Also excluded are close two-body or three-body encounters, where the velocities of the particles would suddenly change. When units like this form a bound system, what is the structure like?

Let the distribution particles in the system be represented by a distribution function f (Section 13.4). Using this function, it is a simple matter to define a **stationary system** by (Equation (13.64))

$$\frac{\partial f}{\partial t} = 0,$$

i.e.,

$$\left(\mathbf{v} \cdot \nabla - \nabla \Phi \cdot \frac{\partial}{\partial \mathbf{v}} \right) f = 0.$$

The zero on the right side means that f is independent of time i.e., f is an integral of motion.

Other integrals of motion also exist, I_m, $m = 1, \dots, n$. Construct the phase space density so that it is a function of the integrals I_m. Then

$$\frac{\partial}{\partial t} f(I_1, \dots, I_n) = \sum_{m=1}^{n} \frac{\partial f}{\partial I_m} \frac{\partial I_m}{\partial t} = 0$$

since $\partial I_m / \partial t = 0$ for all m by the definition of I_m. Therefore the distribution function constructed in this way is a stationary system. This very helpful property of the distribution function will be used extensively in the following. The statement is usually known as **Jeans Theorem**.

We only consider spherically symmetric systems. An integral of motion is the energy per particle

$$E = \frac{1}{2} v^2 + (\Phi - \Phi_0) \tag{13.146}$$

Here we normalize the potential Φ to its value Φ_0 at some radial distance r_0 from the center. It does not depend on time in a stationary system. If the potential is due to the particles which have space density ρ, then the Poisson equation is satisfied

$$\nabla^2 (\Phi - \Phi_0) = 4\pi G \rho. \tag{13.147}$$

The ∇^2 operator has a simple form because of the spherical symmetry which is

$$\frac{1}{r^2} \frac{d}{dr} \left(r^2 \frac{d}{dr} \right) (\Phi - \Phi_0) = 4\pi G \rho. \tag{13.148}$$

Make our distribution function $f(E)$. We start with a function

$$\begin{aligned} f(E) &= \left(2\pi \sigma^2 \right)^{-3/2} \rho_0 \exp\left(-E/\sigma^2 \right), \quad E < 0, \\ f(E) &= 0 \qquad\qquad\qquad\qquad\qquad\qquad E \geq 0. \end{aligned} \tag{13.149}$$

In the above, σ is the velocity dispersion in one dimension and ρ_0 is the density at radial distance r_0. This distribution has the Maxwell–Boltzmann distribution form

$$f(E) \propto \exp(-E/k_B T_e) \tag{13.150}$$

of particle energies in an isothermal gas of temperature T_e. The quantity E is the energy per unit mass. The distribution (Equation (13.149)) is called an **isothermal sphere**. However, there is a fundamental difference between these two. The Maxwell–Boltzmann distribution is a consequence of numerous encounters between gas molecules that lead to the energy per particle of this distribution. The isothermal sphere of dark matter (or stars or galaxies) is a collisionless system. This form of energy distribution may result from violent relaxation during the initial collapse of the system. In spite of this difference, writing

$$\sigma^2 = \frac{k_B T_e}{\mu m_H} \tag{13.150}$$

where μ is the mean molecular weight of the gas or the particles of the system, leads to a formal agreement between Equations (13.149) and (13.150).

The main justification for this functional form comes from its usefulness in describing real stellar systems (based on a variety of observations). Many galaxies have a remarkably constant σ with radius ("flat rotation curve"). For these systems this characteristic may also be expected to arise from violent relaxation.

Now, we will now derive that the equilibrium distribution of density, corresponding to Equation (13.149), is

$$\rho = \rho_0 \frac{r_0^2}{r^2} \tag{13.152}$$

where

$$r_0^2 = \frac{\sigma^2}{2\pi G \rho_0}. \tag{13.153}$$

To do this, we demonstrate that Equation (13.149), integrated over all velocities, leads to Equation (13.152), and that ρ also satisfies Equation (13.148). First

$$\rho = 4\pi \int_0^\infty f(E) v^2 \, dv = \frac{4\pi \rho_0}{(2\pi\sigma^2)^{3/2}} \exp\left\{-\frac{\Phi - \Phi_0}{\sigma^2}\right\} \int_0^\infty \exp\left\{-\frac{v^2}{2\sigma^2}\right\} v^2 \, dv$$

$$\rho = \rho_0 \exp\left\{-\frac{\Phi - \Phi_0}{\sigma^2}\right\} \tag{13.154}$$

since the definite integral has this value

$$\int_0^\infty \cdots = \frac{\left(2\pi\sigma^2\right)^{3/2}}{4\pi} \tag{13.155}$$

The potential corresponding to Equation (13.146) is

$$\Phi = -\int_r^{r_0} \frac{GM(r)}{r^2} dr + \Phi_0 \tag{13.156}$$

where within radial distance r the mass is

$$M(r) = 4\pi \int_0^r \rho r^2 \, dr = 4\pi\rho_0 r_0^2 r. \tag{13.157}$$

Therefore,

$$\Phi - \Phi_0 = 4\pi G\rho_0 r_0^2 \ln\left(\frac{r}{r_0}\right) = \sigma^2 \ln\left(\frac{r}{r_0}\right)^2 \tag{13.158}$$

Substituting Equation (13.153) in Equation (13.154) gives

$$\rho = \rho_0 \left(\frac{r_0}{r}\right)^2, \tag{13.159}$$

which agrees with Equation (13.152).

Substituting $\Phi - \Phi_0$ from Equation (13.158) to the left side of Equation (13.148):

$$\frac{1}{r^2}\frac{d}{dr}\left(r^2\frac{d}{dr}(\Phi - \Phi_0)\right) = \frac{1}{r^2}\frac{d}{dr}\left(4\pi G\rho_0 r_0^2 r\right) = 4\pi G\rho_0\left(\frac{r_0}{r}\right)^2$$

which is equal to the right side of the same equation. This proves that a self-consistent equilibrium model follows from the distribution function of Equation (13.149).

Isothermal sphere models have been used successfully to explain the matter distribution in galaxy halos. Using gravitational lensing Treu and Koopmans (2002) show that in an elliptical galaxy the baryonic and dark matter components together follow the isothermal model, while taken separately they do not! Again, the reason may be that the galaxy halos adopt their distribution functions in a process called **violent relaxation**. In this relaxation, both baryonic and dark matter respond to chaotic changes in the gravitational potential of the galaxy during its formation, leading to an isothermal distribution (Lynden-Bell 1967, Shu 1978). The relaxation process must be able to preserve the segregation of luminous matter and dark matter. Merging of galaxies may accomplish this. The baryonic matter in the form of gas tends to concentrate towards the center of the merger remnant while the dark matter remains in a common extensive halo (Barnes 1992, Barnes and Hernquist 1996). A similar "conspiracy" between baryonic and dark matter results in a flat rotation curve for spiral galaxies, as demonstrated by Van Albada and Sancisi (1986). See Figure 12.7 for an example.

Obvious problems appear using the isothermal sphere as a model of a celestial body: According to Equation (13.157), its mass has no limiting value when $r \to \infty$. Also, we see that, $\rho \to \infty$ as $r \to 0$ density, a singularity as can be seen in Equation (13.159). Because of this behavior, this model is more specifically referred to as a **singular isothermal sphere** (SIS). In order to overcome the former singularity, a modified distribution function called **lowered isothermal model** or **King model** is often used.

$$f(E) = (2\pi\sigma^2)\rho_0\big[\exp\{-(E - E_0)/\sigma^2\} - 1\big], \text{ for } E < E_0$$

$$f(E) = 0, \qquad\qquad\qquad\qquad\qquad\qquad \text{for } E \geq E_0 \tag{13.160}$$

where $E_0 < 0$. A somewhat complicated analytic function represents the resulting density distribution. But this density distribution represents the observed brightness profiles of spherical or nearly spherical stellar systems very well with only a few new observational fitting parameters.

One new parameter E_0 is related to the observational outer radius r_t of the density distribution by

$$E_0 = -\frac{GM(r_t)}{r_t}. \tag{13.161}$$

This fitting parameter radius is called the **tidal radius** since gravitational tides by larger bodies often truncate the density distribution in real astronomical bodies. Both the density and the velocity dispersion go to zero at r_t.

The second observational parameter of a King model is its core radius r_c:

$$r_c^2 = \frac{9\sigma^2}{4\pi G\rho_0}. \tag{13.162}$$

Except for the constant factor, it is identical to the r_0 of the isothermal sphere as shown in Equation (13.152). The significance of r_c for real stellar systems is that the surface brightness falls to roughly ½ of its central value at this projected distance from the center. Inside the core radius the King model is close to isothermal with a rather constant velocity dispersion. This dispersion, is not far from the σ of the isothermal sphere. This dispersion is also an observable quantity. The quantity ρ_0 is the central density. The surface density profiles of King models for different ratios r_t/r_c are shown in Figure 13.5.

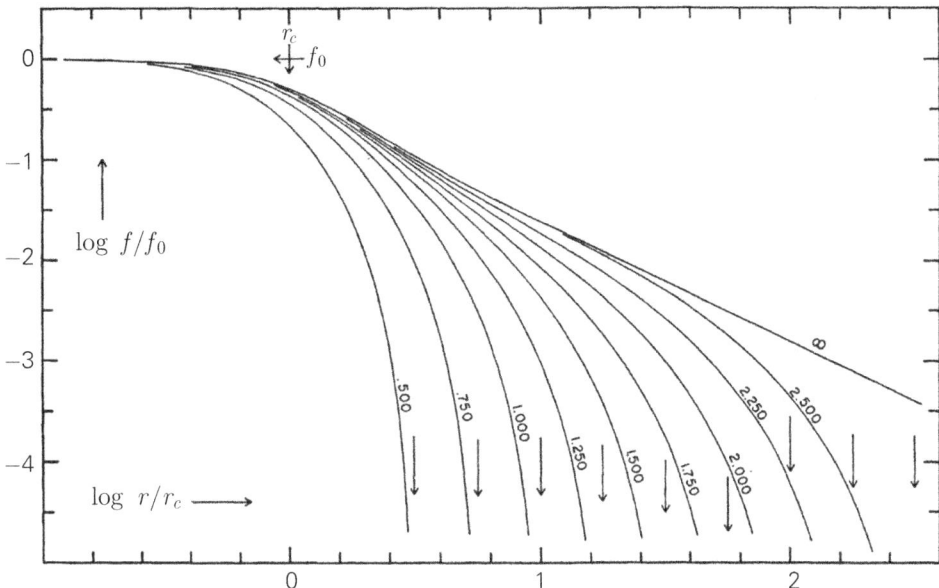

Figure 13.5. Log surface density profiles versus log radius for a variety of King models. These are normalized to f_0, the central surface density giving $\log(f_0/f)$, and r_c, the core radius giving $\log(r/r_c)$. The curves are marked by the concentration parameter $c_{KING} = \log(r_t/r_c)$ where r_t is the tidal radius of the galaxy, as indicated by arrows. NB: The line marked ∞ is incorrect. (From King 1966).

The log ratio $c_{KING} = \log(r_t/r_c)$ is called the concentration parameter. It is typically $\cong 1$ for globular clusters and $\cong 2$ for elliptical galaxies. Figure 13.6 shows a fit to an elliptical galaxy.

Problem. Show that the distribution function whose form is $f(E) \propto (-E)^{5/2}$ gives a self-expression consistent model of the Plummer sphere (Equation (12.11)).

In the case of the hot gas density profile observed in clusters of galaxies a simpler density is often used:

$$\rho = \rho_0 \left(1 + \left(\frac{r}{r_c} \right)^2 \right)^{-\frac{3}{2}\beta}. \qquad (13.163)$$

This is called a **beta model**. At large radii and $\beta = 2/3$, this model behaves like the isothermal model (Equation (13.159)). At small radii there is a constant density core of radius r_c, unlike in the singular isothermal model. The two parameters, r_c and β, are usually enough to describe a wide variety of observed gas profiles.

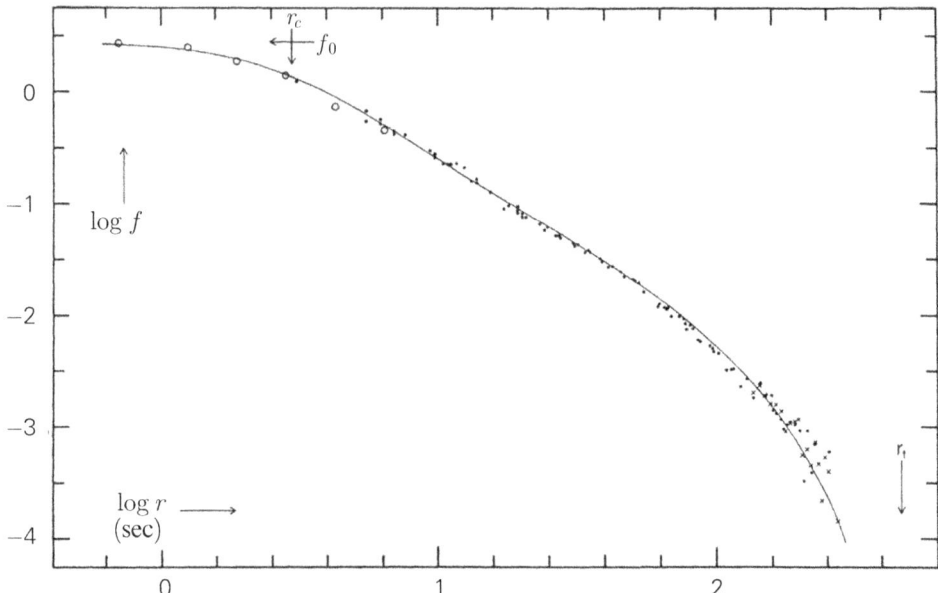

Figure 13.6. Comparison of the observed brightness profile (dots) of the elliptical galaxy NGC 3379 (dots) to a fitted King model (solid line). Scales are the same as Figure 13.5 with a concentration parameter $c_{KING} = 2.20$. The observed core radius (r_c) and the tidal radius (r_t) are indicated by arrows. (From King 1966). NB In contrast to the tidal radius, the core radius is "fictitious" being part of the concentration parameter.

Observations are generally of only the surface density Σ as a function of the projected distance R on the sky. For example, we will carry out the conversion of the $\rho(r)$ profile to the $\Sigma(R)$ profile when $\beta = 1$. The

$$\rho = \rho_0 \left(1 + \left(\frac{r}{r_c} \right)^2 \right)^{-\frac{3}{2}} \tag{13.164}$$

profile is the **modified Hubble law** of galaxy (volume) brightness distribution. This is a very good approximation to an isothermal sphere in its inner parts. When $r \gg r_c$, the modified Hubble law gives $\rho \propto r^{-3}$ rather than the $\rho \propto r^{-2}$ of the isothermal sphere.

For this case, the surface density is easy to integrate:

$$\Sigma(R) = 2 \int_0^\infty \rho \, dz \tag{13.165}$$

where z is the direction along the line of sight from us to the galaxy as defined in Figure 13.7 by

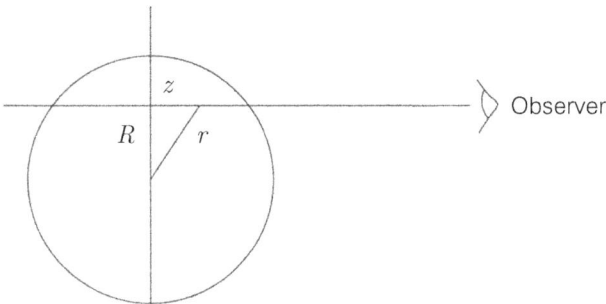

Figure 13.7. Geometry for calculating the surface density from a spherically symmetric mass distribution.

$$r^2 = R^2 + z^2 \tag{13.166}$$

and

$$\Sigma(R) = 2\rho_0 \int_0^\infty \left[1 + \frac{R^2}{r_c^2} + \frac{z^2}{r_c^2} \right]^{-3/2} dz. \tag{13.167}$$

If we make the substitution $y = z(r_c^2 + R^2)^{-1/2}$ the integral is

$$\int_0^\infty \frac{dy}{(1 + y^2)^{3/2}} = \left. \frac{y}{(1 + y^2)^{1/2}} \right|_0^\infty = 1 \tag{13.168}$$

and

$$\Sigma(R) = 2\rho_0 r_c \left[1 + \left(\frac{R}{r_c} \right)^2 \right]^{-1}. \tag{13.169}$$

Now, if we go back to the beta model (Equation (13.163)) and also to observations of hot gas in clusters of galaxies by x-ray emission, the emissivity per unit volume is

$$\varepsilon(r) \propto \rho^2 \propto \left[1 + \left(\frac{r}{r_c} \right)^2 \right]^{-3\beta} \tag{13.170}$$

The surface brightness relative to the central surface brightness S_0 is

$$S = S_0 \left[1 + \left(\frac{r}{r_c} \right)^2 \right]^{-3\beta + 1/2}. \tag{13.171}$$

This last step is not trivial; it can be obtained by calculations similar to the derivation of Equation (13.169), but they are too involved to be presented here.

The origin of β in Equation (13.163), can be seen by assuming that the potential of the cluster of galaxies $\Phi - \Phi_0$ whose gas is emitting the X-rays is such that the cluster is isothermal. The number density of galaxies in the cluster n follows the modified Hubble law

$$\frac{n}{n_0} = \left[1 + \left(\frac{r}{r_c} \right)^2 \right]^{-3/2} \tag{13.172}$$

as a good approximation, relative to the central number density n_0. By Equation (13.154) the same quantity can also be written as

$$\frac{n}{n_0} = \exp\left\{ -\frac{\Phi - \Phi_0}{\langle v^2 \rangle} \right\} \tag{13.173}$$

where we have replaced σ^2 by the velocity dispersion $\langle v^2 \rangle$ of the galaxies.

Now we can presume the cluster gas is relaxed in the same potential well. The shape of the density profile of the gas ρ relative to its central density ρ_0 depends on the gas temperature. We define

$$\beta = \frac{\mu m_H \langle v^2 \rangle}{k_B T_e} \tag{13.174}$$

which is the ratio of the specific (per unit mass) energies of the galaxies and the gas. Therefore (Equation (13.151)),

$$\sigma^2_{gas} = \frac{k_B T_e}{\mu m_H} = \frac{1}{\beta} \langle v^2 \rangle. \tag{13.175}$$

When σ^2_{gas} is substituted for σ^2 in Equation (13.154), we get for the gas

$$\frac{\rho}{\rho_0} = \exp\left\{ -\beta \frac{\Phi - \Phi_0}{\langle v^2 \rangle} \right\} = \left[\exp\left\{ -\frac{\Phi - \Phi_0}{\langle v^2 \rangle} \right\} \right]^\beta = \left(\frac{n}{n_0} \right)^\beta. \tag{13.176}$$

Combining with Equation (13.172) leads to Equation (13.163).

Observations of gravitating matter (dark matter, stars and gas) in clusters of galaxies have shown that the singular isothermal sphere model does not fit observations far from the center of the cluster. For example, the cluster CL 0024 + 1654 has been studied out to 4.5 Mpc from the center by gravitational lensing, and the density fall-off with radius was found to be at least $\rho \propto r^{-2.5}$, and probably is $\rho \propto r^{-3}$, as in the modified Hubble law (Kneib et al. 2003). However, the best fit was obtained using the **NFW density law** (Navarro, Frenk and White 1997), which places the cluster in the context of the universe.

$$\rho(r) = \frac{\rho_c \delta_{NFW}}{(r/r_s)(1 + r/r_s)^2} \tag{13.177}$$

where ρ_c is the critical density of the universe (Equation (8.27)), δ_{NFW} is the dimensionless over-density parameter, and r_s is a scale length. The NFW density law behaves like the modified Hubble law when $r \gg r_s$, but goes to $\rho \propto r^{-1}$ close to the center of the system, rather than to a constant value.

Rather than observations, the NFW density law is primarily based on the numerical simulations of the formation of dark matter halos in an expanding universe. Navarro et al. (1997) find that the NFW density profile arises invariably from of hierarchical mergings of halos, and it should apply in all scales from dwarf galaxy halos to galaxy cluster halos. It is not sensitive to the details of the cosmological model. It is thought that a process akin to violent relaxation is responsible for this density profile.

Rather than using the parameter δ_{NFW} in the equation above, one may use instead a concentration parameter c_{NFW}:

$$\delta_{NFW} = \frac{200}{3} \frac{c_{NFW}^3}{\ln(1 + c_{NFW}) - c_{NFW}/(1 + c_{NFW})}. \qquad (13.178)$$

which is defined to be

$$c_{NFW} = \frac{r_{200}}{r_s} \qquad (13.179)$$

The radial distance r_{200} is from the center of the halo within which the *mean* density of the halo is 200 times greater than the critical density of the universe. Similarly M_{200} is the mass of the halo out to the radial distance r_{200} from the center.

What is the significance of the number 200? We will go through the argument leading to this number. In a spherical perturbation, a region of radius r separates itself from the corresponding background expansion. The radius reaches a maximum value, turns around and starts to go back towards zero. Before zero distance is reached, the system is "virialized" at a radius of about half of the maximum size. This virialization is at a time which is about twice the time of the maximum expansion. At the time of the maximum value, the radius is $(3\pi/4)^{2/3}$ times smaller than the radius of a corresponding freely expanding region of the same mass. At the time of "virialization", the universe has expanded by another additional factor of $2^{2/3}$ while the perturbation has contracted by a factor of 2. Thus, the difference of radii between the freely expanding region and the just formed virialized body is $(3\pi/4)^{2/3} \times 2^{2/3} \times 2$. The difference in mean densities is this number to the third power, i.e., $(3\pi/4)^2 \times 2^2 \times 2^3 = 178$. Since this calculation is very approximate, 178 is typically rounded to 200.

Thus, a system which has been virialized must have an over-density of ~ 200. Observationally, we often do not have a clear idea of the edge of a body, especially in the case of dark matter halos. Then r_{200} can be defined as a reasonable way to describe the radius of the body. Of course, the real radius may be smaller. If the body formed long ago, its radius may not have changed much while the scale of the universe increased. Therefore, in reality the over-density should generally be greater than 200.

As an example of the kind of parameter values that we meet in clusters of galaxies, the main body of CL $0024 + 1654$ has a mass $M_{200} = 5.6 \times 10^{14} M_{\odot}$. It has a radius $r_{200} = 1.7$ Mpc. And, finally, its the scale length $r_s = 76$ kpc, thus the concentration parameter $c_{NFW} = 22$ (Kneib et al. 2003). For the Coma cluster $M_{200} = 11.4 \times 10^{14} M_{\odot}$, $r_{200} = 2.1$ Mpc, $r_s = 240$ kpc and $c_{NFW} = 8.8$ (Rines et al. 2001). For individual galaxies plus their halos, the parameter values are $M_{200} \cong 10^{12} M_{\odot}$, $r_{200} \cong 200$ kpc and $r_s \cong 20$ kpc have been determined from gravitational lensing data (Hoekstra et al. 2004). A typical galaxy luminosity $2 \times 10^{10} L_{\odot}$ so that the mass to light ratio is 50, much more than for a typical star in the neighborhood of our Sun.

In the theory of Galaxy halo formation by Navarro et al. (1997), δ_{NFW} is proportional to the universe's mean density at the time of the halo's formation. Smaller halos form first as indicated by the fact that their mean densities today are greater than the mean densities of large halos. The main properties of halos of different masses are thus explained. How well the theory explains details like the density profiles of Galaxy sized halos is still controversial.

Chapter 14

Dark energy and gravitating matter from structure in the universe

14.1 Introduction

Throughout this book, it has become increasingly evident that the universe is fairly simple in its differential geometry. As Einstein assumed in his first paper (1917) on relativistic cosmology, the assumption of uniformity and isotropy provides quite an appropriate approximation to the real universe. Furthermore, observational data indicate that the co-moving space of the universe is nearly flat.

On the other hand, the energy content of the universe appears to be rather complex. As we have discussed earlier, the majority of gravitating matter in the universe is dark matter. As we discuss in this chapter, the current view is that at the present cosmic epoch dark energy dominates over gravitating matter globally (in Chapter 15 we show that it is also important on certain scales locally). On the whole, the major cosmic components are dark energy and dark matter which comprise together more than 95% of the total cosmic energy. As we have seen, the origin of "ordinary" (baryonic) matter is almost as unclear as the nature of the dark matter and dark energy. It is embarrassing that only the cosmic microwave background (CMB) radiation that contributes 0.005% to cosmic energy is well understood in its nature and origin. We first discuss how the angular structure in the CMB radiation gives information about the cosmological parameters, including the amounts of baryonic and dark matter and dark energy.

Recent precise data are provided by the WMAP (Wilkinson Microwave Anisotropy Probe) observations. Together with the data on cosmological supernovae, of Galaxy clustering in large scale from the Sloan Digital Sky Survey (SDSS) and of other results of modern cosmological observations, the WMAP data give the concordance dataset (Ostriker and Steinhardt 1995) that is the empirical basis of the current standard cosmological model.

14.2 Describing structure in the CMB radiation

Initial "seed" perturbations are needed for the formation of structure in the universe. The physical nature of the initial seeds remains essentially unknown. There are no direct indications from first principles on the initial amplitude or the possible dependence of the amplitude on the spatial scales. One of interesting hypotheses treats the origin of the seed perturbations as resulting from quantum fluctuations at a very early (perhaps, pre-Friedmann) epoch of the expansion. If so, the generation of the perturbations is essentially random. The generation may be described with the use of statistical methods assuming that the central quantity of the process— the perturbation amplitude—is a random variable.

Suitable perturbations after their appearance may be amplified and can develop into cosmic structures of different scales. Rather than just looking at the distribution of galaxies and other structures today, initial stages of the process are actually imprinted in the CMB which we discuss here. Galaxies and systems of galaxies appear later. J. Silk in 1967 first recognized that the CMB photons come to us from the "last-scattering surface" carrying information about the relic radiation from the era of recombination when radiation decoupled from baryons. The spatial distribution of the emitted radiation was not perfectly isotropic because radiation was involved in the gravitational instability prior to recombination. Since adiabatic perturbations are associated with temperature fluctuations in the baryon-photon plasma, these fluctuations imprinted in the radiation distribution are preserved after the recombination. The imprints are observed now as angular variations of the CMB temperature.

In the same year, Rainer K. Sachs and Arthur M. Wolfe (1967) considered observational horizon-scale fluctuations at the recombination era and predicted temperature fluctuations in the observed CMB and the corresponding angular scale. In both cases, the initial perturbations were fixed at the early epoch of the redshift $z \cong 1\,100$ when they were weak and linear.

The first observational detection of CMB imprints left by initial cosmic perturbations was made in 1992 by the Russian orbital observatory RELICT and the American space laboratory COBE (COsmic Background Explorer). The COBE observations were of a higher accuracy and reliability but both missions agreed with each other, providing a direct observational test—and confirmation—for the above random quantum physics of gravitational instability.

A strong anisotropy (a dipole) in the CMB arises from our motion through the local cosmological rest frame. Taking into account the motion of the Earth relative to the Sun, the motion of the Sun relative to the center of our Galaxy, as well as the motion of the Galaxy relative to the Local Group, we find that the direction and the magnitude of the CMB dipole implies a Local Group motion towards the Galactic longitude $l = 268°$ and latitude $b = 27°$ with a speed of ≈ 600 km/s (Smoot et al. 1991). In our local sky this direction is in the constellation of Hydra.

Such a high velocity of the Local Group, when considered together with the rather regular Hubble law around us (with velocity dispersion $\ll 600$ km/s), means that some rather large volume of the local universe of galaxies has to be moving as a whole relative to the CMB and sufficiently distant galaxies. Thus one has attempted to understand such a bulk motion with this speed and direction as arising from the combined anisotropic gravitational forces of all the galaxies (and dark matter) within some large distance of ~ 100 Mpc from us. The question is still somewhat open, but astronomers have been able to identify mass concentrations which are responsible at least to a significant part of the observed peculiar velocity. The nearest one is the nearby Virgo cluster, generating a ~ 220 km/s infall velocity within the Local Supergalaxy (Jerjen and Tammann 1993; Theureau et al. 1997) which accounts for almost 30% of the 630 km/s velocity. For some time it was thought that the remaining velocity vector (a flow towards the Hydra-Centaurus supercluster) could be explained by a hypothetical "Great Attractor" (GA) lying behind Centaurus at a distance of about 60 Mpc and having a mass of a few times of $10^{16} M_\odot$ (Lynden-Bell et al. 1988; Strauss et al. 1992).

However, subsequent studies of the possible mass over-density or the expected infall velocity field around the GA have not been able to find a single high mass attractor. For instance, Kocevski and Ebeling (2006) concluded from a large all-sky sample of (X-ray-selected) clusters of galaxies that

> "44% of the Local Group's peculiar velocity is due to infall into the Great Attractor region, while 56% is in the form of large-scale flow induced by more distant over-densities between 130 and 180 h^{-1} Mpc away".

One such over-density could be the Shapley supercluster at about 160 h^{-1} Mpc (or 230 Mpc for $h = 0.7$).

The fact that the local bulk flow extends beyond 200 Mpc or more is an important piece of information; it tells us that we have to average the galaxy space over at least the 200 Mpc scale in order to find the cosmological co-moving rest frame.

Historically, the large scale streaming motions of galaxies were first suggested by Rubin et al. (1976) who inferred from a large sample of spiral galaxies (ScI) that our Local Group is participating in a migration with large numbers of other galaxies with the speed of about 450 km/s. However, with its direction of $l = 163°$, $b = -11°$ it much differs from the speed relative to the CMB, and it is generally thought that the "Rubin–Ford effect" was influenced by some systematic error either caused by a selection bias in the data set (Fall and Jones 1976) or by anisotropic light extinction in our Galaxy (Teerikorpi 1978; note that the derived velocity vector lies close to the dusty Galactic plane) or both.

Cosmological simulations in the concordance model values have shown that large scale motions of the observed size are induced, and **in this sense** the CMB dipole component is well accounted for in the ΛCDM model as an effect. Therefore we can move on to consider higher order multipole components which carry early "primary source" information from the time of origin of the CMB at recombination.

COBE identified temperature variations at a level $\delta T = (30 \pm 5) \times 10^{-6}$ K (Smoot et al. 1992). The relative temperature fluctuation is thus

$$|\delta T / T|_{COBE} \cong 10^{-5}. \tag{14.1}$$

The measurements were performed at the angular scale $\theta = 10°$. This angular scale corresponds to a linear scale at recombination of

$$R_{COBE}(t_{rec}) \cong 3.5ct_0\theta(1 + z_{rec})^{-1} \cong 7 \times 10^{24}\,\text{cm}. \tag{14.2}$$

Problem. Using general relations in Section 9.2, prove that the angular size θ of an object with a proper size R at the redshift z is $\theta = R\beta(1 + z)/(ct_0)$, where $\beta \cong 1/z$ for $z \ll 1$ and $\beta \cong 1/3.5$ for $z \gg 1$. Assume the co-moving space is flat and $\Omega_M = 0.27$. Make use of the approximation of Equation (9.6) between $z = 0$ and $z = 10$.

The Equation (14.2) result is 13 times larger than the Hubble radius at the same epoch:

$$R_H(t_{rec}) = 1.5ct_{rec} \cong 5.4 \times 10^{23}\,\text{cm}. \tag{14.3}$$

Figure 14.1 shows WMAP CMB temperature fluctuations over the full sky with temperature variations represented by color variations (red is warmer and blue cooler). The mean is quite uniform at 2.728 K again with the fluctuations in the fourth decimal. We want to describe these fluctuations in terms of strength and their relative positions (in angle on the sky).

Figure 14.1. WMAP CMB temperature fluctuations over the full sky (5 years of data). Intensity represents variations of the remnant glow from the young universe. White is warmer and dark cooler. The mean is 2.728 K with the fluctuations of 200 µK. The foreground Milky Way emission and the dipole component due to our motion relative to the CMB has been subtracted from the map. Credit WMAP science team. http://www.nasa.gov/images/content/216398main_fullsky.jpg

A fundamental way of analyzing the CMB temperature fluctuation map is to calculate the power spectrum, a plot of the correlation power in different multipole components as a function of the "multipole number". The multipole method of looking at the CMB map is explained in Box 12. Figure 14.4 shows the resulting power spectrum, displayed such that the horizontal line represents the Harrison–Zeldovich initial spectrum, and the bumps in the spectrum give other details of the cosmological model. In the next few sections we describe the basic model, and then consider different details that we may read from these data.

Box 12. Multipole expansion.

A function of an angle θ (say, the polar angle) on the sky is conveniently represented by **multipole** components. A dipole distribution is a function that has its maxima and minima at opposite sides of the sky. The CMB dipole radiation is warmer in the forward direction and cooler in the opposite direction with respect to the motion of the Solar System, our Galaxy and the Local Group which arises from the Doppler shift. This component is taken out from the CMB maps, and the remaining higher multipoles are studied. A quadrupole component is strong if there are warmer patches in the CMB sky preferentially 180 degrees apart, say, at the poles, while the equatorial region is cooler (or vice versa). It is obvious that the foreground Milky Way makes a strong contribution to this component and information on the quadrupole component of the CMB of distant origin is thus less certain than higher multipoles components. Studies of the original CMB fluctuations usually concentrate on higher multipoles.

In order to see the appearance of multipole components, consider the problem of expressing the electric potential at point $S(r, \theta)$ of a charge q placed at point $(a, 0)$ in spherical polar coordinates. Due to symmetry, the ϕ coordinate is ignored. The charge lies on the z-axis at distance a $(a < r)$ from the origin O (see Figure 14.2).

Taking the distance between the charge and point S is r_1, the potential

$$\Phi = \frac{q}{r_1}.$$
(B12.1)

Using triangle OSq in Figure 14.2 the potential may be written

$$\Phi = \frac{q}{\sqrt{r^2 + a^2 - 2ar\cos\theta}}.$$
(B12.2)

Expanding the denominator in powers of a/r.

$$\Phi = \frac{q}{r} \sum_{l=0}^{\infty} P_l(\cos\theta) \left(\frac{a}{r}\right)^l$$
(B12.3)

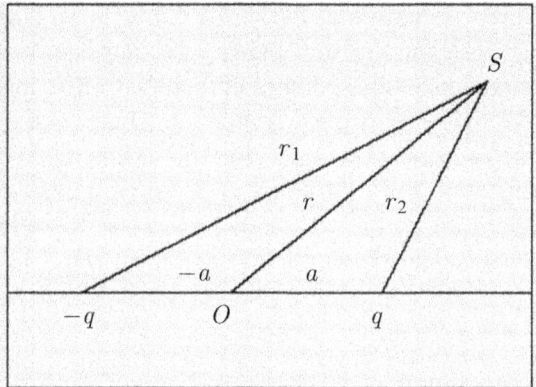

Figure 14.2. Calculation of the potential Φ at point S, arising from charges q at $z = a$ and $-q$ at $z = -a$. The z-axis is horizontal, and the origin of the coordinate system r, θ is at O.

The coefficients P_l are called **Legendre polynomials**. The first few are

$$P_0 = 1$$

$$P_1 = \cos\theta$$

$$P_2 = \frac{3}{2}\cos^2\theta - \frac{1}{2}$$

(B12.4)

The general formula is

$$P_l(x) = \sum_{k=0}^{[l/2]} (-1)^k \frac{(2l - 2k)!}{2^l k!(l - k)!(l - 2k)!} x^{l-2k}$$

(B12.5)

where $x = \cos\theta$ and $[l/2] = l/2$ if l is even and $[l/2] = (l - 1)/2$ if l is odd. For more detail see e.g., Arfken and Weber (1995). Figure 14.3 plots some Legendre polynomials.

Adding a charge $-q$ at $z = -a$ (Figure 14.2) resulting in the potential

$$\Phi = q\left(\frac{1}{r_1} - \frac{1}{r_2}\right) = \frac{q}{r}\left[\sum_{l=0}^{\infty} P_l(\cos\theta)\left(\frac{a}{r}\right)^l - \sum_{l=0}^{\infty} P_l(\cos\theta)(-1)^l\left(\frac{a}{r}\right)^l\right]$$

$$= \frac{2q}{r}\left[P_1(\cos\theta)\frac{a}{r} + P_3(\cos\theta)\left(\frac{a}{r}\right)^3 + \cdots\right]$$

. (B12.6)

Now $2aq$ is called the dipole moment. The first (and dominant) term is the electric dipole potential.

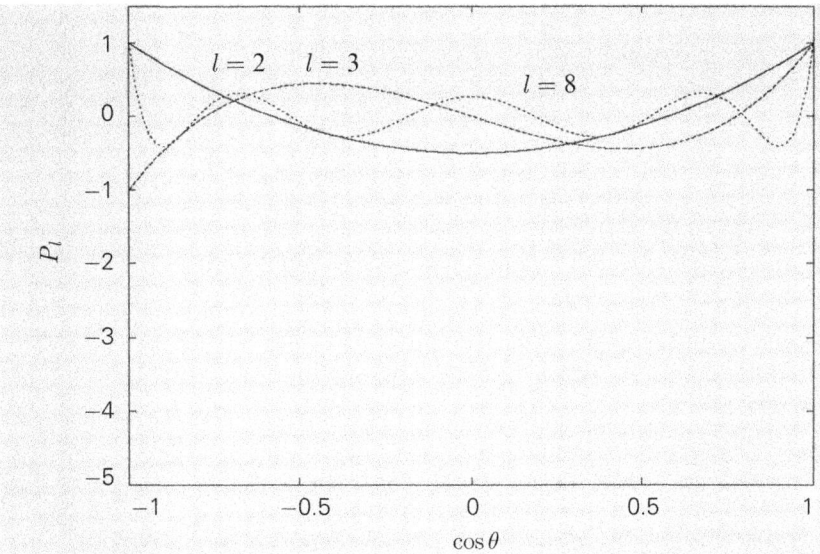

Figure 14.3. Legendre polynomials $P_2(\cos\theta)$, $P_3(\cos\theta)$, and $P_8(\cos\theta)$.

By adding more charges on the z-axis, one can generate higher linear multipole systems: quadrupoles whose portion of the expansion starts with the $P_2(\cos\theta)$ term, octupoles whose dominant term is the $P_3(\cos\theta)$ term, etc.

Problem. Assume the linear charge system with $+q$ at $z = -a$ and at $z = +a$, and $-2q$ at $z = 0$. Show that it generates a series with a leading quadrupole term. Show also that a system of charges with $-q$ at $z = -2a$, $+2q$ at $z = -a$, $-2q$ at $z = +a$ and $+q$ at $z = +2a$ generates a series with a leading octupole term.

The Legendre polynomials satisfy an orthonormality condition

$$\int_{-1}^{+1} P_l(x) P_m(x)\, dx = \frac{2}{2l+1}\, \delta_{m,l} \tag{B12.7}$$

where the Kronecker delta $\delta_{l,m}$ equals 0 if $l \neq m$ and $\delta_{l,m} = 1$ if $l = m$. This property leads to a function $f(x)$ as a sum of the Legendre polynomials:

$$f(x) = \sum_{l=0}^{\infty} a_l P_l(x), \tag{B12.8}$$

which is called the **Legendre series**. Its coefficients are

$$a_l = \frac{2l+1}{2} \int_{-1}^{+1} f(x) P_l(x) \, dx, \qquad (B12.9)$$

as can be easily verified using the orthonormal condition.

Problem. Prove the above statement.

CMB radiation shows relative temperature variations $\delta T/T$ in a direction ψ and in a nearby direction $\psi + \theta$. We calculate a two-point correlation function (to be discussed later)

$$C(\theta) = \left\langle \frac{\delta T}{T}(\psi) \frac{\delta T}{T}(\psi + \theta) \right\rangle \qquad (B12.10)$$

and expand it in the Legendre series with coefficients C_l. Usually $l(l+1) C_l/2\pi$ is plotted as a function of l. The scale-invariant spectrum whose physical origin we discussed earlier, appears as a horizontal line. There are good reasons to believe that this is the primordial dark matter fluctuation spectrum that is directly translated into temperature fluctuations. In a plot like this, it is easy to see how the primordial spectrum has been modified via amplification.

14.3   The power spectrum

The CMB power spectrum is associated with the underlying power spectrum of density perturbations at the time of origin of the CMB photons. Let us study such a spectrum in a little more detail.

The amplitude of weak linear density perturbations $\delta(t, \mathbf{r})$ is positive for over-densities and negative for under-densities. It is as function of time and spatial coordinates. The amplitude is zero when it is averaged over a sufficiently large volume V occupied by many both over-densities and under-densities of various shapes and sizes. So at any moment of time:

$$\langle \delta(t, \mathbf{r}) \rangle = 0. \qquad (14.4)$$

The amplitude can be represented by its Fourier transform in the volume V:

$$\delta(t, \mathbf{r}) = \sum_k \delta_{\mathbf{k}} e^{i \mathbf{k} \cdot \mathbf{r}} \qquad (14.5)$$

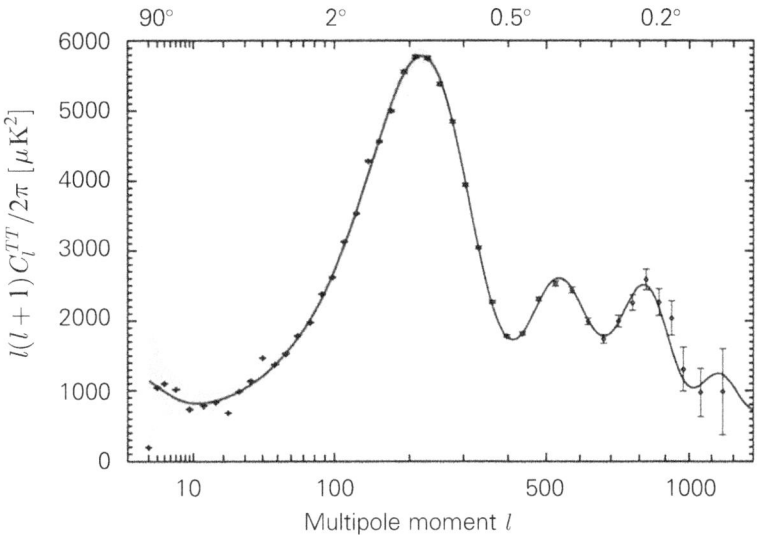

Figure 14.4. Comparison of the best fit concordance model to the WMAP observed temperature angular power spectrum. Dots are the unbinned data (Spergel et al. 2003, Komatsu 2011). The multipole moment l values (upper and lower scale tic marks) are related angular values $\theta \approx 100°/l$. Approximate angular values are given on top. Especially note that the peaks in the angular structure appear on scales equal to or less than about one degree not the COBE 10 degrees. Credit NASA WMAP science team 7 year results. http://wmap.gsfc.nasa.gov/resources/camb_tool/index.html shows the seven year WMAP observations versus the various models. http://lambda.gsfc.nasa.gov/toolbox/education/Exploring_CMB_with_CAMB.pdf. http://icosmo.org/cgi-bin/Interactive_Step1.cgi.

where:

$$\delta_{\mathbf{k}}(t) = (V)^{-1} \int_V \delta(t, \mathbf{r}) e^{-i\mathbf{k}\cdot\mathbf{r}} \mathrm{d}\mathbf{r}. \tag{14.6}$$

The wave vector $\mathbf{k}$ has discrete values, so that, for instance, in a cubic volume $V = L^3$, L being the length of the side of the cube,

$$k_x = 2\pi n_1/L, \quad k_y = 2\pi n_2/L, \quad k_z = 2\pi n_3/L \tag{14.7}$$

where n_1, n_2, n_3 are integers (see Box 13).

Box 13. The Fourier Transform.

For oscillations we often approximate functions by their **Fourier series**, using sine and cosine terms. This approximation can be extended to any function that is sufficiently regular, but it is most useful for periodic functions.

Consider a periodic function $\delta(x)$ with the period 2π. It may be represented by the series

$$\delta(x) = \frac{1}{2}a_0 + a_1 \cos x + a_2 \cos 2x + a_3 \cos 3x + \cdots + b_1 \sin x + b_2 \sin 2x$$
$$+ b_3 \sin 3x + \cdots \tag{B13.1}$$

The coefficients are given by

$$a_k = \frac{1}{\pi}\int_{-\pi}^{\pi} \delta(x) \cos kx\, dx, \quad k = 0, 1 \ldots \tag{B13.2}$$

$$b_k = \frac{1}{\pi}\int_{-\pi}^{\pi} \delta(x) \sin kx\, dx, \quad k = 1, 2, \ldots$$

Problem. Show that the Fourier series of the function

$$f(x) = 0, \quad -\pi \leq x < 0$$
$$= 1, \quad 0 < x \leq \pi \tag{B13.3}$$

is

$$f(x) = \frac{1}{2} + \frac{2}{\pi}\left(\sin x + \frac{1}{3}\sin 3x + \frac{1}{5}\sin 5x + \ldots\right). \tag{B13.4}$$

The series can be written more compactly with the complex exponential:

$$\sin kx = \frac{1}{2i}\left(e^{ikx} - e^{-ikx}\right)$$

$$\cos kx = \frac{1}{2}\left(e^{ikx} + e^{-ikx}\right). \tag{B13.5}$$

We get

$$\delta(x) = \sum_{k=-\infty}^{\infty} \delta_k e^{ikx}$$
$$\tag{B13.6}$$

$$\delta_k = \frac{1}{2\pi}\int_{-\pi}^{\pi} \delta(x) e^{-ikx}\, dx, \quad k = \cdots - 1, 0, 1, \ldots$$

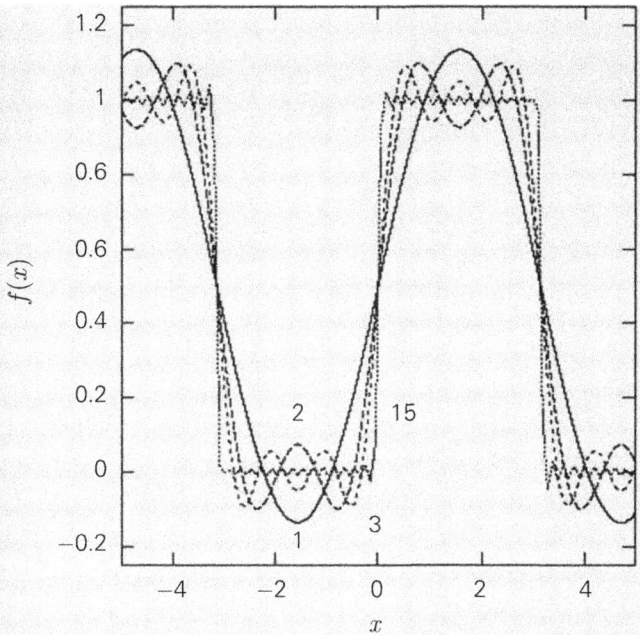

Figure 14.5. The Fourier series presentation of the function given in Equation (B13.3) is shown using 1, 2, 3 and 15 terms from the series given in Equation (B13.4), after the constant term.

The average value of $|\delta(x)|^2$ is connected to the Fourier coefficients δ_k by **Parceval's Theorem**

$$\langle |\delta(x)|^2 \rangle = \sum_{k=-\infty}^{\infty} \delta_k^2. \tag{B13.7}$$

Derivations of the above are found in standard textbooks (e.g., Arfken and Weber 1995).

Figure 14.5 is an illustration of how the approximation of $f(x)$ is improved with the increasing number of terms in the series.

Next assume a function that is periodic in the interval $(-L/2, L/2)$ with the period L instead of 2π. The function $\sin n \frac{2\pi}{L} x$, $n = 1, 2 \ldots$ has the period of L since

$$\sin\left(n\frac{2\pi}{L}(x + L)\right) = \sin\left(n\frac{2\pi}{L}x + 2\pi n\right) = \sin n\frac{2\pi}{L}x.$$

Similarly, $\cos n\frac{2\pi}{L}x$ and $e^{in\frac{2\pi}{L}x}$ have the same period L. Defining

$$k \equiv \frac{2\pi}{L} n, \tag{B13.8}$$

Equations (B13.6) are generalized to

$$\delta(x) = \sum_{k=-\infty}^{\infty} \delta_k e^{ikx}$$

$$\tag{B13.9}$$

$$\delta_k = \frac{1}{L} \int_{-L/2}^{L/2} \delta(x) e^{-ikx} dx.$$

For a three-dimensional function $\delta(\mathbf{r}) \equiv \delta(x)\delta(y)\delta(z)$ we may write

$$\delta(\mathbf{r}) = \sum_{k_x=-\infty}^{\infty} \delta_{k_x} e^{ik_x x} \sum_{k_y=-\infty}^{\infty} \delta_{k_y} e^{ik_y y} \sum_{k_z=-\infty}^{\infty} \delta_{k_z} e^{ik_z z} = \sum_k \sum_k \sum_k \delta_{\mathbf{k}} e^{i\mathbf{k}\cdot\mathbf{r}} \tag{B13.10}$$

where

$$\delta_{\mathbf{k}} = \delta_{k_x}\delta_{k_y}\delta_{k_z} = \frac{1}{L^3} \int \int_{-L/2}^{L/2} \int \delta(x)\delta(y)\delta(z) e^{-ik_x x} e^{-ik_y y} e^{-ik_z z} dx\, dy\, dz$$

$$\tag{B13.11}$$

$$= \frac{1}{V} \int_V \delta(\mathbf{r}) e^{-i\mathbf{k}\cdot\mathbf{r}} d\mathbf{r}$$

and the volume $V = L^3$.

We will now generalize the Fourier series to **Fourier integral**. Rewriting Equations (B13.1) and (B13.2) for a function $\delta(x)$ of period L and combine:

$$\delta(x) = \frac{1}{L} \int_{-L/2}^{L/2} \delta(t)\, dt + \frac{2}{L} \sum_{k=1}^{\infty} \cos kx \int_{-L/2}^{L/2} \delta(t) \cos kt\, dt + \frac{2}{L} \sum_{k=1}^{\infty} \sin kx \int_{-L/2}^{L/2} \delta(t) \sin kt\, dt$$

$$= \frac{1}{L} \int_{-L/2}^{L/2} \delta(t)\, dt + \frac{2}{L} \sum_{k=1}^{\infty} \int_{-L/2}^{L/2} \delta(t) \cos k(t-x)\, dt \tag{B13.12}$$

Let L go to infinity so that the interval $(-L/2, L/2)$ becomes $(-\infty, \infty)$, and define the change of k corresponding to one digit jump in n in Equation (B13.8) by $\Delta k = \frac{2\pi}{L}$. Then

$$\delta(x) = \frac{1}{\pi} \sum_{k=1}^{\infty} \Delta k \int_{-\infty}^{\infty} \delta(t) \cos k(t-x)\, dt. \tag{B13.13}$$

This leads to the Fourier integral in the limit.

$$\delta(x) = \frac{1}{\pi} \int_0^\infty dk \int_{-\infty}^\infty \delta(t) \cos k(t-x)\, dt = \frac{1}{2\pi} \int_{-\infty}^\infty dk \int_{-\infty}^\infty \delta(t) \cos k(t-x)\, dt$$

$$(\text{B13.14})$$

The last formulation is based on the fact that $\cos k(t-x)$ is an even function of k. Since $\sin k(t-x)$ is an odd function of k,

$$\frac{i}{2\pi} \int_{-\infty}^\infty dk \int_{-\infty}^\infty \delta(t) \sin k(t-x)\, dt = 0. \qquad (\text{B13.15})$$

Adding Equations (B13.14) and (B13.15) we get

$$\delta(x) = \frac{1}{2\pi} \int_{-\infty}^\infty e^{-ikx} dk \int_{-\infty}^\infty \delta(t) e^{ikt}\, dt. \qquad (\text{B13.16})$$

This is equivalent to the **Fourier transform pair**

$$\delta(x) = \frac{1}{2\pi} \int_{-\infty}^\infty \delta_k e^{-ikx}\, dk$$

$$(\text{B13.17})$$

$$\delta_k = \int_{-\infty}^\infty \delta(t) e^{ikt}\, dt.$$

In order that Equation (B13.14) holds for any function $\delta(x)$, we define a Dirac delta function

$$\delta_D(x - x') = \frac{1}{2\pi} \int_{-\infty}^\infty e^{ik(x-x')}\, dk. \qquad (\text{B13.18})$$

As usual, the Dirac delta function has the property

$$\delta(x) = \int_{-\infty}^\infty \delta(t)\delta_D(x-t)\, dt. \qquad (\text{B13.19})$$

Generalization to three dimensions is obtained by using the three-dimensional delta function

$$\delta_D(\mathbf{r} - \mathbf{r}') = \delta_D(x-x')\delta_D(y-y')\delta_D(z-z')$$

$$= \frac{1}{2\pi} \int_{-\infty}^\infty e^{ik_x(x-x')}\, dk_x \frac{1}{2\pi} \int_{-\infty}^\infty e^{ik_y(y-y')}\, dk_y \frac{1}{2\pi} \int_{-\infty}^\infty e^{ik_z(z-z')}\, dk_z$$

$$= \frac{1}{(2\pi)^3} \int_{-\infty}^\infty \int_{-\infty}^\infty \int_{-\infty}^\infty e^{i\mathbf{k} \cdot (\mathbf{r}-\mathbf{r}')}\, d\mathbf{k}. \qquad (\text{B13.20})$$

Using this function it is easily verified that

$$\delta(\mathbf{r}) = \frac{V}{(2\pi)^3} \int \delta_{\mathbf{k}} e^{-i\mathbf{k}\cdot\mathbf{r}} d\mathbf{k}$$

$$\delta_{\mathbf{k}} = \frac{1}{V} \int \delta(\mathbf{r}) e^{i\mathbf{k}\cdot\mathbf{r}} d\mathbf{r}$$

(B13.21)

is a Fourier transform pair.

Usually the random quantities (like the Fourier transforms $\delta_{\mathbf{k}}$) are assumed to be distributed in accordance with the normal (Gaussian) distribution law:

$$f(y) = (2\pi\sigma^2)^{-1/2} \exp(-y^2/2\sigma^2)$$

(14.8)

where y is a random variable and $\sigma^2 = \langle y^2 \rangle - \langle y \rangle^2$ and the average values are found using of the same function $f(y)$. Also assume isotropy in space so that the $\mathbf{k}$ vector has no preferred direction. Then we use the scalar k instead of the vector $\mathbf{k}$. The differential $d\mathbf{k}$ becomes $4\pi k^2 dk$.

In the statistical description, the most important statistical characteristic of density perturbations is the **power spectrum**, which is defined as

$$P(t, k) = \langle |\delta_{\mathbf{k}}(t)|^2 \rangle.$$

(14.9)

Here the brackets $\langle \ldots \rangle$ mean to average over the normal distribution. The power spectrum gives a quantitative measure of the density amplitude for different wave vectors. Rephrasing, $P(t, k)$ shows how the amplitude depends on the spatial scale of the perturbations $r = 2\pi/k$. Like in Chapter 13 we measure r as well as k in the co-moving frame of reference.

14.4   Perturbations of the gravitational potential

Let us now consider the perturbations of the gravitational potential that accompany density perturbations. Define density perturbation amplitude as $\delta = \delta\rho/\rho$. In Newtonian mechanics, the gravitational potential as a function of time and spatial coordinates follows from the equation of motion, the continuity equation, and the Poisson equation in a self-consistent way together with the density and velocity of matter. When a solution for the density perturbation is known, the perturbation of the potential, $\delta\Phi$, is easily calculated. Approximately:

$$\delta\Phi \approx \frac{G\delta M}{R},$$

(14.10)

where $\delta M \approx \delta \rho R^3$ and R is the size of the perturbed region. Using the density amplitude δ, one has

$$\delta \Phi \approx G\rho R^2 \delta. \tag{14.11}$$

This relation enables calculation of the behavior of the potential perturbations in various epochs of the perturbation evolution.

During the radiation domination era, the growing mode of the density perturbations is described by $\delta \propto t$ (Equation (13.110)), the density as $\rho \propto t^{-2}$ (Equation 13.108)), and, finally, the size scale $R(t) \propto t^{1/2}$ (see Equations (13.108)). Substituting, the perturbation of the potential turns out to be time-independent with the exponents summing to zero, $-2 + 1 + 1 = 0$, resulting in a constant:

$$\delta \Phi \propto \rho R^2 \delta \propto \text{Const.} \tag{14.12}$$

During the radiation domination era, for the falling or weakening mode of perturbation, $\delta \propto t^{-1}$, Equation (13.108), with the density and size scale varying as before, the exponents sum to $-2 + 1 - 1 = -2$, Thus the perturbation of the potential is falling:

$$\delta \Phi \propto \rho R^2 \delta \propto t^{-2}. \tag{14.13}$$

During the matter domination epoch the growing mode of the density perturbations is given by $\delta \propto t^{2/3}$ (Equation (13.106)) and the density and size scale with time as $\rho \propto t^{-2}$, $R(t) \propto t^{2/3}$ (Equation (13.105)). Then the perturbation of the potential does not depend on time:

$$\delta \Phi \propto \rho R^2 \delta \propto \text{Const.} \tag{14.14}$$

The falling mode of perturbations at the same matter epoch is characterized by $\delta \propto t^{-1}$, again with the same density and size variation. Thus the perturbation of the potential is also falling:

$$\delta \Phi \propto \rho R^2 \delta \propto t^{-5/3}. \tag{14.15}$$

To summarize, the perturbation of the gravitational potential accompanying a weak linear density perturbations can only be constant or falling with time, while the density amplitude of perturbations can be growing.

This result is in complete agreement with the General Relativity treatment of linear perturbations performed by Lifshitz (1946) for radiation domination and matter domination epochs. Remarkably, the Newtonian treatment gives the same results for linear perturbations as the General Relativity treatment does. This is even true in the completely unjustified case of the growing mode at radiation domination where the spatial size of these perturbations is larger than the Hubble radius: $R > ct$. For this condition, Newtonian mechanics is completely inapplicable because—in Newtonian terms—the velocity of expansion approaches the speed of light and the gravitational potential goes to c^2, as the distance goes to the Hubble radius. However, the similarity of the mathematical descriptions of the linear perturbations is so far-reaching in General Relativity and Newtonian mechanics that the final formulae are actually the same in both. This similarity exists not only for the case of weak linear perturbations. It also appears in nonlinear spherically symmetrical density perturbations, as may be seen from the Tolman solution and its Newtonian analog which follow.

14.5   Harrison–Zeldovich spectrum of density perturbations

We now turn to initial adiabatic perturbations and introduce their spectrum. The spectrum is a description of the dependence of the perturbation amplitude on the spatial scale. In the simplest case, this dependence may be a power-law:

$$\delta(t) \propto R(t)^q \tag{14.16}$$

Here q is a constant. $R(t)$ is the size of the perturbation area, its spatial scale at a given moment of time t. Perturbations are presumed to have this spectrum "from the beginning", that is from the epoch of their generation in the very early universe. Using a Lagrangian size for the perturbation region r to characterize its time independent co-moving spatial scale for the early radiation domination epoch,

$$R(t) = r(t/t_1)^{1/2} \tag{14.17}$$

Here r is the perturbation's arbitrary size at t_1. The non-relativistic matter (dark matter and baryons of total density ρ_M) mass is conserved in the perturbation, therefore the scale r is directly associated with:

$$M(t, r) = \frac{4\pi}{3} \rho_M(t) R(t)^3 = \frac{4\pi}{3} \rho_M(t_1) r^3. \tag{14.18}$$

The perturbation spectrum is then

$$\delta \propto r^q \propto M^{q/3}. \tag{14.19}$$

We use this spectrum to consider the gravitational potential perturbation under adiabatic perturbations. According to general relations for the perturbation of the potential $\delta\Phi$ in Equation (14.11),

$$\delta\Phi \approx G\rho_M R^2 \delta, \tag{14.20}$$

This quantity is time-independent for the rising mode of the gravitational instability. This quantity may also be scale-independent, for a particular perturbation spectrum power law index. In terms of the Lagrangian scale,

$$\delta\Phi \cong G\rho_1 r^2 \delta_1 \propto r^2 \delta_1 \propto r^{q+2}, \tag{14.21}$$

where ρ_1 and δ_1 are respectively the density and the density perturbation at $t = t_1$. The gravitational potential perturbation is scale independent, if

$$q = -2. \tag{14.22}$$

The scale-invariant (for gravity perturbations) spectrum

$$\delta \propto r^{-2} \propto M^{-2/3} \tag{14.23}$$

is the **Harrison–Zeldovich** spectrum which was suggested independently by both Edward R. Harrison (1970) and Zeldovich (1972).

Perturbations with the Harrison–Zeldovich spectrum at any time have all the same value of the gravitational potential perturbation. This value cannot yet be obtained theoretically from fundamental physics. However, it may be simply estimated. Focusing on a specific moment of time, calculate the potential perturbation $\delta\Phi$ at the epoch $z = z_N$. This is when the largest cosmic structures, large clusters of galaxies (of mass M_L) become nonlinear. At this nonlinear moment, $\delta \cong 1$, and the perturbed potential is thus

$$\delta\Phi \cong G\rho_M(z_N)R_L(z_N)^2, \tag{14.24}$$

Here

$$R_L \cong \left[\frac{3}{4\pi}M_L/\rho_M(z_N)\right]^{1/3} \tag{14.25}$$

is the typical size of a perturbation with the mass M_L at the nonlinear epoch. Then

$$\delta\Phi \cong G[\rho_M(z_N)]^{1/3}\left[\frac{3}{4\pi}M_L\right]^{2/3}. \tag{14.26}$$

A typical mass of the largest clusters of galaxies $M_L \cong 3 \times 10^{15}\,M_\odot$ which is mostly dark matter. These clusters are observed to already exist at the cosmic time corresponding to $z \approx 1$ (Bahcall and Fan 1998, Tran et al. 1999, Mullis et al. 2005). For them to exist at this time, these clusters would have become nonlinear at roughly half this time, or at $z_N \approx 2.25$. Taking $z_N \cong 2.25$, the gravitating matter density $\rho_M(2.25)$ is $(1 + 2.25)^3$ times greater than today, $\rho_M(0) = 3.5 \times 10^{-8}\,M_\odot\,\text{pc}^{-3}$.

The quantity Δ is a basic universal constant in the theory of gravitational instability which is defined to be the ratio of the potential energy associated with the gravitational fluctuations of mass δM to the rest mass energy $\delta M c^2$ of the same element of matter. With the Harrison–Zeldovich spectrum, this is a scale-invariant time-independent measure of the initial cosmic structure. Substituting these numbers in the dimensionless quantity:

$\Delta \equiv \delta\Phi/c^2$

$\cong 4.3 \cdot 10^{-3} \, (\mathrm{km/s})^2 \, pc/M_\odot (1.2 \cdot 10^{-6})^{1/3} (M_\odot/\mathrm{pc}^3)^{1/3} (3/4\pi)^{2/3} (3 \cdot 10^{15})^{2/3} \, M_\odot^{2/3} / [(3 \cdot 10^5)^2 (\mathrm{km/s})^2]$

$= 0.4 \times 10^{-5} \approx 10^{-5}.$ (14.27)

The amplitude of the growing adiabatic perturbations at time t and at the spatial scale R in terms of Δ for the Harrison–Zeldovich spectrum is

$$\delta \cong \Delta \frac{c^2}{G\rho_M R^2} \cong 6\pi\Delta (ct/R)^2 \qquad (14.28)$$

where Equations (8.18) and (14.19) have been used: $G\rho_M = \left(6\pi t^2\right)^{-1}$.

If the scale of the perturbations is always greater than the Jeans length, i.e., the mass scale $M > M_J(t_{rec}) \approx 10 M_J \, (t_*)$, (Equation (13.135) with $t_{rec} \cong 10 t_*$), the relation describes the evolution of the perturbations "from the very beginning" to the termination of gravitational instability (at $z \cong z_V$—see Section 13.6) or up to the nonlinear stage of their growth.

Computer simulations of the formation of the large-scale structure of the universe usually assume the Harrison–Zeldovich spectrum. The results of such simulations are in general qualitative agreement with what is actually observed on the scales of groups, clusters and superclusters.

14.6 Perturbations and CMB

In this section, we present the basic physical processes that produce temperature fluctuations, or temperature anisotropy, in the CMB.

Initial perturbations that exist in the form of peculiar common motions of radiation and baryonic plasma prior to the recombination generate the CMB anisotropy. The generation is due to three different physical effects. The first of them is the Doppler effect produced by the peculiar velocity of the perturbations. To evaluate the perturbation Doppler effect, consider a perturbation region in which plasma and radiation move together with the velocity v relative to the unperturbed cosmological expansion. Inside the region, photons are in thermodynamic equilibrium, and their frequency distribution is described by the Planck universal law (Equation (10.23)):

$$n_R(\omega) = \omega^2/(\pi^2 c^3)\left[\exp\left(\frac{\hbar\omega}{k_B T}\right) - 1\right]^{-1}, \qquad (14.29)$$

where $\hbar$ is the Planck constant, k_B is the Boltzmann constant, $n_R(\omega)\,d\omega$ is the number density of photons with the frequency between ω and $\omega + d\omega$ and T is the photon temperature. The photon frequency is Doppler-shifted in the reference frame of the isotropic background, so that instead of the unshifted ω one will have

$$\bar{\omega} = \omega[1 + (v/c)\cos\theta], \qquad (14.30)$$

where θ is the angle between the velocity of the bulk motion and the wave vector of the photon. Since the perturbations are weak, and $(v/c)\cos\theta$ is much less than 1, the exponent of the Planck distribution has:

$$\hbar\bar{\omega}/(k_B T)\left(1 + \frac{v}{c}\cos\theta\right)^{-1} = \hbar\bar{\omega}\left/\left[k_B T\left(1 + \frac{\delta T}{T}\right)\right]\right. \qquad (14.31)$$

The relative temperature variation is

$$(\delta T/T)_1 = \frac{v}{c}\cos\theta. \qquad (14.32)$$

The second mechanism of CMB anisotropy is density variation in adiabatic perturbations. In these the radiation density ρ_R is slightly larger (in over-densities) or smaller (in under-densities) compared to the unperturbed radiation density. This is called the Silk mechanism. One has for the corresponding temperature fluctuations:

$$\left(\frac{\delta T}{T}\right)_2 = \frac{1}{4}\frac{\delta\rho_R}{\rho_R} = \frac{1}{3}\frac{\delta\rho_B}{\rho_B}. \qquad (14.33)$$

since $\rho_R \propto T^4$. In this relation, we make use of the fact that the number density of photons, $n_R \propto T^3$, and that the ratio n_R/ρ_B is the same in the adiabatic perturbation and in the unperturbed photon plasma fluid. This relation links the amplitude of perturbations in baryonic matter (at recombination) with CMB temperature fluctuations.

Lastly, for the third mechanism, perturbations of the gravitational potential or the metric perturbations generate the CMB anisotropy. This mechanism is the already mentioned Sachs–Wolfe effect. Metric perturbations affect the propagation of photons. The photon frequency changes when the photon travels in a changing gravitational potential. This is the gravitational shift (see Equation (3.13)), and photons may be redshifted by various amounts, on the last-scattering surface. As we discussed above, density perturbations $\rho\delta$ at a spatial scale R produce perturbations of the gravitational potential in over-dense or under-dense regions with the amplitude (Equation (14.11)):

$$\delta\Phi \cong -G\rho R^2\delta. \tag{14.34}$$

The density perturbation $\rho\delta$ is related to all gravitating non-vacuum matter. The minus means over-densities produce potential valleys and under-densities produce potential hills in the gravitational field. At recombination, photons that leave a potential well come to us redshifted, while photons from a potential hill are blue-shifted relative to the mean redshift level. For weak effects, there is a linear relation between the frequency change $\delta\omega$ and the amplitude of the potential perturbation. In this case, $\delta\omega/\omega = \delta\nu/\nu = 1 - \gamma^{-1}$ by Equation (3.14) where $\gamma^{-1} = (1 - 2\delta\Phi/c^2)^{1/2}$ if we put $\delta\Phi = GM/r$ in Equation (3.13). Then to first order

$$\delta\omega/\omega \cong \delta\Phi/c^2. \tag{14.35}$$

This change in frequency affects all the photons from the perturbation region, so that all frequencies in the radiation spectrum have equal fractional shifts, and the photon distribution remains the Planck equilibrium distribution. In particular, the frequency ω_{max} that corresponds to the maximum in the Planck distribution is shifted in accordance with Equation (14.35). However, from Equation (10.25), frequency ω_{max} is taken to be proportional to the radiation temperature. Because of this, the original emission temperature is perturbed with the relative amplitude

$$\left(\frac{\delta T}{T}\right)_3 \cong \frac{\delta\Phi}{c^2} \cong -\frac{G\rho}{c^2}R^2\delta. \tag{14.36}$$

Since $G\rho \cong (6\pi t^2)^{-1}$ (Equation (8.18)), finally:

$$\left(\frac{\delta T}{T}\right)_3 \cong -\left(\frac{R}{ct}\right)^2\frac{\delta}{6\pi}. \tag{14.37}$$

It is seen from the first factor on the right side of Equation (14.37) that the Sachs–Wolfe effect is most important for the largest scales that are comparable to the horizon scale ct or exceed it.

The cosmic age t and the characteristics of the perturbations are a "snapshot" at the last scattering moment near the recombination and decoupling time $t_{rec} \cong 10^{13}$ sec. These three effects described above are the basic physics of the CMB anisotropy generation by protogalactic perturbations. The CMB temperature variations arise from different amounts of gravitational redshift which photons suffer when they escape from dark matter concentrations.

14.7   The cosmic horizon at CMB emission

We will describe qualitatively how some of the cosmological parameter values affect the theoretical line in the amplitude versus multipole number plot in Figure 14.4

above. In the very largest scales, corresponding to low multipole numbers, the choice of the vertical scale is such that the line is expected to be practically horizontal. The angular scale of these horizontal portions is larger than the size of the cosmic horizon diameter or the Hubble radius at the time when the photons are decoupled and emitted. This flat spectrum is the result of the previously discussed scale-independent initial density perturbation spectrum in the dark matter. Note that the third of the effects is directly associated with perturbations of the gravitational potential which are scale-invariant and time-independent for the Harrison–Zeldovich spectrum. In this spectrum, one has for the CMB temperature fluctuations:

$$\left(\frac{\delta T}{T}\right)_3 \cong \Delta \approx 10^{-5}. \tag{14.38}$$

This is the fractional level of temperature fluctuations expected for the larger mass scales $M > 10 M_J(t_*)$ at which the initial spectrum is preserved.

The distances observable today with modern ground-based and orbital telescopes extend almost up to today's observational horizon. This cosmic horizon radius is the distance in the co-moving space that light can cover during the existence of the universe. It is also called the particle horizon ("particle" referring especially to photons as messengers). We see in Friedmann's models that the present-day horizon radius is approximately $R_0(t_0) \cong ct_0 \cong 1.30 \times 10^{28}$ cm. Here $t_0 = 13.7 \pm 0.2$ Gyr is the current age of the universe. The most remote observed objects, which are quasars and first galaxies with redshifts around 8, are seen at distances exceeding ten billion light years, or $\approx 10^{28}$ cm.

Now let us calculate roughly the size of the particle horizon at the time of recombination and then the corresponding angular size in our sky. We ask how long a distance in the co-moving frame a photon has been able to traverse during the time t_0. In spherical coordinates the distance element has the familiar form $ds^2 = c^2 dt^2 - R(t)^2 dr^2 = 0$. In the time interval $t = 0 \to t = t_0$, the photon's radial co-moving coordinate changes by the amount $r_{particle} = c \int_0^{t_0} dt/R(t)$. By multiplying by the scale factor $R(t_0)$ one obtains the corresponding proper distance (in cm):

$$L_{particle} = R(t_0) c \int_0^{t_0} dt/R(t) \tag{14.39}$$

For instance, for the Einstein–de Sitter model $R(t) = R_0 r (t/t_0)^{2/3}$ and the present age $t_0 = (2/3)/H_0$. By integrating we derive that the radius of the particle horizon is two times the Hubble distance:

$$L_{part} = 2c/H_0 \quad \text{(Einstein-de Sitter model)}. \tag{14.40}$$

Let us now estimate the size of the particle horizon as it was at the epoch corresponding to the recombination (at $z = 1100$). For simplicity, we use here the Einstein–

de Sitter model. During radiation domination, the scale factor grows as $R(t) \sim t^{1/2}$, hence

$$L_{rec} = cR(t_{rec}) \int_0^{t_{rec}} dt/R(t) = 2t_{rec}c.$$

Recombination occurred about 360 000 years ($= t_{rec}$) after the big bang, hence $L_{rec} \approx 0.2$ Mpc. Today we see such a length at the angle θ which we can calculate using the angular size distance $D_A = 2c/H_0[((1 + z)^{1/2} - 1)/(1 + z)^{3/2}] \approx 2(c/H_0)(1 + z)^{-1}$: $\theta = L_{rc}/D_A \approx 0.5(0.2/3000\,h^{-1})(1 + z)$. When $z = 1\,100$ and $h = 0.7$, in degrees $\theta = (360/2\pi)0.032 = 1.5°$.

A more accurate calculation using the standard Lambda model does not essentially change the result that CMB photons arriving from directions only a few degrees apart in the sky could not have been causally connected. In spite of this, the background radiation is remarkably isotropic all over the sky. An explanation of this famous causality problem has been given by the inflation theory of the early universe (Chapter 17).

In scales less than the Hubble radius or the particle horizon, it is possible that different regions have "communicated" with each other to form density concentrations. The interaction increases correlation power and produces the peaks in the power spectrum on angular scales less than $\sim$1 degree. These are called **acoustic peaks** since an important form of interaction between different regions in the early universe is via sound waves.

The angular scale in the observed multipole spectrum where the horizontal line ends and peaks begin is important since it tells us the angular size of the Hubble radius at the time of the recombination. Assume that the angular size of the Hubble radius is known. Its angular size in the sky depends on the overall curvature of the universe. Qualitatively, for a universe which is not flat, we may deduce from Figure 4.4 that the angular size of the Hubble radius is greater if $K > 0$ than if $K < 0$. The flat space with $K = 0$ will be intermediate between the two. The value of the curvature (dependent on the total density parameter Ω) shifts the theoretical curve in Figure 14.4 horizontally. By matching the theory with observations the curvature has been found to be very close to zero; in fact the fit with the data in Figure 14.4 is based on putting $K = 0$ exactly with $\Omega = 1$.

14.8   The origin of peaks in CMB angular size spectrum

The baryonic gas before recombination is a hot gas of light atomic nuclei, electrons, and photons. The existence and amplitudes of the CMB peaks reflect processes in the early universe before and up to recombination at t_{rec} in the sub-horizon scale (see review and references in Eisenstein and Bennett 2008). The breakdown of the universe's matter components can be studied using the amplitudes of the peaks in the CMB angular size frequency spectrum.

With $\sim 10^{12}$ photons per cubic centimeter versus only 10^4 nuclei in the same volume and at the same temperature, radiation pressure from photons dominates compared to pressure from the baryonic nuclei. Dark matter acts via gravitation and not by pressure (just as it does today). The photons provide a medium for the propagation of "sound" waves which travel at the speed $c_s \approx 0.4c$ at the time of the recombination. Pressure changes are transmitted via these acoustic waves. Rather than being governed solely by pressure as in the case of sound in air, the acoustic waves in the early universe were modified by the force of gravity mainly from dark matter and to a much lesser degree from baryonic matter.

In the scales $> cH(t_{rec})$ there is no time to transport energy from one region to another. Fluctuations in the sub-horizon scale $\lambda = 2\pi/k$ may be described as a combination of sine waves of different wave numbers k. Consider a single quasi-spherical over-density fluctuation as representing the effect of such perturbations in the scale of π/k. The time for a sound wave to cross the perturbation is $t_{cross} \sim (kc_s/\pi)^{-1}$. If $t_{cross} > t_{rec}$ there is no time for pressure to act. In the opposite case the photon pressure causes oscillations which lead to temperature fluctuations. The decoupling of matter and radiation catches oscillations at different phases hence leading to different amplitudes depending on scale.

The first ($l \approx 220$) peak in Figure 14.4 is connected to the largest oscillations coinciding with positive density perturbations in dark matter. Baryonic matter and the photons fall inward to dark matter potential wells to create compressed hot density enhancements at the time of recombination. These compressed areas of photons and baryonic matter create regions of enhanced thermal emission in the enhancements. This is schematically shown in Figure 14.6.

The second peak is of shorter wavelength so the dark matter fell inward to create concentrations of photons and baryons earlier than for the first peak. This region contains dark matter, baryons and photons. The pressure results in a spherical sound wave of both baryons and photons moving with a speed about half the speed of light outwards from the over-density. The dark matter only interacts gravitationally and so it stays at the center of the sound wave, the origin of the over-density. Before decoupling, the photons and baryons move outwards together. After decoupling the photons are no longer interacting with the baryonic matter so they diffuse away. This relieves the pressure on the system, leaving a shell of baryonic matter at a fixed radius. This radius is often referred to as the sound horizon. Without the photo-baryon pressure driving the system outwards, the only remaining force on the baryons is gravitational. Therefore, the baryons and dark matter (still at the center of the perturbation) form a configuration which includes over-densities of gravitating matter both at the original site of the anisotropy and in a shell at the sound horizon.

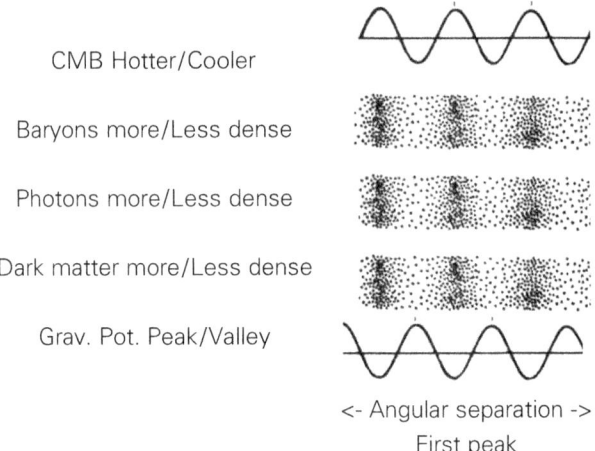

CMB Hotter/Cooler

Baryons more/Less dense

Photons more/Less dense

Dark matter more/Less dense

Grav. Pot. Peak/Valley

<- Angular separation ->
First peak

Figure 14.6. Schematic diagram of origin of the first peak in the CMB temperature angular power spectrum. (Redrawn from part of a figure in Hu and White 2004).

In the second $l \approx 550$ peak dark matter and the baryon-photon condensations are out of phase relative to each other at the time of recombination as shown in Figure 14.7. This results in a peak of smaller amplitude in the CMB power spectrum. In the creation of the density fluctuation of the second peak, dark matter gravity compresses the baryon-photon gas into denser regions in the gravitational potential valleys, just like in Figure 14.6. However, this was not at the time of recombination, but earlier when t_{cross} was much less than t_{rec}. By the time of recombination, the increased photon gas radiation pressure caused the photons and baryons to flow out from potential valleys toward the adjoining peaks. This results in a weaker enhancement of emission there at recombination (when transparency happens).

The ripples in the density of space continue to attract matter and eventually galaxies formed in a similar pattern, therefore one would expect to see a greater number of galaxies separated by the sound horizon than by nearby length scales.

The peaks with higher l values follow the same pattern: the third peak forms like the first one, but at smaller scale, the fourth peak like the second one at smaller scale, etc. However, going to higher peak numbers, a new feature comes in, damping on small scales. This is because the last scattering surface of the CMB photons is not exactly a two-dimensional surface but a layer of finite width (in redshift range, the thickness of the layer is $\Delta z \approx 80$). The temperature fluctuations are damped on scales smaller than this since photons from positive and negative fluctuations are mixed; this happens in the scale of a few arc-minutes or less.

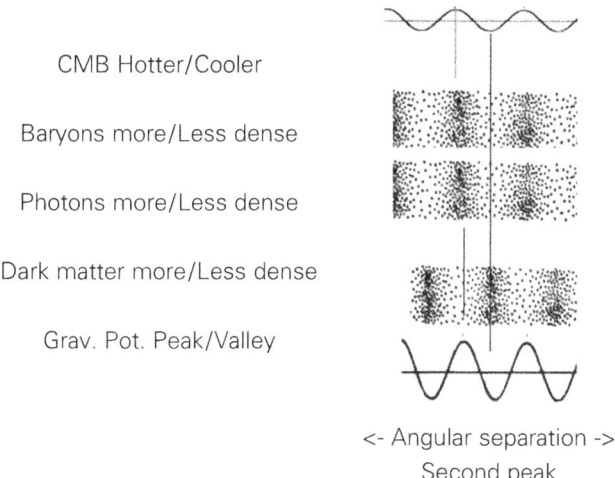

CMB Hotter/Cooler

Baryons more/Less dense

Photons more/Less dense

Dark matter more/Less dense

Grav. Pot. Peak/Valley

<- Angular separation ->
Second peak

Figure 14.7. Schematic diagram of origin of the second peak in the CMB temperature angular power spectrum. Figure is drawn larger but CMB "wavelength", is shorter with smaller amplitude than the first peak. Note the phase difference between dark matter and baryonic matter and photons. (Redrawn from part of a figure in Hu and White 2004).

In approximate terms, we may say that the first peak shows the universe is close to spatially flat. The second peak indicates substantial amounts of baryons and the third peak measures the physical density of the dark matter. The damping tail provides consistency checks of underlying assumptions. However, as we explain in the next section, CMB anisotropies are sensitive to a host of cosmological parameters. There are parameter degeneracies where different combinations of parameters conspire to produce the same CMB spectrum within observational error limits.

Other information, CMB polarization, and secondary anisotropies can break these degeneracies. But there are some fairly good indicators, such as the ratio of the first two peaks which can be used to estimate the density of dark matter relative to baryonic matter. The estimated value from this ratio is in agreement with the relative amounts of dark matter and baryonic matter via studies of systems of galaxies and the Big Bang nucleosynthesis discussed previously.

14.9   Cosmological parameters from CMB peaks: additional primary observations

The modeling the CMB power spectrum is not simple. One reason is that the photons of the CMB were generated over a period of 120,000 yr, which represents a fair

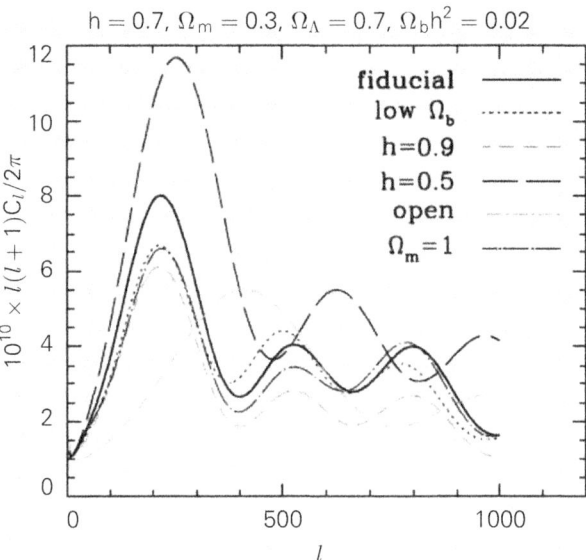

Figure 14.8. The effect of varying one cosmological parameter at the time away from the concordance values (curve labelled "fiducial" whose values are listed at the top).

fraction of the age of the universe (380,000 yr) at CMB emission. Thus the earliest photons of the CMB arise in different conditions from the later photons. The observed CMB spectrum is thus a combination from different radiating layers, and the calculation of the integrated CMB requires computer modeling. In practice, a large number of model CMB spectra are calculated using different parameter values. By comparing with observations the best parameter values are found together with the uncertainties. The optimal parameter values for WMAP given in this text refer to the Figure 14.4 fit.

Figure 14.8 shows how the model is changed from the Concordance model, if one of the cosmological parameters is varied. The "fiducial" model fits with WMAP observations extremely well (Figure 14.4). Thus a major alteration of the model by tweaking just one parameter is not possible. However, tweaking several parameters at the same time may still produce an acceptable result (this situation is usually called a degeneracy). For example, $\Omega_M = 1.3$ and $\Omega_\Lambda = 0$ produces a fit as good as the flat universe $\Omega_M = 0.2404$ and $\Omega_\Lambda = 0.7596$. Note that when going along a Ω_Λ versus Ω_M plot the total density parameter changes from $\Omega = 0.92$ (open space) to $\Omega = 1.3$ (closed space). The fact that we keep rather close to the flat space $\Omega = 1$ reflects the fact that the angular position of the first big peak of the CMB spectrum is especially sensitive to the curvature (i.e., $\Omega = \Omega_\Lambda + \Omega_M$) and less sensitive to the relative contributions from dark matter and dark energy. However, varying Ω_M is accompanied by a variation of the Hubble constant, too, because in the analysis of the CMB spectrum these two parameters are degenerate, appearing together as a primary fitting parameter $\Omega_M h^2 (H_0 = 100\, h$ km/s/Mpc). Another important fitting parameter is the combination $\Omega_B h^2$.

Rather than 71 km/s/Mpc, the $\Omega_M = 1.3$ and $\Omega_\Lambda = 0$ fit requires a Hubble constant of 32 km/s/Mpc which is well beyond the observational error. The Hubble constant is a primary "outside" (coming from other observations than the CMB) parameter important in interpreting the WMAP data. This importance is one reason for our extensive discussion of the Hubble constant's value earlier in this text. The present-day Hubble constant as determined from measurements in the relatively local galaxy universe (distance < 1000 Mpc), is now usually taken to be $H_0 = 73.8 \pm 2.4$ km/s/Mpc. In the review of the Hubble constant in Section 5.5 we mentioned that it is still possible that the true value of H_0 could be somewhat lower; for instance, if actually $H_0 = 63$ km/s/Mpc, then the CMB fit would give $\Omega_M = 0.31$ instead of 0.24.

The new "precision cosmology" from the CMB analysis has not made the classical cosmology of the galaxy universe obsolete! Also, the age of the universe for the mentioned alternative pair of parameters (1.3,0), 19.3 billion years, is too long compared to the locally determined values (from white dwarf stars etc.). Again, these age estimates coming from outside serve to constrain the WMAP values.

But if we are considering only acceptable fits to the CMB data in the $(\Omega_M, \Omega_\Lambda)$—plane, and not worrying what happens to the other parameters, we obtain a narrow strip of solutions (Figure 14.9, labeled CMB). Note that if extended down and to the left, it would intersect the horizontal axis at (1.3,0) which is the solution discussed above that, despite fitting the Figure 14.4 CMB amplitude plot, has too small a Hubble constant and too long an age to be valid.

We note that from the WMAP (five-year) data alone (without any outside constraints) the best fit gives $h = 0.72 \pm 0.03$ (Hinshaw et al. 2009). (see Section 10.2).

We will now consider the other constraints. The supernova constraint (labeled SNe) comes from the following considerations. In Figure 14.10 we show the plot of Figure 9.5 again, but rather than comparing it with the Concordance model directly, we ask what Friedmann model would give a best fit to the data, and with what error bounds.

Now we use the supernova observations in a simple and transparent manner to obtain complementary information on the density parameters. As in the earlier Figure 9.5, Figure 14.9 plots the observations of binned supernova (type Ia) distance moduli at different redshift intervals, in comparison with those for a Milne "coasting model" i.e., a model with no deceleration or acceleration ($\Delta(m - M) = 0$ horizontal line). In Figure 14.10, we compare the data with a simple polynomial fit (solid line). We note that the fitted curve becomes horizontal, i.e., the deceleration changes to acceleration at $z \approx 0.6$. We now ask what this observation implies in the (Ω_M, Ω_V) plane.

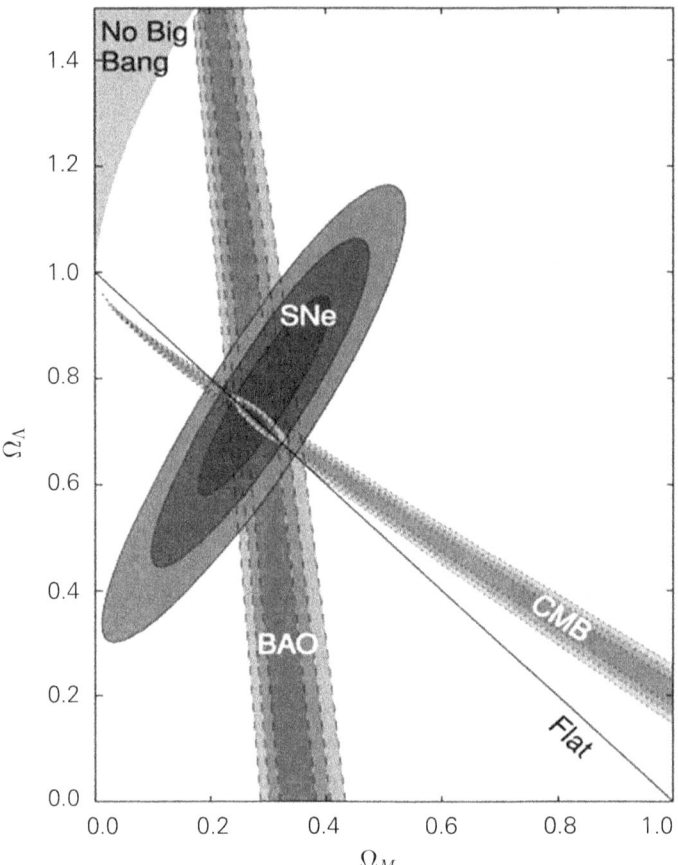

Figure 14.9. The Ω_M and Ω_Λ values corresponding to theory and observational data. The single black line corresponds to values for a flat critical density universe (Flat). The CMB angular frequency data form the narrow diverging line (CMB). The values deduced from supernovae are shown as a set of nested elongated ellipses (SNe). The broad nearly vertical line corresponds to baryon acoustic observations (BAO). In all these from the outer edges, boundaries of progressively darker tones correspond to 68.3%, 95.4%, and 99.7% confidence regions. A cosmological constant dark energy ($w = -1$) has been assumed. The very small ellipses at the intersection of these are the net result (Amanullah, 2010 by permission). arXiv:1004.1711 http://supernova.lbl.gov/Union/.

Equations (8.25) and (8.26) for the case of $\rho_V > 0$ become

$$-2\rho_V + \rho = \frac{3}{4\pi G}\left(-\frac{R(t_0)\ddot{R}(t_0)}{\dot{R}(t_0)^2} \cdot \frac{\dot{R}(t_0)^2}{R(t_0)^2}\right) = \frac{3q_0 H_0^2}{4\pi G}. \tag{14.41}$$

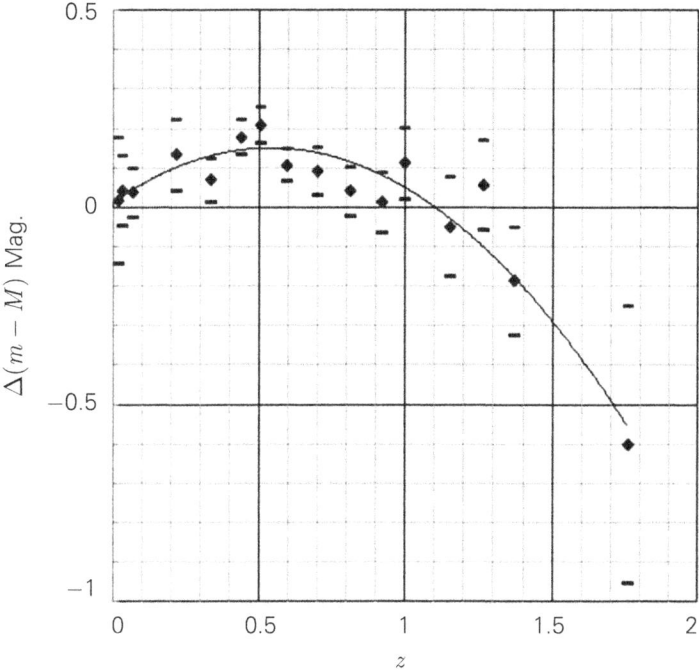

Figure 14.10. Binned supernova (type Ia) distance moduli at different redshift intervals, in comparison with those for a Milne "coasting model" i.e., a model with no deceleration or acceleration ($\Delta(m - M) = 0$ horizontal line). In contrast to Figure 9.5. We show a simple polynomial fit trend line (solid line) together with observational points and their associated error bars. The data points in the figure are adapted from Riess et al. (2007) for 206 supernovae with the most accurate distances. Also see http://www.astro.ucla.edu/~wright/sne_cosmology.html for data tables and later references.

Dividing Equation (14.41) by 2 and also by the critical density, $\rho_c = 3H_0^2/(8\pi G)$, we can obtain a relation between both density parameters relative to critical density and q_0,

$$\Omega_V = \Omega_M/2 - q_0. \tag{14.42}$$

Here we use the definition (Equation (8.28))

$$\Omega_M = \rho_0/\rho_c = \frac{8\pi G}{3H_0^2}\rho_0.$$

The definition of Ω_V is similar (Equation (8.30)). As noted earlier, the present-day observed value for q_0 is -0.6 which is a net acceleration. Equation (14.42) is one of two observational conditions to estimate the Omegas.

From Equation (14.41) we see that, when $\rho \leq 2\rho_V$, the deceleration changes into acceleration. Let z_V be the transition redshift, estimated earlier in the Concordance model to be 0.75 (Equation (8.37)). What happens if we replace it by $z_V = 0.6$?

Since the mean density evolves as (Equations (6.8) and (8.12))

$$\rho = (\Omega_M \rho_c)[1 + z]^3$$

we have

$$\Omega_M \rho_c [1 + z_V]^3 = 2\Omega_V \rho_c$$

at the time when $\rho = 2\rho_V$. From the equation above we obtain

$$(1 + z_V)^3 = 2\Omega_V / \Omega_M.$$

This equation is the second of two observational conditions to estimate the Omegas.

Since $z = 0.6$ for the polynomial fit,

$$\Omega_V / \Omega_M = 2.048.$$

Including the first observational condition from above,

$$-\Omega_V + \Omega_M/2 = q_0 = -0.6, \quad \text{we obtain}$$

$$\Omega_V = 0.793$$

$$\Omega_M = 0.387.$$

Note that these two do not add up to one i.e., we are simply taking the supernova observations at face value and *not* assuming a flat universe. The SN result is a bit above and to the right of the Concordance point which is the intersection of the flat universe and CMB lines in Figure 14.9. The difference between the fitting result and the Concordance result is not statistically significant, considering the observational errors, as shown by the error contours in Figure 14.9. In agreement with our simple approximate calculations, more sophisticated analysis of the data also indicates that the "best choice" for the supernovae is actually near the center of the nested error ellipses above and to the right of the Concordance intersection. More data will probably reduce the difference still more.

14.10   Spatial Correlations of Galaxies

There is another constraint in Figure 14.9, corresponding to baryon acoustic oscillations (BAO). We will discuss this constraint below. Before that, we have to take a look at galaxy spatial correlations. Galaxies tend to associate with each other: a small number together makes a group of galaxies; a larger number of galaxies in a single system is called a cluster. Clusters of galaxies together with groups sometimes make a supercluster. Galaxies are also found in filaments and in sheets that often leave a void (a low galaxy density region) between them (Jõeveer et al. 1978, Jõeveer and Einasto 1978, Tully and Fisher 1978, Tifft and Gregory 1976, 1978, Chincarini 1978, Tarenghi et al. 1978). Recall that the Friedmann model is based on a uniformly distributed cosmic matter—it is evidently important that the visible matter has, on the contrary, a very lumpy distribution. We already mentioned that the structures now seen in the galaxy universe were formed by gravitational clustering from initial small seeds. In this section we describe how the present distribution of galaxies is characterized by statistical correlation functions, allowing one to compare theoretical predictions with the real galaxy universe.

Until the mid-1970's, our view of the galaxy distribution was mostly 2-dimensional: there existed a number of catalogs of galaxies which gave good basic material for the study of the galaxy distribution in the sky (e.g., Zwicky et al. 1961–1968, Nilson 1973). But for the most part, the galaxy redshifts were unknown, and thus we did not have detailed knowledge of the galaxy distances. Thus the 3rd dimension, the depth of the galaxy field, was missing. In this respect, a landmark event in the study of clustering was the IAU Symposium 79 *The large scale structure of the Universe* held in Tallinn, Estonia, September 1977. There the redshift entered the scene as a distance indicator (via the Hubble law) and started to reveal the complex spatial distribution of galaxies.

In the Center for Astrophysics Redshift Survey John P. Huchra and Margaret J. Geller complemented the Zwicky and Nilson catalogues. They obtained the red-shifts and other information of about 15 000 nearby galaxies (Davis and Peebles 1983, Geller and Huchra 1989, Huchra et al. 1999). This survey confirmed that galaxies lie preferentially on surfaces surrounded by low-density regions (Kirshner et al. 1981, de Lapparent et al. 1986, 1988). These low density regions have typically only 20% of the mean density, with the typical diameters of about 70 Mpc. One of the largest sheets of galaxies near to us is called "the Great Wall", with dimensions of $5 \times 85 \times 240$ Mpc! It is, however, not unique; similar "walls" have been observed elsewhere. The picture of the galaxy distribution resembles cells of a honeycomb as can be seen in the cross-section view of Figure 14.11 from Shectman et al. (1996).

Many additional redshift surveys have been carried out. More than 26,000 red-shifts were measured in the Las Campanas Redshift Survey, while the Anglo-Australian Two Degree Field (2dF) Galaxy Survey reached about 220,000 galaxies. The Sloan Digital Sky Survey (SDSS) has more than a million galaxy redshifts. This survey was carried out by an international consortium with members from the United States, United Kingdom, Germany, Japan, and Korea. It used a rather

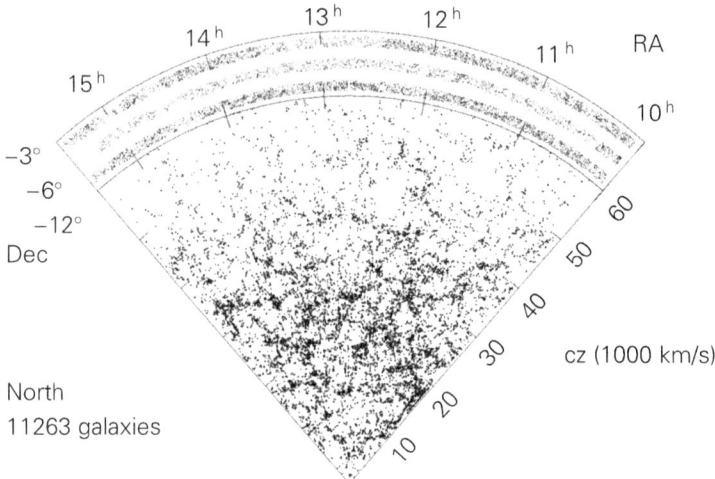

Figure 14.11. The distribution of galaxies in a slice of the universe, observed in the Las Campanas Redshift Survey. Redshift is used as the radial distance measure. Shectman et al. 1996.

modest sized 2.5 meter telescope equipped with a state-of-the-art spectrograph which could measure simultaneously the redshifts of 640 galaxies in one exposure. Also other surveys have provided unprecedented new data for analyzing galaxy distributions.

To quantify the clustering we need a quantitative parameter. Most common is the **two-point correlation function** ξ. Assume two small volumes ΔV_1 and ΔV_2 separated by a distance r_{12}. If the average number density of galaxies is n, the probability of finding a galaxy in ΔV_1 is $n\Delta V_1$. Similarly, the probability of finding a galaxy in ΔV_2 is $n\Delta V_2$. If the galaxy positions are uncorrelated, the joint probability of finding a galaxy in ΔV_1 and another galaxy in ΔV_2 is the product $n^2\Delta V_1\Delta V_2$. However, if the galaxy positions are correlated, the joint probability is

$$\Delta P = n^2[1 + \xi(r_{12})]\Delta V_1\Delta V_2. \tag{14.43}$$

where $\xi(r_{12})$ is the two-point correlation function. If $\xi(r_{12}) > 0$, there is an excess of neighbors around volumes ΔV_1 and ΔV_2 at separation r_{12}. In the reverse, $\xi(r_{12}) < 0$ would indicate that galaxies would tend to avoid each other at that distance. Thus $\xi(r)$ is a quantitative measure of clustering.

Another way of defining correlation is by using the fluctuations $\bar{\rho}\delta$ in the number of galaxies about the mean value $\bar{\rho}$. The number per unit volume is then

$$\rho = \bar{\rho}(1 + \delta)$$

By definition $\langle \bar{\rho}\delta \rangle = 0$.

The two-point correlation function is effective for detecting spherical clusters and voids but not so efficient for sheets or filaments of galaxies (Gott et al. 1979, Dekel and Aarseth 1984). A better description of correlations in this type of structures requires considering the displacement r_{12} in three perpendicular directions separately. For example in sheets of galaxies $\xi(r_{12}) > 0$ when r_{12} lies along the sheet but $\xi(r_{12}) < 0$ is a possibility for r_{12} perpendicular to the sheet. Instead of the magnitude separation r_{12}, we should use the separation vector $\mathbf{r} - \mathbf{r}'$.

The correlation function $\xi(\mathbf{r} - \mathbf{r}')$ is now defined (for $\mathbf{r} \neq \mathbf{r}'$) by

$$\langle \bar{\rho}\delta(\mathbf{r})\bar{\rho}\delta(\mathbf{r}') \rangle = \bar{\rho}^2 \xi(\mathbf{r} - \mathbf{r}') \tag{14.44}$$

where the two volumes are at positions $\mathbf{r}$ and $\mathbf{r}'$, separated by $\mathbf{r} - \mathbf{r}'$.

Another alternative way of describing the correlation region is by the use of wavelength with a wave vector $\mathbf{k}$. Substitute $\delta(t, \mathbf{r})$ from Equation (14.5) into Equation (14.44):

$$\xi(\mathbf{r} - \mathbf{r}') = \langle \delta(\mathbf{r})\delta^*(\mathbf{r}') \rangle = \left\langle \sum_k \sum_{k'} \delta_\mathbf{k}\delta_{\mathbf{k}'}^* e^{-i\mathbf{k}'\cdot\mathbf{r}'} e^{i\mathbf{k}\cdot\mathbf{r}} \right\rangle$$

The density excess $\delta(\mathbf{r}')$ must be a real number. Thus we have replaced $\delta(\mathbf{r}')$ by its complex conjugate $\delta^*(\mathbf{r}')$. We may rearrange the exponents by using the identity

$$\mathbf{k}\cdot\mathbf{r} - \mathbf{k}'\cdot\mathbf{r}' = (\mathbf{k} - \mathbf{k}')\cdot\mathbf{r} + \mathbf{k}'\cdot(\mathbf{r} - \mathbf{r}').$$

Then

$$\xi(\mathbf{r} - \mathbf{r}') = \left\langle \sum_k \sum_{k'} \delta_\mathbf{k}\delta_{\mathbf{k}'}^* e^{i(\mathbf{k}-\mathbf{k}')\cdot\mathbf{r}} e^{i\mathbf{k}'\cdot(\mathbf{r}-\mathbf{r}')} \right\rangle$$

To calculate the volume average, we integrate every term of the above double summation over the volume, taking account of the periodic boundary conditions. First fix k' and take the terms with different k, and considering only integration over one of the spatial directions, say, the x-axis. We get terms of the type

$$\text{Re}(e^{i(k-k')x}) = \cos(k - k')x = \cos kx \cos k'x + \sin kx \sin k'x.$$

After integration the averaged terms become

$$\frac{1}{2\pi}\left[\int_{-\pi}^{\pi} \cos kx \cos k'x\,dx + \int_{-\pi}^{\pi} \sin kx \sin k'x\,dx \right] = \delta_{k,k'}$$

where $\delta_{k,k'}$ is the Kronecker delta function: $\delta_{k,k'} = 0$ if $k \neq k'$ and $\delta_{k,k'} = 1$ if $k = k'$.

More generally, terms contributing to the sum are those with $\mathbf{k} = \mathbf{k}'$. Thus only one term remains from each k summation. Write the remaining summation over k' as:

$$\xi(\mathbf{r} - \mathbf{r}') = \sum_{k'} \langle |\delta_{\mathbf{k}'}|^2 \rangle e^{-i\mathbf{k}' \cdot (\mathbf{r} - \mathbf{r}')}$$

Now let the volume become arbitrarily large so the summation can be replaced by an integral over all k states (see Box 13, noting that we drop the primes here, and simply use $\mathbf{r}$ in place of $\mathbf{r} - \mathbf{r}'$):

$$\xi(\mathbf{r}) = \frac{V}{(2\pi)^3} \int \langle |\delta_{\mathbf{k}}|^2 \rangle e^{-i\mathbf{k} \cdot \mathbf{r}} d\mathbf{k}$$

Comparing the above with Equation (B13.21), the correlation function is the Fourier transform of the power spectrum $P(t, k)$, as defined in Equation (14.9):

$$P(t, k) = \langle |\delta_{\mathbf{k}}|^2 \rangle = \frac{1}{V} \int \xi(\mathbf{r}) e^{i\mathbf{k} \cdot \mathbf{r}} d\mathbf{r}$$

Usually one can assume that there is no preferred direction in space, i.e., $P(k)$ is isotropic with spherical symmetry. The integral is then most conveniently evaluated in spherical polar coordinates. Since $\xi(r)$ is a real number, we take the real part of $\exp(i\mathbf{k} \cdot \mathbf{r}) \to \cos(kr \cos \theta)$; the angle θ is measured from the direction of the $\mathbf{k}$ vector. Thus

$$\xi(r) = \frac{V}{(2\pi)^3} \int P(k) k^2 dk \int_0^{2\pi} d\phi \int_0^{\pi} \cos(kr \cos \theta) \sin \theta d\theta. \tag{14.45}$$

Changing to the variable $y = kr \cos \theta$ we get

$$\xi(r) = \frac{V}{(2\pi)^2} \int P(k) k^2 dk \int_{-kr}^{kr} \cos y \frac{dy}{kr}$$

$$= \frac{V}{(2\pi)^2} \int P(k) k^2 dk \frac{\sin y}{kr} \bigg|_{-kr}^{kr} = \frac{2V}{(2\pi)^2} \int P(k) \frac{\sin kr}{kr} k^2 dk. \tag{14.46}$$

It is found in observations of galaxies that typically

$$\xi(r_{12}) \approx \left(\frac{r_{12}}{r_0} \right)^{-\gamma} \tag{14.47}$$

where $\gamma > 0$. As long as $r_{12} < r_0$ there is strong correlation or clustering. Thus r_0 is called the correlation length. In galaxy surveys typically $r_0 \cong 7$ Mpc and $\gamma \cong 1.8$.

It has been hoped that on large scales (say >50 Mpc) the correlation function can detect the cross-over to homogeneity (the size of the "homogeneity cell"), and indeed there have been results in this direction. For example, when calculated from the Las Campanas redshift survey (LCRS) from the 1990s, for $r_{12} \geq 40$ Mpc the correlation function oscillates weakly around zero, suggesting that there are only weak signs of galaxy clustering and voids at these large scales (see Figure 14.11). However, the LCRS survey covers a quite narrow slice of the sky (1% of it), which makes it difficult to measure the correlation function on large scales. The modern SDSS survey covers about ¼ of the sky and has the deepness of about 1 000 Mpc, making it a much better data base for studying the spatial distribution.

Using another kind of correlation function, the so-called conditional density $\Gamma(r)$ function (which was introduced to extragalactic studies by Pietronero (1987) and measures the behavior of number density as a function of distance as properly averaged over all sample galaxies), Hogg et al. (2005) found for the deep SDSS Luminous Red galaxy sample that $\Gamma(r)$ is a power-law $r^{-\gamma}$ corresponding to $\gamma \approx 1$ in the interval 1–35 Mpc. On scales 35–100 Mpc there is a deviation from the power law, and within 100–140 Mpc the $\Gamma(r)$ function (i.e., density) achieves a constant value. This was interpreted as a detection of the homogeneity scale. However, even with the big SDSS survey it may be difficult to detect reliably the homogeneity scale. Sylos-Labini et al. (2009b) noted that while up to 40 Mpc galaxy structures have well-defined power-law correlations, on larger scales it is not possible to consider whole sample averages as useful statistical descriptors. This means that the density fluctuations are too large in amplitude and too extended in space to be self-averaging on such large scales inside the sample volumes. Sylos-Labini et al. concluded that the galaxy distribution is inhomogeneous up to the largest scales ($r \approx 140$ Mpc) probed by the SDSS samples.

In a *Nature* review, Wu, Lahav and Rees (1999) gave an apt summary of the situation:

"The Universe is inhomogeneous—and essentially fractal—on the scale of galaxies and clusters of galaxies, but most cosmologists believe that on larger scales it becomes isotropic and homogeneous."

It is also true that even now we are still not quite certain about the large scale beyond which the lumpy galaxy universe looks really smooth.

Within the integral of Equation (14.46) the function $(\sin kr)/kr$ oscillates with decreasing amplitude as r increases. At the same time for k at small scales, $P(k)$ decreases. Thus the main contribution to the integral from the function $(\sin kr)/kr$ comes when kr is small; i.e., $(\sin kr)/kr \to 1$. The limit of integration may be taken as $k = 0$ and $k = r^{-1}$. With these approximations the power spectrum

$$P(k) \propto k^n, \quad n < 0, \ n \neq -3 \tag{14.48}$$

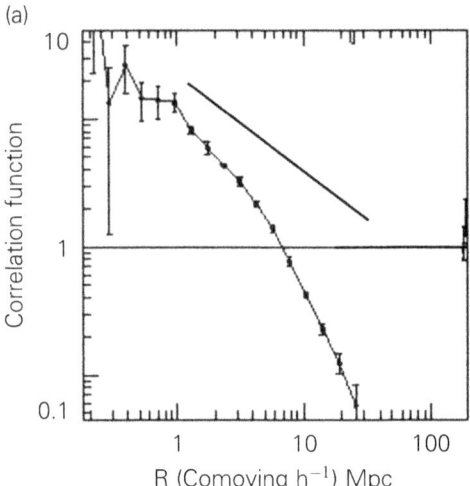

(a)

Correlation function

R (Comoving h^{-1}) Mpc

Figure 14.12. The observed correlation function and power spectrum in the Las Campanas redshift survey (LCRS). The top figure (a) shows the observed LCRS ξ for intermediate-to-large scales. The dots and error bars denote ξ for the combined North and South Cap sample, the dashed line for the Northern Cap sample alone and the dotted line for the Southern Cap sample alone. Finally, an approximate power law straight line is shown (Tucker et al. 1997). The bottom figure (b) shows the deconvolved LCRS power spectrum compared to the power spectra of three samples drawn from the combined SSRS2 and CfA2 redshift surveys. (Lin et al. 1995).

which corresponds to the correlation function

$$\xi(r) \propto r^{-(3+n)}. \tag{14.49}$$

In observations $n \approx -1.5$ and $\gamma \approx 1.5$, confirming the approximate correspondence (see Figure 14.12).

It is usual to express the power spectrum in another form. Differentiating the correlation function with respect to the logarithmic interval of k:

$$\frac{d\xi}{d \ln k} = k \frac{d\xi}{dk} \approx \frac{V}{2\pi^2} P(k) k^3 \equiv \Delta_k^2$$

when $(\sin kr)/kr \approx 1$ in Equation (14.46). The newly defined quantity Δ_k^2 represents the relative strength of density fluctuations in a logarithmic bin around the wavenumber k. We see the significance of this quantity Δ_k^2 from the following considerations.

Considering a region of radius $R \cong k^{-1}$ with a relative over-density δ_R:

$$\rho(\mathbf{r}) = \rho_0(1 + \delta_R) \tag{14.50}$$

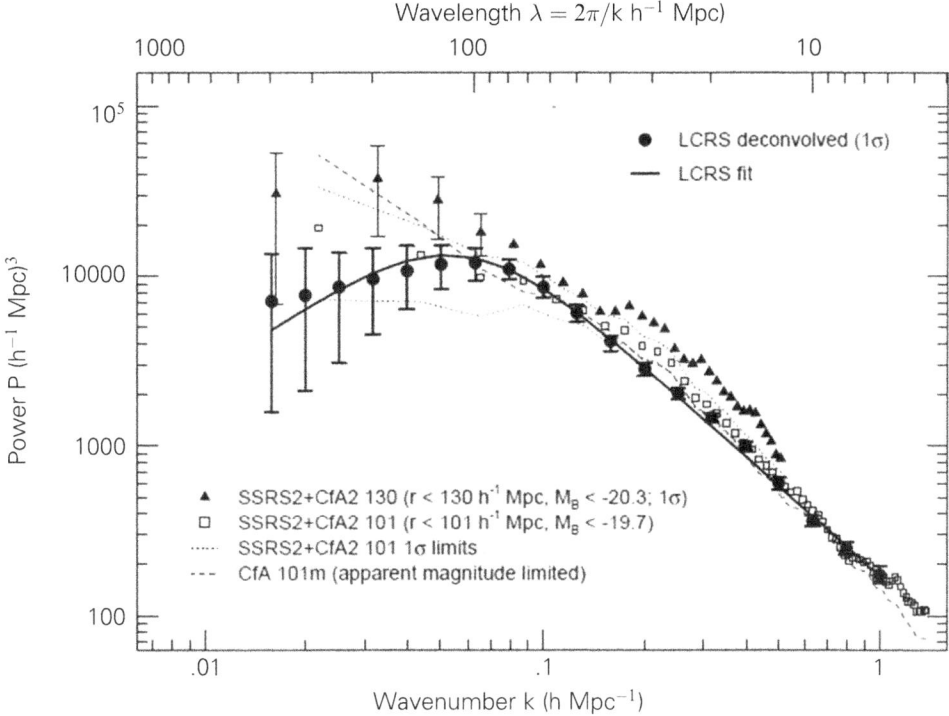

Figure 14.12. (Continued)

where $\rho(\mathbf{r})$ is the mean density at point $\mathbf{r}$, averaged over spheres of radius R, and ρ_0 is the overall average density of galaxies. If we draw such a sphere at different places $\mathbf{r}$ and calculate the average, we must get $\langle\delta_R\rangle = 0$. However, the variance $\langle\delta_R^2\rangle$ is different from zero and measures the lumpiness of the galaxy distribution in scale R. The quantity Δ_k^2 is a measure of lumpiness in the scale k^{-1}, i.e., it is essentially the same as $\langle\delta_R^2\rangle$. Thus we may write

$$\langle\delta_R^2\rangle \approx \Delta_k^2. \tag{14.51}$$

Customarily one compares observational samples as well as theoretical calculations at a definite scale R. This has been chosen $R \cong 8h^{-1}$ Mpc when $H_0 = 100\,h$ km s^{-1} Mpc^{-1}; for $h = 0.71$, $R = 11$ Mpc. Then define

$$\sigma_8 = \langle\delta_R^2\rangle^{1/2}. \tag{14.52}$$

We find typically in observations $\sigma_8 \cong 0.84$. This means that structures in this scale are just about 1.84 times the mean density. In contrast, collapsed bodies have radii that are one half of the maximum radii, i.e., the mean density is 8 times greater than when the expansion of the body came to a halt. At that time the mean density of the body was already something like 5 times greater than the mean density. This makes a density contrast of $5 \times 8 = 40$. However, the contrast increases as the universe expands, and the value is about 200 by the time the body has virialized and has become a bound system. Recall our earlier discussion of the factor of 200.

The density excess $\delta_R \cong 0.84$ (i.e., factor of 1.84 contrast) in the scale $R \cong 11$ Mpc signifies that structures of this scale will probably never become virialized. A detailed discussion of the Fourier analysis of density fluctuations is given in the book by Peebles (1980, 1993) and Peacock (1999) and Martinez and Saar (2002).

The above discussion was mainly based on the 2-point correlation function which is the usual approach to the spatial fluctuations in the number density of galaxies, since it was pioneered by Totsuji and Kihara (1969) and then developed by Peebles and his collaborators. We mentioned above also the (conditional density) $\Gamma(r)$ function which can give complementary information on the galaxy distribution in situations and for samples where the $\xi(r)$ function encounters problems (e.g., when the spatial distribution cannot be described as fluctuations around a well-defined average density). A review of the theory and use of the $\xi(r)$ and $\Gamma(r)$ functions, and how these two functions are connected, may be found in Chapters 10 and 11 of Baryshev and Teerikorpi (2012). For an extensive treatment of questions related to statistical physics of cosmic structures, see the book by Gabrielli et al. (2005).

14.11   The CMB and the baryon acoustic oscillation (BAO) spectrum

As was mentioned in Section 14.4, we will now discuss how the image of the acoustic sound crossing scale at the time of matter-radiation separation is imprinted in the matter density, and later shows up in the distribution of galaxy density. Even though most galaxies are expected to form in dark matter centers which attract baryons, also baryon shells around these centers should show up in galaxy distributions as an approximately 1% excess over the mean background density of the corresponding distance scale and with about 150 Mpc radius at the present epoch.

The initial length scale of the baryon acoustic oscillations is basically obtained from the sound horizon at the recombination epoch ($z = 1\,100$). The size of the corresponding structure in the galaxy distribution increases along with expanding space.

The BAO imprints in the matter distribution should be present in the galaxy correlation function. Because the prediction is rather straightforward and gives a well-defined spatial scale, the search for the BAO is an important test of the standard model. In addition, this length scale serves as a standard ruler which can be used e.g., to derive the value of the Hubble constant (Beutler et al. (2011) derived $H_0 = 67 \pm 3$ km/s/Mpc).

Indeed, this feature is seen both in the power spectrum analysis and in the correlation function of the largest galaxy survey samples (Figures 14.12, 14.13 and 14.14). The existence of the preferred distance scale, a standard ruler $AB(z = 0)$, which in this case has increased along with expanding space, allows us to predict its angular size in the sky θ_{AB} as a function of redshift once the angular size distance $d_A(z)$ is known (Equation (9.3)):

$$\theta_{AB}(z) = [AB/(1+z)]/d_A(z) = AB/d_M(z), \quad \text{where } d_M(z) = d_A(z)(1+z). \quad (14.53)$$

Here the quantity d_M is called the co-moving distance of the standard ruler. When the usual angular size distance d_A allows us to calculate the angular size of a true unchanging ruler as seen at different redshifts, the co-moving distance does the same for a ruler whose size varies along with the universe (hence the factor $1 + z$). Especially, if we find a co-moving structure at low redshifts, we can calculate its expected angular size at the recombination as $\theta_{AB}(z_{rec}) = AB/d_M(z_{rec})$.

The co-moving distance is calculated as an integral over the light path from the ruler to us (Equation (9.4)):

$$d_M = \int_0^z \frac{dz'}{H(z)'} \quad (14.54)$$

and the Hubble parameter is obtained by dividing Equation (8.14) by $R(t)^2$:

$$H(z)^2 = H_0^2 \left[\Omega_M(1+z)^3 + \Omega_K(1+z)^2 + \Omega_V\right]. \quad (14.55)$$

Here we have defined the curvature density parameter

$$\Omega_K(t) = -\frac{c^2 k}{H^2(t)R^2(t)}$$

which could also be defined by

$$1 - \Omega_K = \Omega_M + \Omega_V. \quad (14.56)$$

Thus it measures the deviation from flatness in the three spatial dimensions.

Thus using the baryon acoustic oscillations as standard rulers we may measure cosmological distances (Equation (14.53)). These distances trace the expansion history of the universe (Equation (14.54)), which in turn depends on the dark energy properties of the cosmological model (Equation (14.55)). Altogether this is a very promising method of establishing the values of Ω_M and Ω_V.

Although we emphasize the angular size of the standard ruler in the sky in our discussion, note from Figure 14.13 that the value of Ω_M may also be determined from the size of the acoustic bump. Eisenstein et al. (2005) calculated the large-scale correlation function measured from spectroscopic observations of 46,748 luminous red galaxies from the Sloan Digital Sky Survey. The survey covers over 3816 square degrees and $0.16 \ll 0.47$ to study large-scale structure. They found a well-detected peak in the correlation function at a separation which is an excellent match to the expected shape and location of the imprint of the recombination-epoch baryon acoustic oscillations (BAO) on low-redshift clustering of matter (Figure 14.14). The detection confirms a firm prediction of standard cosmological theory. The physical size of the BAO feature in co-moving coordinates (in today's measure) is 154.7 Mpc. According to Eisenstein et al 2005, this physical size plus omegas near the concordance values ($\Omega_M = 0.3$ and $\Omega_\Lambda = 0.7$, a flat universe with $h = 0.7$) produce features at the angular scales seen in the CMB observations. The BAO is consistent with a flat universe with dark energy and gravitating matter corresponding to the Concordance values. However, it is fair to mention that detecting and measuring features of the correlation function on such very large scales can have problems causing systematic errors in the results; it is good to make checks using larger samples (Sylos-Labini et al. 2009a).

We may now place the band of possible $(\Omega_M, \Omega_\Lambda)$ values from BAO observations in Figure 14.9. We notice that altogether only a rather small region of the $(\Omega_M, \Omega_\Lambda)$ space is allowed if all the different restrictions are satisfied as specified in the small set of nested ellipses around the intersection point. In Figure 14.9, which shows the Ω_M and Ω_Λ values corresponding to theory and observational data, the single black line is for a theoretical flat critical density universe. As noted earlier, the CMB data form the narrow diverging observational orange line (CMB). Also as previously noted, the observed values for 557 supernovae provide the set of nested elongated ellipses (SNe). Now, the broad almost vertical line corresponds to observed baryon acoustic oscillations (BAO). We see that the BAOs constrain the value of the pair of omegas via its intersection with the CMB line. The actual omegas should be within the intersection. Indeed, the BAO serves to constrain the omega values better than the supernovae, particularly after systematic errors for the supernovae are taken into consideration. The small dark gray ellipses are a fit to all these (assuming a flat universe) which is close to the concordance values used earlier ($\Omega_M = 0.27$ and $\Omega_\Lambda = 0.73$). For the BAO, from the outer edges, boundaries of progressively less

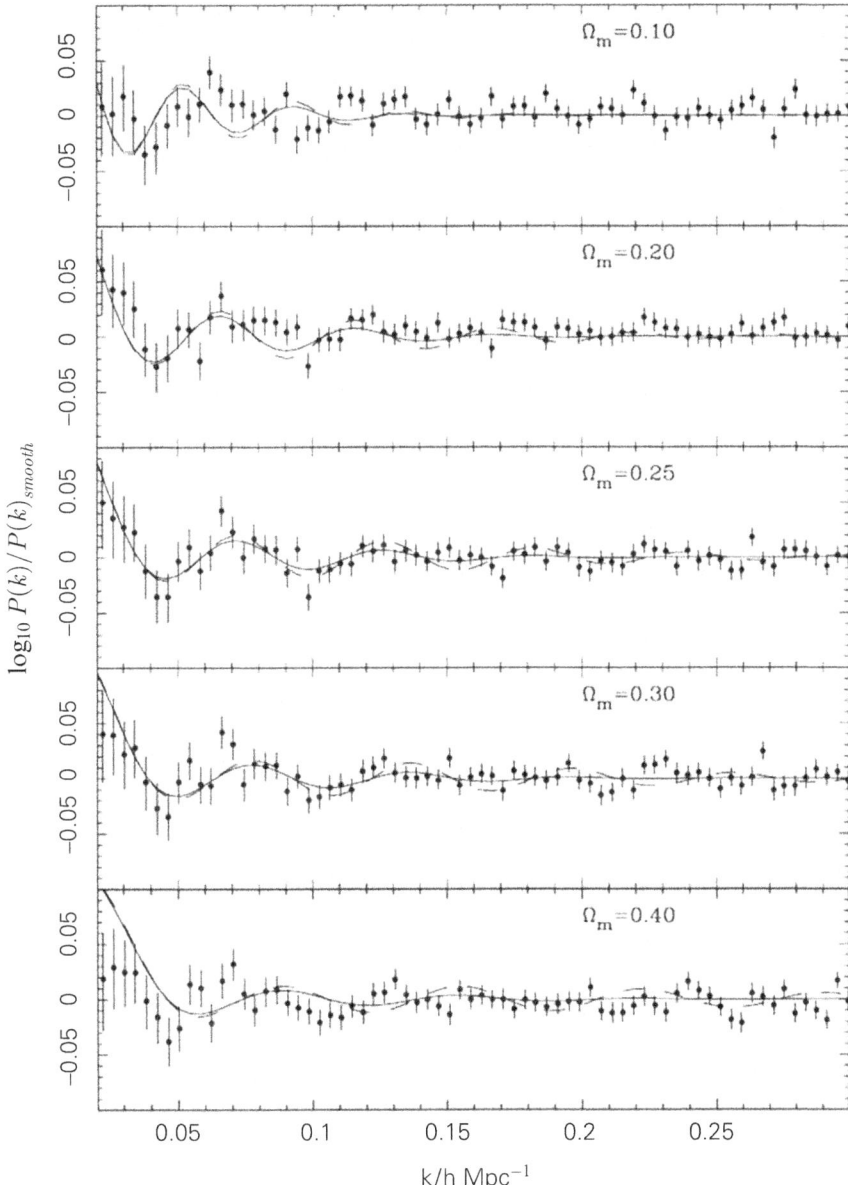

Figure 14.13. Baryonic acoustic oscillations in a galaxy survey. Ratio of the power spectra from the SDSS to the smooth cubic spline fit used to model the overall shape of the measured power spectra (filled circles with errors). Data are plotted using five flat models to convert from redshift to co-moving distance, with matter densities shown in each panel. Also in each panel are plotted the BAOs predicted by a CDM model with the same matter density, $h = 0.73$, and a 17% baryon fraction (solid lines). The observed oscillations approximate those predicted by this model for $\Omega_M = 0.2$ to $\Omega_M = 0.3$. (Percival et al. 2007a, b).

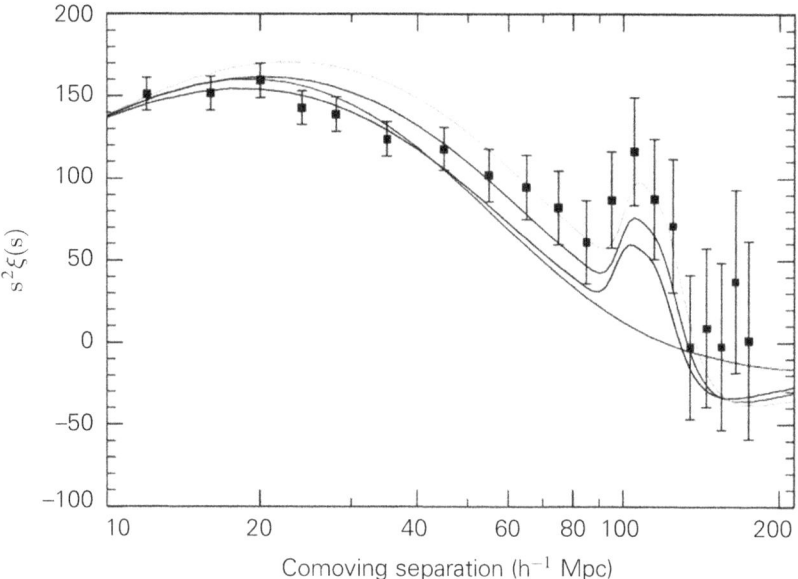

Figure 14.14. Correlation function in the SDSS galaxy survey where the galaxies are typically at the redshift $z = 0.35$ (Eisenstein et al. 2005). The lines from top to bottom model the 100 h^{-1} Mpc baryon acoustic oscillation peak at different assumed levels of total matter density: $\Omega_M h^2 = 0.12$, 0.13, 0.14, 0.024, while the baryonic matter level is taken to be $\Omega_B h^2 = 0.024$. Note that the line without dark matter does not have an acoustic peak.

dark tones correspond to 68.3%, 95.4%, and 99.7% confidence regions. A cosmologically constant dark energy ($w = -1$) has been assumed. (Amanullah 2010).

14.12   The dark energy equation of state: is dark energy density a function of time?

We now explain the significance of the parameter w. Earlier in Chapter 10.3 we described the differing equations of state for matter and radiation. These can be characterized by the ratio of pressure to energy density, $w = P/\rho c^2$. Dark energy can be similarly described. Here we follow Howell's (2011) review for surveys, values and uncertainties describing dark energy w estimates. In the present-day universe, matter between galaxies is so thin that it exerts no pressure on average. Matter has $w = 0$; it simply dilutes with volume as space expands.

As we have seen, dark energy exerts a repulsive force, so w for dark energy must be negative. If $R(t)$ is the length scale in the universe, then the time evolution of density of a component of the universe can be described by

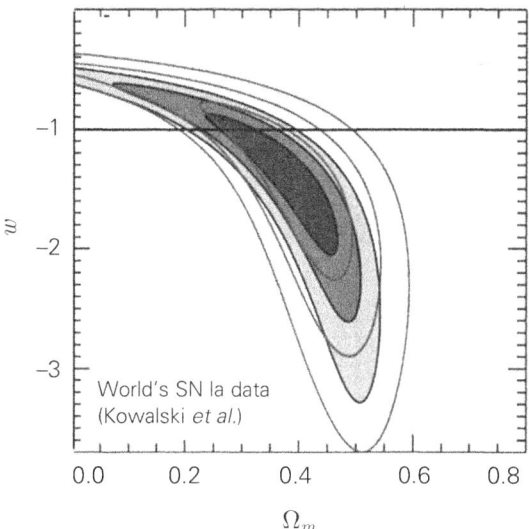

Figure 14.15. Dark energy equation of state constraints from Type Ia supernovae. The panel shows models of w made using SNe Ia (assuming a flat universe). $\Omega M - w$ statistical-only constraints are nested gray shaded lines from Kowalski et al., (2008) combining many data sets. Kowalski et al. (2008) showed that systematic uncertainties (unshaded lines) are significant. The inner to outer contours are 68.3%. 95.4%, and 99.7% confidence regions. Figure courtesy of Alex Conley adapted from Howell (2011).

$$\rho \propto R(t)^{-3(1+w)}. \tag{14.57}$$

If dark energy has $w = -1$, its energy density does not dilute with the Universe's expansion. On the other hand, if $w < -1$ the dark energy density ρ grows with time, and will ultimately create a "big rip", destroying galaxies and even atoms. Alternatively, w can also be a scalar field with $w > -1$ and then dark energy can weaken with time. This alternative is often termed "quintessence". Measurements of $\langle w \rangle$ or $w(z)$ are obtained by building a map of the history of the expansion of the Universe using SN Ia as standardized candles. To determine cosmological parameters, a Hubble diagram (distance or distance modulus versus redshift) is constructed. Cosmological model parameters are varied and compared to the observations. The comparison can be used to see if w varies with time. The results for different surveys and errors are shown in Figure 14.15 and in Table 14.1. A constant $w = -1$ seems to be consistent with the observations but errors should be reduced with more supernovae and better understanding of this and other standard candles. Results are summarized in the Table 14.1.

Table 14.1. Best fit values of $\langle w \rangle$ and error results. Table excerpted from Howell (2011). Note: For recent results using more supernovae see (http://arxiv.org/abs/1105.3470v1).

	SNSL3	CfA/ESSENCE	SDSSII	SCP
Systematic (assuming flatness)	−1.061	−0.987	−0.96	−0.997
Statistical + Systematic error	0.069	0.13	0.13	0.08
Sources	Conley, A. et al. 2011; Sullivan, M. et al. 2011	Hicken, M. et al. 2009; Wood-Vasey et al. 2007	Kessler et al. 2009	Amanullah, R. et al. 2010

14.13   Dark energy determined from gravitational lensing of the CMB

One of the important properties of the cosmic microwave background fluctuations is that they are Gaussian: the temperature values at any n points in the sky have a probability distribution which follows an n-dimensional Gaussian distribution. It means that the temperature correlation between any two points on the sky is only a function of their angular distance. This has been found to be true to a great degree, and it is likely to have been exactly true when the CMB photons start their long journey to us.

However, something may happen to the photons on the way. They travel through a universe which is not exactly homogeneous, but is full of density perturbations in different stages of clustering evolution. When photons pass by the denser regions, their light paths are bent by weak and strong gravitational lensing. Mostly it is the weak lensing that is important. Lensing deflects the CMB photon by an average of 3 arc minutes, and the deflections are coherent in the degree scales. Recently the Atacama Cosmology Telescope, a 6-meter telescope in the Atacama Desert in Chile, has observed the statistical signature of these coherent deflections. They appear as extra correlation power in the CMB power spectrum at the range $l \approx 100$ to $l \approx 1\,000$, currently seen at 4σ significance level.

The importance of the coherent deflections of the CMB photons lies in the fact that they arise from the whole travel path of the photon, and thus carry a message of the distribution of matter in difference size structures at different stages of evolution of the universe (Sherwin et al. 2011). The structure evolution depends on the dark energy parameter $\Omega_V = 1 - \Omega_M$ via Equations (9.5) and (13.90). Through detailed simulations, it has been shown that dark energy is necessary in the models that reproduce the observations. Even though the result is not as constraining as other methods described above, it is still important as a completely independent confirmation of dark energy.

14.14   Cosmic 3D space: finite or infinite?

Here we discuss recent work on the question of whether our universe is finite or infinite. It shows the importance of study of the CMB. We emphasize that we cannot simply assume that our 3D space is flat when analyzing the CMB along with other data such as the Hubble constant, the BAO, supernovae etc. Relativity and modern cosmology has clarified the problem, but not solved it yet. As we discussed earlier in this book, the true reality of nature is the 4D space-time, not space and time separately. There are many various ways to separate a 3D space from the space-time. Technically, it is made by the choice of a reference frame and a coordinate system. Any 3D space is a section of the 4D space-time, or its 3D hypersurface, and the 3D geometry depends on how space was identified. In the Friedmann theory, the 3D space is the co-moving space, which means that moving cosmic matter serves as a reference frame. This matter is uniformly distributed and isotropically expands in this space, so that the space is uniform and isotropic. The corresponding time is the proper time of the expanding matter. Co-moving space and the proper time form together the 4D space-time of the Friedmann model which is covariant, i.e., independent of the reference frame.

The most interesting example of possible spaces is the "observable space" which is the 3D section of the 4D space-time by the past light-cone. This space is associated with the real procedure of observation. In this space, matter distribution is not uniform: its density is lower nearby and higher at the remote distances which are seen at earlier stages of the cosmological expansion. Since the life-time of the universe is finite, light (or any other signal) was emitted not earlier than about 14 Gyr ago, covering a finite distance before reaching the observer at present. Therefore the distances that can, in principle be reached in observations, are finite, and the total volume of observable space is finite as well. The largest distances available for observations are near the horizon radius, ct_0, where $t_0 = 13.7$ Gyr is the age of the Universe.

Thus, we have one answer to the question in the title of this section. The 3D observable space of the universe is finite. It contains a finite amount of energy, and a finite total number of stars and galaxies as well. The largest and only observable object is usually called the MetaGalaxy.

This answer is not, however, complete. It says nothing about other 3D spaces of our space-time, and, about the co-moving space widely used in cosmological theory. Recall that Einstein's cosmological model in his 1917 paper describes the universe as an isotropic 3D co-moving space of constant positive curvature. He considered this space as similar to a 2D sphere: not only the metric, but also the structure of the space as a whole, that is its topology is an extension of the 2D model. The total volume of such a spherical space should be finite, just as the area of the surface of the 2D ball is finite. A spherical finite space is in good correspondence with the cosmological tradition that dates back to the first attempts to imagine the whole shape and size of the universe by ancient civilizations.

The Moscow theoreticians Lev D. Landau (1962 Physics Nobel Prize) and Eugeny M. Lifshitz in the now famous and widely translated *The Course of Theoretical Physics* present the finiteness of the positive-curvature space as obvious and natural. They calculate—as an application to the Friedmann solution for a positive spatial curvature expanding universe—the total volume of the space:

$$V = 2\pi^2 a_K^3, \tag{14.58}$$

where a_K is the curvature radius. They conclude that

"the space of positive curvature proves to be closed in itself, i.e., finite in volume, but, of course, not having boundaries".

However, Friedmann wrote in the popular science book *The World as Space and Time*, (Petrograd, Academia, 1923) that

"... distorted ideas have spread about the finiteness, closeness, curvature and other properties of our space which are supposedly established by the Relativity Principle".

He mentioned

"... misunderstanding repeated not only in popular-science papers and books, but also in more serious and specialist works on the Relativity Principle. I mean the notorious question of the finiteness of the Universe, i.e., of the finiteness of our physical space filled with luminous stars. It is claimed that having found a constant positive curvature of the space one can conclude that it is finite, and above all that a straight line in the Universe has a finite length, that the volume of the Universe is also finite, etc. This statement may be based either on misunderstanding or on additional hypotheses."

General Relativity uses differential geometry to describe the "local" properties of the world given by the space-time metric. The key quantities of the differential geometry are the metric tensor and—in the case of an isotropic 3D space—the curvature radius of the space. However, differential geometry says nothing about the overall shape and volume of the space, the topology of the 3D space and space-time. While there are some relations between the overall design of the space

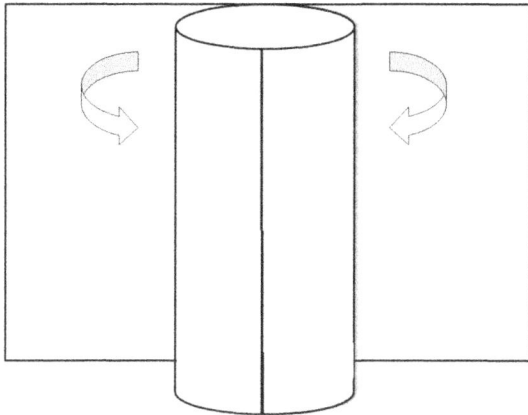

Figure 14.16. A cylinder can be made from a flat rectangular sheet. Assembling the cylinder, the two edges of the rectangle coincide.

and its curvature, these are not simple one-to-one relations. Anyway, topology of the space cannot be derived from the General Relativity theory.

To clarify, Friedmann in the above cited *The World as Space and Time*, gives a pedagogical example. The metric of the surface of a 2D infinite cylinder and the metric of a 2D plane are identical: this is the metric of the 2D Euclidean space. A cylinder can be assembled from a plane as shown in Figure 14.16. If one draws a triangle on the plane, nothing special will happen to the triangle when the sheet edges are glued together, as in Figure 14.16. The sum of the angles of the triangle will remain equal to two right angles, and the Pythagorean Theorem which holds for the Euclidean plane does not lose its validity for the cylinder. Examining the triangle, a two dimensional creature could not distinguish between inhabiting a flat sheet or a cylinder.

However, on the cylinder, there are "straight lines of finite length" (like a circle at the end of the cylinder on Figure 14.16), while there are no such lines in a plane. The cylinder, although infinite in one direction, has a finite size in the directions normal to its axis, so thus it is finite and closed in these directions.

Using the example of a plane and a cylinder, Friedmann concludes:

"Thus, the metric of the world alone does not enable us to solve the problem of the finiteness of the Universe. To solve it, we need additional theoretical and experimental investigations."

It is known from topology that 18 different versions of topology are possible for Euclidean flat metrics. For a cylinder, there is one special direction. Moving in this direction, as you exit to the left you enter from the right. In an extension of this idea, a toroidal flat finite space is possible. As you exit to the left you enter from the right, in any direction in this space. For curved spaces, the number of allowable topological realizations is infinite. For example, a spherical space may be infinite, as Friedmann explains in his second cosmological paper (1924). He writes:

"We call a space finite, if the distance between two non-coincident points in it does not exceed a certain positive number whatever this pair of points may be. Therefore, before considering the problem of finiteness of the space, we should make a convention about what points of this space will be considered different. For example, if we consider the surface of a sphere embedded in a three-dimensional Euclidean space, the points laying on the same latitude and having a longitude difference of 360 degrees may be considered as coincident; if on the contrary we regard these points different, we should get a multi-sheet spherical surface in the Euclidean space".

In this last case,

"the distance may be made infinitely large, if one attributes the points to different sheets in an appropriate way."

As for Friedmann's "additional theoretical and experimental investigations" being needed to solve the problem, now, 90 years later, no general theory has been suggested that would relate the topology of space-time or 3D space topology to its matter content in the manner that General Relativity relates the differential geometry to the energy-momentum tensor. Some aspects of this problem have been studied in a constructive way. Recall for the toroidal flat finite space that as you exit to the left you enter from the right, in any direction. Extending this to any light emitting object, a topological "ghost effect" is possible, in which multiple images of the same object may be observed in the sky. This effect happens in a topologically closed space of a finite volume. The light from a given luminous object may reach the observer not from one direction as in ordinary space but in various ways. For instance, it may be seen by rays in the opposite direction in space which passed around the space. The observer shall see the same object in two opposite directions. Needless to say, the light rays cannot be traveling for longer than the age of the universe. The problem was studied both theoretically and observationally in the 1970s by the mathematician Dmitry D. Sokoloff of Moscow University and the astrophysicist Victoriy F. Shvartsman of SAO of the Russian Academy of Sciences (1974). See also the works by Zeldovich and Novikov (1967), Sokoloff (1971), Sokoloff and Starobinsky (1975).

No such effect has yet been definitely found in optical observations. Most recently topological effects have been first searched for in the Cosmic Microwave Background (CMB) by the French theoretician Jean-Pierre Luminet and his co-workers in a 2003 article in *Nature*. Rather than discrete sources, indications of the Universe's spatial topology can be searched for in the power spectrum of the

CMB anisotropies. The CMB's anisotropy is produced by initial adiabatic perturbations which seed large scale cosmic structure. These are observed as weak variations of the CMB temperature on the sky. The "CMB sky" with these temperature
variations may be described by a sum of spherical harmonics (discussed earlier in
this book). We have already discussed how the relative strengths of the harmonics
described by the CMB power spectrum carry information about the physical properties of the universe. In addition to the relative amounts of matter and dark matter
etc., the harmonics may contain information about the geometry—both differential
and global—of the co-moving space. Topological information would be contained
in the low end of the power spectrum which corresponds to the largest angular and
spatial scales.

Recapitulating our earlier discussion in this text, the lowest harmonic is the
dipole with wave number $l = 1$ and an angular scale of $180°$. This component
from the universe itself is unobservable because of the Doppler effect dipole due
to the Solar System's motion relative the CMB. The dipole anisotropy is 100 times
stronger than the cosmological dipole so the latter is hidden by the Doppler effect.
The lowest harmonic which is observable is the quadrupole with wave number $l = 2$
and the angle scale of $90°$. Luminet interprets the WMAP data on the CMB power
spectrum to show that a quadrupole is only about one-seventh as strong as it would
be expected in an infinite flat space. The octopole with the wave number $l = 3$ and
the angle $60°$ is also weak at about 70% of the expected (in an infinite space) value.
Larger wave numbers up to $l = 900$ correspond to relatively small spatial scales
where no weakening effect was found.

Significant power loss on angular scales wider than $60°$ implies that the broadest
spatial scales are missing in the CMB. Luminet et al. (2003) suggest in their *Nature*
article that this weakening is because space itself is not big enough to support these
scales since the universe's volume is finite i.e., it is relatively small.

A quantitative model of a finite-size universe can create the peculiarities of the
WMAP data. This is a model with a 3D co-moving space which is the Poincaré
dodecahedral space (a construct long studied in topology). To mentally imagine
this space, first note that any ordinary 2D sphere may be completely covered by
12 regular spherical pentagons which fit snugly. The corner angles are all exactly
$120°$. Each is a pentagonal part of the sphere. Note that the ordinary Euclidean
pentagonal dodecahedron of solid geometry is a solid with 12 equal flat faces; but
here the faces are pieces of a spherical surface (Figure 14.17).

Now consider a 3D sphere, or a hypersphere, the 3D surface of a 4D ball. The
hypersphere is covered by 120 regular spherical dodecahedra. They fit together
snugly, and this is because their edge angles are exactly $120°$. Each of them is a
dodecahedral piece of the hypersphere. Spherical dodecahedra make up the Poincaré
dodecahedral space. It is not so easy to imagine such a space. One can use of a
special visualization software for spherical dodecahedra and the Poincaré space
which is available at http://www.geometrygames.org/CurvedSpaces

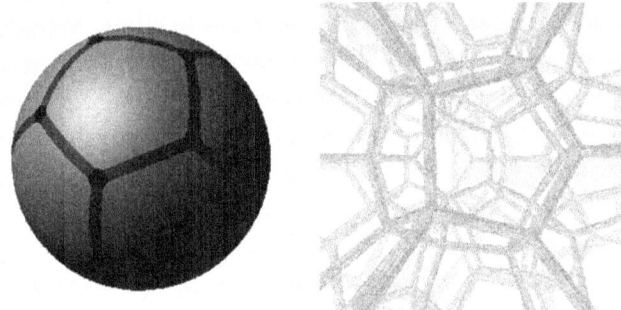

Figure 14.17. (Luminet et al. (2003). Spherical pentagons and dodecahedra fit snugly unlike their Euclidean counterparts. (a) 12 spherical pentagons tile an ordinary sphere. (b) 120 spherical dodecahedra tile the surface of a hypersphere.

The Poincaré space is a positively curved space. It has a multiply connected topology whose volume is 120 times smaller than the volume of a simply connected hypersphere of the same curvature radius. Now opposite faces of a dodecahedral block are abstractly connected to each other. When light goes out through one face, it re-enters from the face on the opposite side. Thus we are aware only of one block of the dodecahedral space.

Remarkably the dodecahedral space is globally uniform. Because of this, its geometry (as well as the CMB anisotropy) looks statistically the same to any observer wherever the observer is located in this space. This conforms to the cosmological principle. Note that "simpler" finite spaces are typically non-uniform and look different to observers at different locations.

For the finite model, since the density parameter Ω_o exceeds unity slightly, the differential geometry of the space is described by the metric of positive spatial curvature (as in Einstein's model of 1917 and Friedmann's solution of 1922). The CMB anisotropy power spectrum for the model by Luminet et al. (calculated with a special complex computer program) produces the power spectrum pattern at both low and higher harmonics, if the present-day cosmological density parameter has a special value $\Omega_0 = 1.013$. With $\Omega_0 = 1.013$, the maximal radial distances $R_U(t_0)$ at the present epoch are within the narrow interval

$$0.82 R_0(t_0) \leq R_U(t_0) \leq 1.03 R_0(t_0), \tag{14.59}$$

where $R_0(t_0) = ct_0 \approx 1.3 \times 10^{28}$ cm is the present-day value of the horizon radius, and $t_0 = 13.7$ Gyr is the cosmic age. The constraints here are the observed CMB spectrum and the cosmic age.

Briefly, Luminet et al. (2003) suggest that the co-moving space of our universe is finite, and its present-day radial size is near the present-day horizon radius. Commenting on this result, the journal *Nature* concludes:

"If confirmed, this is a major discovery about the nature of the Universe" (Ellis 2003).

Interpretation of the WMAP data searching for expected angular patterns on the sky has proven ambiguous (e.g., Cornish et al. 2004 compared to Roukema et al. 2004).

However, the Luminet picture of a finite co-moving space can be tested with other observations. What may definitely be tested is the rigid prediction of the value for the density parameter $\Omega_0 = 1.013$. Here we emphasize that the value must conform with the WMAP CMB data but also including the Hubble constant, BAO and supernova data as we have discussed earlier in this chapter.

The value of the present-day standard density parameter (summing all the cosmic energies at the present day) from the WMAP seventh year release from Equation (10.2) is $\Omega_0 = 1.0023 \pm 0.0056/0.0054$ (Jarosik, et al. 2011), for the best estimate giving 1.0079 for the one standard deviation above and 1.0135 two deviations above the best value.

We see that the 1.013 requirement is almost two standard deviations away from the nearly flat best value. Further analysis of data from the WMAP laboratory and from its successor, the Planck satellite or additional supernova data may well move the value of the parameter still closer toward unity with smaller error bars. In such a case, the concrete model with the dodecahedral space may definitely be ruled out. Right now the data do not favor the dodecahedral space model. Perhaps some other, even more sophisticated, topologies of a finite space may be true for the universe and still be consistent the CMB anisotropy spectrum.

Chapter 15

The local path to dark energy

The ΛCDM cosmology views dark energy as having constant density everywhere so that it fills the voids as well as the interiors of galaxy systems, consistent with the conception of uniform and constant vacuum density (Einstein's Λ-term). Thus it may represent, together with other relativistic components of the universe, the naturally homogeneous substance on which the Friedmann model is based.

If this is true, its local density should be identical to that inferred from the cosmological recession of distant galaxies using Ia supernovae as distance indicators and from the analysis of the anisotropies of the microwave background radiation, i.e., $\rho_V \approx 7 \times 10^{-30}$ g cm^{-3}.

Possible alternative explanations of the SNIa Hubble diagram (Chapter 9), requiring no dark energy, typically refer to phenomena on large scales. If any of them is valid and there is no dark energy (or $\Lambda = 0$), then one does not expect any true dark energy effect on much smaller scales either. For instance, if some dust extinction (unlikely) causes the extra dimming of distant supernovae, this would have no effect on local observations. But, if dark energy exists like suggested by the large-scale observations and is smooth and constant like the vacuum energy (with the equation of state $\rho_V = -P/c^2$, then it should exist also locally. According to General Relativity, gravity depends on pressure as well as density: the effective gravitating density $\rho_{eff} = \rho + 3P/c^2$ is negative for a vacuum $(= -2\rho_V)$, and this leads to repulsion ("antigravity").

Hence the study of the antigravity in our neighbourhood, and on short scales in general, can be seen as an important test of the standard cosmological model, somewhat analogous to the attempts to detect non-baryonic dark matter particles "down here" in laboratories.

15.1 A gravitating system within dark energy: the zero-gravity radius

Soon after the discovery of the universal acceleration from observations of distant supernovae, a simple question was asked (Chernin et al. 2000; Chernin 2001): at what distance from the Local Group do the gravity of its mass and the antigravity of the dark energy (if its local density equals its globally measured value) balance each other? This question, which was inspired by the properties of the expansion flow around the Local Group, was a first step towards the local study of dark energy. As a result, during the last decade several local dark energy effects have been discussed, including accelerating outflows close to and around galaxy groups. In addition to important insight obtained on the local value of dark energy and its effect on dynamical mass determinations, an older puzzling observation, the asymmetry of redshifts of galaxy group members (Byrd and Valtonen 1985) has now been explained as caused by dark energy-driven expansion of the outer parts of galaxy groups (Section 15.6).

Treating the Local Group (or generally a mass concentration) as a point mass M on the background of the antigraviting dark energy, its gravity produces the radial force $-GM/r^2$, where r is the distance from the group barycenter. The antigravity of the vacuum produces the radial force $G2\rho_V(4\pi/3)r^3/r^2 = G(8\pi/3)\rho_V r$. Here $-2\rho_V$ is the effective gravitating density of vacuum.

The radial component of motion in this gravity/antigravity force field obeys the Newtonian equation

$$\ddot{r}(t, \chi) = -GM/r^2 + r/A^2, \tag{15.1}$$

where $r(t, \chi)$ is the distance of a particle to the barycenter of the mass concentration and χ is the Lagrangian coordinate of the particle. The important constant

$$A = [(8\pi G/3)\rho_V]^{-1/2} \tag{15.2}$$

is the characteristic vacuum time and has the value $= 5 \times 10^{17}$ s (or 16 Gyr) for the Concordance value $\rho_V = 7 \times 10^{-30}$ g cm^{-3}.

It is seen from Equation (15.1) that the gravity force ($\propto 1/r^2$) dominates the antigravity force ($\propto r$) at small distances, and the acceleration is negative there. At the distance

$$R_V = \left(GMA^2\right)^{1/3} = [(3/8\pi)(M/\rho_V)]^{1/3} \tag{15.3}$$

the gravity and antigravity balance each other, so the acceleration is zero at the "zero-gravity sphere" of the radius R_V. At larger distances, $r > R_V$, antigravity dominates, and the acceleration is positive. For instance, for the Local Group mass $M \approx 2 \times 10^{12} M_\odot$ and the global dark energy density the zero-gravity distance is $R_V \approx 1.3$ Mpc.

These expressions for the antigravity force and the equation of motion are given here in terms of Newtonian mechanics. Its structure looks natural, but it actually needs justification from General Relativity. A static GR space-time for a mass embedded in the uniform vacuum gives such a justification.

The above set of the Newtonian spherically symmetrical equations has an exact analogue (a spherically symmetric metric) in General Relativity (e.g., Chernin et al. 2006). There is a well-known solution for a point-like body on the vacuum background:

$$ds^2 = c^2 F(r)dt^2 - r^2 d\Omega^2 - F(r)^{-1}dr^2, \tag{15.4}$$

where

$$F(r) = 1 - 2GM/(c^2r) - (r/A)^2. \tag{15.5}$$

This time-independent metric describes a spherically-symmetrical static space-time (also see Box 14). Identifying the constant parameters M and A here with the mass of the central body and the vacuum time/length, respectively, we may use this metric for the local cosmology. The local flow particles move along geodesics in the space-time of Equations (15.4) and (15.5).

In the limit of small deviations from the Newtonian (Galilean) space-time, the metric (Equation (15.5)) takes the form:

$$(F(r))^{1/2} \approx 1 - GM/(c^2r) - 1/2(r/cA)^2 = 1 + U(r)/c^2, \tag{15.6}$$

where $U(r)$ is the gravitational potential of the Newtonian theory with vacuum background.

Characteristics of zero-gravity spheres: If we try to calculate the zero-gravity radius around a point in a **homogenous** distribution (a zero-gravity sphere), this radius will increase directly proportional to the radius of the considered matter sphere. This means that one cannot ascribe physical significance to the zero-gravity radius within a fully uniform universe—every point is as it were on the surfaces of a great number of zero-gravity spheres of arbitrarily different sizes, not feeling any force. But the things change once some matter structure appears in the universe. A density enhancement does not appear alone, but together with the zero-gravity sphere around it.

The zero-gravity sphere for a point mass M has special significance in an expanding universe. A light test particle at $r > R_V$ experiences an acceleration outwards relative to the point mass. If it has even a small recession velocity away from M, it participates in an accelerated expansion (Teerikorpi et al. 2005).

The total amount of dark energy within the zero-gravity sphere is one half of the rest energy of the central mass concentration as can be seen from the formula for R_V (Equation (15.3)).

We may also consider two identical isolated point masses M. For this system the zero-gravity distance, where the two masses have zero-acceleration relative to their center-of-mass, is $R_{EE} = 2^{1/3}R_V \approx 1.26\,R_V$, i.e., somewhat larger than the zero-gravity distance for one particle. This radius also defines an "equal energy" sphere—in such a sphere around the mass M the total energy associated with the central mass and the corresponding dark energy are equal. The significance of this radius for two identical point masses is the same as the zero-gravity radius for a test particle. Separated by the distance $r > R_{EE}$ they experience outward acceleration. We may add other mass points sparsely enough so that their equal energy spheres do not intersect. Then if the spheres are originally at rest relative to each other or are recessing, the system will scatter with accelerating expansion (Teerikorpi et al. 2005).

These examples illustrate in simple situations the general result that in vacuum-dominated expanding regions perturbations do not grow. They also lead one to

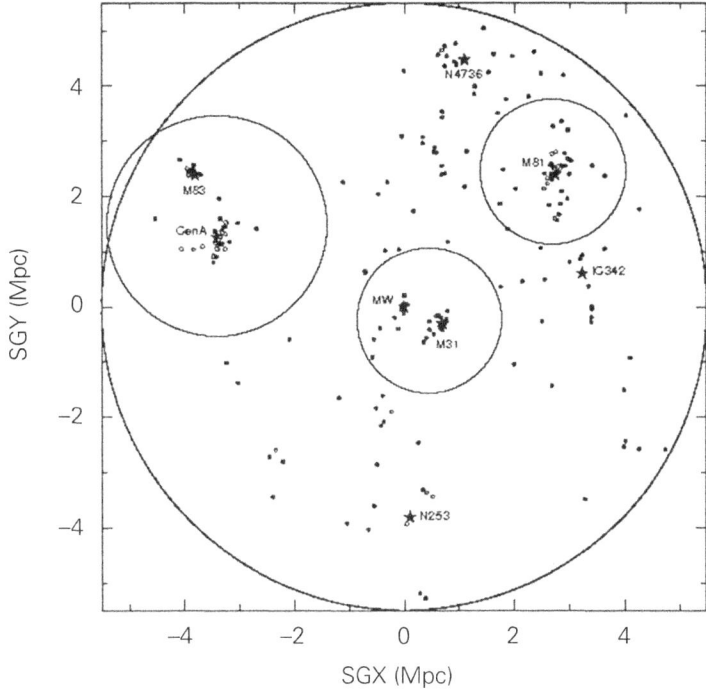

Figure 15.1. Approximate zero-gravity spheres around the Local Group (at the center) and two nearby galaxy groups. The radii have been calculated using the masses $2 \times 10^{12}\, M_\odot$ for the LG and the M81/M82 Group and $7 \times 10^{12}\, M_\odot$ for the CenA/M83 Group (the underlying map presenting the local environment up to about 5 Mpc as projected onto the supergalactic plane is from Karachentsev et al. 2003).

consider the relation of the zero-gravity spheres to the exact Einstein–Straus (1945) solution where any local region may be described as a spherical "vacuole" embedded within the uniform distribution of matter (Chernin et al. 2006). The size of the vacuole $R_{ES}(t, M)$ increases with time in accordance with the Friedmann expansion solution $R_{ES} \propto R(t)$, and the central mass M is equal to the mass of the (non-relativistic) matter within the sphere of radius R_{ES} in the smooth matter distribution of the Friedmann universe: $M = (4\pi/3)\rho_M(t)R_{ES}(t)^3 = \text{const.}$

The metric inside the vacuole is static and the Einstein–Straus (1945) solution demonstrates how such a local non-Friedmann static cosmological model may be compatible with the Friedmann metric (and dynamics) of the global cosmological expansion.

It is interesting to see that the zero-gravity radius and the (present-time) Einstein–Straus vacuole radius are simply related by

$$R_{ES}(t_0) = (2\rho_V/\rho_M)^{1/3} R_V. \tag{15.7}$$

Table 15.1. Size-parameters for cosmic systems of different masses (the approximate radius of the optical system, the zero-gravity radius, the Einstein–Straus vacuole radius) for standard values of ρ_V and ρ_V/ρ_M.

$M/M_\odot$	R_{syst}	R_V	R_{ES}	Example
1	0.0002 pc	100 pc	–	the Solar System
2×10^6	0.1 kpc	10 kpc	–	a globular cluster
1×10^{12}	0.1 Mpc	1.0 Mpc	1.7 Mpc	a compact binary galaxy
2×10^{12}	1 Mpc	1.3 Mpc	2.2 Mpc	the Local Group
1×10^{15}	2 Mpc	10.3 Mpc	17.5 Mpc	the Virgo cluster
2×10^{15}	5 Mpc	13.0 Mpc	21.1 Mpc	the Coma cluster

For instance, for the ratio $\rho_V/\rho_M = 0.7/0.3$ one obtains $R_{ES}(t_0) = 1.67 R_V$ and for $\rho_V/\rho_M = 0.8/0.2$ the relation is $R_{ES}(t_0) = 2R_V$. Physically, the Einstein–Straus vacuole can be seen as the volume from which the gravitational instability has gathered the matter to form the central mass concentration.

A picture is theoretically possible in which vacuoles of various sizes and masses fill almost all cosmic space without intersecting each other. This highly non-uniform cosmological model could describe the regular global expansion in terms of the relative motions of discrete masses that are in the centers of the vacuoles, while the contribution of the uniform matter is negligible. The masses move apart from each other on the uniform dark energy background, in accordance with the Hubble law and the expansion factor given by the Friedmann theory. Because the zero-gravity radius is well inside the vacuole in such a model, each vacuole is "repelled" from every other one and the expansion is generally accelerating.

Figure 15.1 shows the map of our local extragalactic environment up to about 5 Mpc, together with the approximate zero-gravity spheres[1] drawn around the Local Group and two nearby galaxy groups. The spheres do not intersect. This suggests that the groups are presently receding from each other with acceleration. We can calculate an interesting lower limit. If the local dark energy density is, say, 10% of the global one, then each radius is $10^{1/3} \approx 2.5$ times larger and there would be deceleration instead.

Table 15.1 gives the zero-gravity and the Einstein–Straus radii for a few astronomical systems of different masses, for the standard values of ρ_V and ρ_V/ρ_M. We see that for stars, star clusters and individual galaxies (even for tight binary galaxies)

---

1   Chernin et al. (2004) calculated the zero-gravity surface around the Local Group, dominated by the Milky Way and M31 pair, and found that it is almost spherical and remains nearly unchanged during a 12.5 Gyr history of the LG.

the zero-gravity radius is much larger than the size of the system which is located deep in the gravity-dominated region.

15.2   Dynamical structure of a gravitating system within dark energy

For galaxy groups and clusters, the zero-gravity radius is near or within the region where the outflow of galaxies begins to be observed (e.g., around 1.5 Mpc for the Local Group). It is especially on such scales where the dynamical structure of the system and its close neighborhood can shed light on the local density of dark energy. In order to study this dynamical structure we ask what happens to the test particles (dwarf galaxies) that have left the central region of the system?

The particles move radially practically as predicted by the Newtonian equation of motion, where the forces are the gravity of the central mass and the antigravity of the dark energy. It was given as Equation (15.1).

The first integral of Equation (15.1) expresses, as usual, the mechanical energy conservation:

$$\frac{1}{2}\dot{r}^2 = \frac{GM}{r} + \frac{1}{2}\left(\frac{r}{A}\right)^2 + E(\chi), \tag{15.8}$$

where $E(\chi)$ is the total mechanical energy of a particle with the Lagrangian coordinate χ (per unit mass). Here the potential energy

$$U(r) = -\frac{GM}{r} - \frac{1}{2}\left(\frac{r}{A}\right)^2 \tag{15.9}$$

is always negative. Because of the vacuum, the gravitational potential cannot be normalized to $U = 0$ at $r = \infty$, and $U(r)$ goes to $-\infty$ in both limits of $r \to 0$ and $r \to \infty$. This means, in particular, that the trajectories with $E < 0$ are not necessarily finite. Such behavior of the potential has a clear analogue in General Relativity applied to the same problem.

The test particles of the local expansion flow move along approximately radial trajectories in the potential $U(r)$. The total energy of each particle that has escaped from the gravity potential well of the central concentration must exceed the maximal value of the potential U:

$$E > U_{\max} = -\frac{3}{2}\frac{GM}{R_V}. \tag{15.10}$$

It is quite convenient to normalize our equations to the zero-gravity distance R_V and consider the Hubble diagram with normalized distance and velocity (x- and y-axes): $x = r/R_V$ and $y = V/H_V R_V = V/V_V$ (the quantity V_V is formally the speed of the vacuum flow at the distance R_V). Then radially moving test particles will move along curves, which depend only on the constant total mechanical energy E of the particle (Teerikorpi et al. 2008):

$$y = x\left(1 + 2x^{-3} - 2\alpha x^{-2}\right)^{1/2}. \tag{15.11}$$

Here the quantity $H_V \equiv 1/A$ is a physical constant directly related to the dark energy density ρ_V (or the cosmological constant), the "vacuum Hubble constant" (see Section 8.3 and Equation (15.14)). The factor α parameterizes the energy, so that

$$E = -\alpha GM/R_V. \tag{15.12}$$

The present position of the particle on its own curve depends on the initial conditions (distance and time), but each curve has a velocity minimum at $x = 1$ (i.e., $r = R_V$).

The energy with $\alpha = 3/2$ is special: it is the minimum energy that still allows a particle initially below $x = 1$ to reach this zero-gravity border (and if the energy is slightly larger) to continue to the dark energy-dominated region $x > 1$, where it starts accelerating. In the ideal case one does not expect particles with $x > 1$ below this minimum velocity curve. If one changes the values of the parameters M and ρ_V (or A), the $y - x$ curves do not change, but the normalized positions of observed test particles do change as R_V and H_V change.

Figure 15.2 shows different regions in the normalized Hubble diagram. Below $r = R_V$, we have indicated the positive minimum velocity curve and its negative symmetric counterpart. It is between these curves where galaxies have energies below the escape energy $-(3/2)\, GM/R_V$, slowly moving outwards ($V > 0$) or falling back ($V < 0$) towards the group. The interior of this region defines the group as a bound entity: a galaxy will not escape beyond R_V unless it obtains sufficient energy from an interaction.

At $R = R_V$, the minimum velocity curve reaches zero after which it starts increasing. It defines the lower envelope of the permitted locations of outflowing galaxies in the Hubble diagram. In practice it seems that the isoenergy curve $E = 0 = \alpha$ is an approximate upper envelope, characterizing the highest-velocity tail of galaxies escaping from the group.

In this normalized representation the straight diagonal line $y = x$ gives the "vacuum" flow when dark energy is fully dominating.[2] It corresponds to the "vacuum" Hubble constant H_V, and is asymptotically approached by the out-flying particles beyond $x = 1$.

---

2   This limit is described by de Sitter's static solution. As is well-known, this solution has the metric of Equation (15.4) with $F(r) = 1 - (r/(cA))^2$. The space-time of de Sitter's solution is determined by the vacuum alone, which is always static itself, and the vacuum completely controls both space-time metric and any geodesic in this space-time.

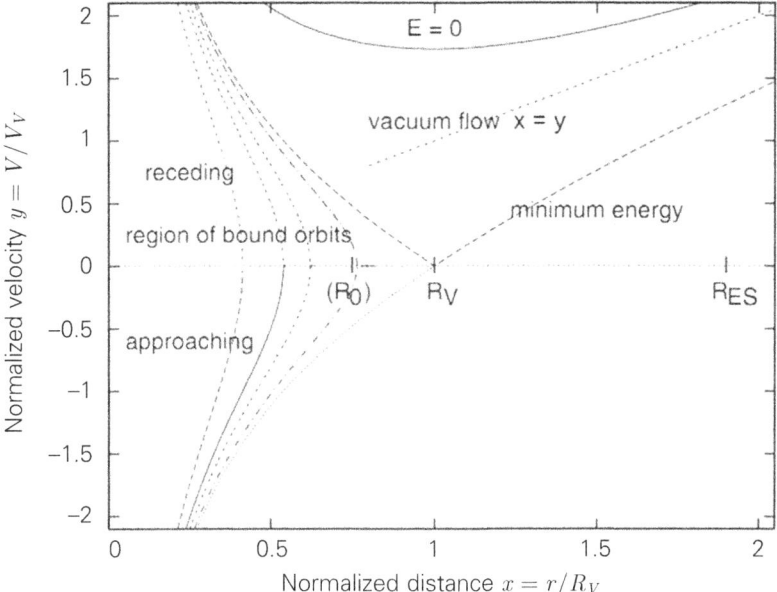

Figure 15.2. Diagram showing different regions in the normalized Hubble diagram around a mass concentration in the vacuum. In the region of bound orbits a dwarf galaxy cannot move into the region of the vacuum flow unless it receives extra energy as a result of an interaction with other galaxies. Adapted from Teerikorpi et al. 2008.

Indeed, asymptotically, when r goes to infinity and the vacuum dominates entirely, the solution of Equation (15.8) is exponential (see Section 8.3)

$$r(t, \chi) \propto e^{(t/A)}, \tag{15.13}$$

so that the linear velocity-distance law, $\dot{r} = r/A$, appears with the constant expansion rate

$$H_V \equiv 1/A = (8\pi G \rho_V/3)^{1/2} = 61.0 \times (\rho_V/7 \times 10^{-30}\,\mathrm{g/cm^3})^{1/2}\,\mathrm{km/s/Mpc}. \tag{15.14}$$

The global Hubble relation is given by the line with the somewhat steeper slope $H_0/H_V = (1 + \rho_M/\rho_V)^{1/2} \approx 1.2$. Correspondingly, the vacuum Hubble time scale $T_V = 1/H_V$ is larger than the global Hubble time by the same factor $(1 + \rho_M/\rho_V)^{1/2} = (\Omega_V)^{-1/2}$ for a flat universe and for ρ_V = the global dark energy density. Thus T_V is a natural upper limit for the age. In the standard model $T_V = 16 \times 10^9$ yr and the age of the universe (13.6×10^9 yrs) is about $0.85 \times T_V$.

In fact, the galaxies presently located in the range $1 \lesssim r/R_V \lesssim 1.7$ have spent in the past more than half of their time within the gravity dominated region. For the standard model, it is only at $r/R_V > 1.8$ where the galaxies have spent about equal

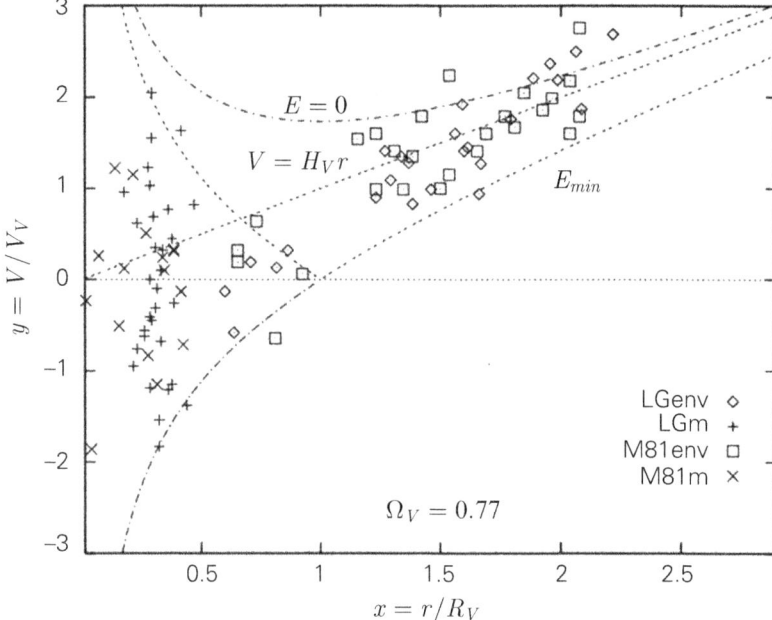

Figure 15.3. The normalized Hubble diagram for the galaxies in the environments of the LG and M81 groups, for the near standard vacuum density. The velocity-distance relation for the vacuum flow $y = x$ ($V = H_V x r$) is shown. The curve for the lower limit velocity is given below and above $x = 1$; below $x = 1$ its negative counterpart is shown. The members of the groups are found within these curves. Adapted from Teerikorpi et al. 2008.

or longer time in the dark energy dominated region. This is why the current Hubble relation has not had time to approach quite close to the vacuum flow, even though the static zero-energy border has existed since the formation of the group. Only in the distant future will it approach that relation for the dark energy dominated cosmos.

We will discuss the Local Group separately in the next section. Here we show a combined normalized Hubble diagram for the Local Group, the nearby M81 group, and their environments (Figure 15.3). The distance data (Chernin et al. 2007a, b) come mainly from high accuracy observations of the TRGB (Tip of the Red Giant Branch) using the Hubble Space Telescope as reported by Karachentsev and his team in a series of papers. The M81 group centered on the M81–M82 pair in Ursa Major is at the small distance of about 4 Mpc and is about equally massive as the Local Group. The distances and velocities refer to the center of the group. We also indicate the galaxies usually regarded as members of the groups (referred to by "$-m$", in the symbol lists of the diagrams).

We have used in this diagram the same mass $2 \times 10^{12} M_\odot$ for both the LG and the M81 group. Here the vacuum density $\rho_V = 7 \times 10^{-30}$ g/cm^3 corresponds to $\Omega_V = 0.77$ (for $H_0 = 70$) and $H_V = 60$ km/s/Mpc. The galaxy positions are displayed in the units of $R_V = 1.3$ Mpc and $H_V R_V = 78$ km/s on the x- and y-axis, respectively. This near standard vacuum density describes well the observed outflow around both groups, and the energy condition to overcome the potential well ($E > -3/2 \; GM/R_V$) is not violated in the relevant range $x = r/R_V > 1$. A rather similar normalized diagram results from the more massive Centaurus A group (its location is shown Figure 15.1).

In fact, extending the study of the outflow pattern to the much more massive Virgo cluster ($M \approx 1 \times 10^{12} M_\odot$, $R_V \approx 10$ Mpc) Chernin et al. (2010) have suggested that different galaxy systems (groups and larger) may have a universal two-part design: a quasi-stationary bound central component with an expanding outflow around it on the scales up to what would roughly correspond to the Einstein–Straus radius. The dynamic structure of such systems would reflect the gravity-antigravity interplay, and the zero-gravity radius is the main physical parameter. It forms a natural upper limit to the size of a gravitating bound system within dark energy.

15.3   Dark energy and determination of mass in systems of galaxies

The "lost gravity" effect of dark energy influences determination of mass of systems of galaxies. Especially, we consider here the modifications of the Kahn–Woltjer model which was discussed in classical terms in Section 12.7 and the Virial Theorem, following the treatment in Chernin et al. (2009). Table 15.1 shows that significant effects are expected only for systems larger than compact binary galaxies.

As we described in Section 12.7 Kahn and Woltjer (1959) used a simple linear two body dynamics to describe the relative motion of the Milky Way and M31 galaxies. The motion of the galaxies was described (in the reference frame of the binary's center of mass) by the equation of motion whose first integral is $(1/2)v^2 = GM/r + E(M)$. Here $E(M)$ is the total energy per unit mass ($M = M_1 + M_2$ is the total mass).

The energy is negative for a gravitationally bound system, $E(M) < 0$. This inequality and the energy conservation equation above with the observed values $r = 0.7$ Mpc and $v = -120$ km/s lead to an absolute lower limit for the estimated binary mass: $M > 1 \times 10^{12} M_\odot$. With the "timing argument" additionally used, the mass is $M \approx 4.5 \times 10^{12} M_\odot$, if the maximum separation was about 4.4 Gyr ago (corresponding to the calculation in Section 12.7 with $T = 13.2$ Gyr).

Now we will see how inclusion of dark energy can modify the dynamics of the Local Group and other systems of galaxies. With a minimal modification of the original KW method, including the dark energy background ρ_V, the equation of motion and its first integral become:

$$\frac{dv}{dt} = -\frac{GM}{r^2} + \left(\frac{G8\pi}{3}\right)\rho_V r, \qquad (15.15)$$

$$\frac{1}{2}v^2 = \frac{GM}{r} + \left(\frac{G4\pi}{3}\right)\rho_V r^2 + E(M). \qquad (15.16)$$

Now the total energy for a bound system embedded in the dark energy background must be smaller than an upper limit which depends on the mass and the dark energy density:

$$E(M) < -\frac{3}{2}GM^{2/3}\left[\left(\frac{8\pi}{3}\right)\rho_V\right]^{1/3}. \qquad (15.17)$$

The limiting value corresponds to the case where the distance between the component galaxies could just reach the zero-gravity distance.

With the same values as above for the separation and relative velocity of the two galaxies, the absolute lower mass limit increases to $M > 3.2 \times 10^{12}\, M_\odot$ (when the local dark energy density = its global value). Also the timing argument leads to an increased mass, $M \approx 5.3 \times 10^{12}\, M_\odot$ (Chernin et al. 2009; Binney and Tremaine 2008). Actually such a value should be viewed as an upper limit, because we do not know the details of the forming process of the Milky Way/M31 binary. For instance, if the maximal separation of the galaxies was 6 Gyr ago (as might be motivated by the ΛCDM cosmology: gravitational instability was terminated in the linear regime about 6 Gyr ago with the start of the dark energy dominated epoch (e.g., Chernin et al. 2003) $M \approx 4.3 \times 10^{12}\, M_\odot$. The effect is illustrated in Figure 15.4 where the dependence of the inferred mass on the look-back time to the maximum separation is shown for both the classical and modified Kahn–Woltjer models. Though the uncertainty due to the imperfectly known kinematical situation, the dynamical history, and the values of the relevant parameters (especially the mutual velocity today) is rather large, these examples show that dark energy has a considerable effect on the mass determination using the Kahn–Woltjer method.

We discussed the classical Virial Theorem in Section 12.2. For a system within dark energy, the Virial Theorem needs an extra term due to the contribution of the particle-dark energy interaction to the total potential energy (Forman 1970; Jackson 1970). When positive, the inclusion of the cosmological constant leads to a correction upwards for the mass estimates (Chernin et al. 2009; Chernin et al. 2011).

In the presence of dark energy, the total potential energy of a quasi-stationary gravitationally bound many-particle system includes in addition to the sum U_1 of

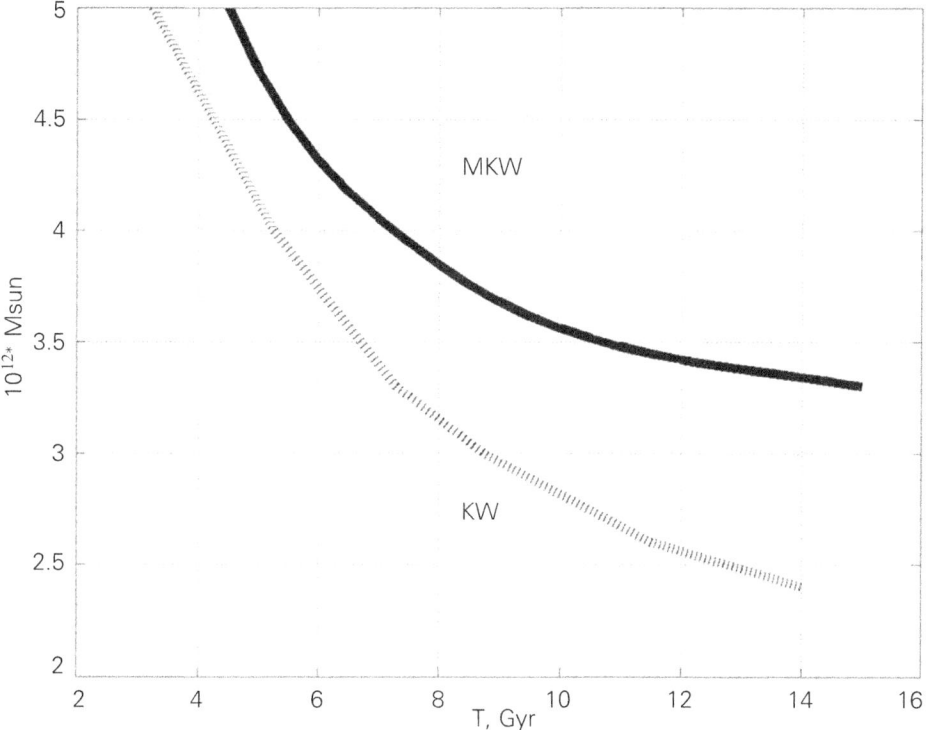

Figure 15.4. Difference between the Local Group mass predictions from the classical Kahn–Woltjer estimator (KW) and its modified form (MKW). The x-axis gives the assumed look-back time of maximum separation and the y-axis gives the calculated total mass of the Milky Way and M31 binary. The local dark energy density is taken to be equal to its global value. From Chernin et al. 2009.

the mutual potential energies of the member particles, also the sum U_2 of the potential energies of these particles in the force field of dark energy:

$$U_1 = -\frac{1}{2} \sum \frac{G m_i m_j}{|r_i - r_j|};$$

$$U_2 = -\frac{4\pi \rho_V}{3} G \sum m_i r_i^2. \tag{15.18}$$

Here r_i is the radius-vector of a particle in the frame of the system's barycenter; the summation in U_1 is over all particle pairs $(i \neq j)$. The summation in U_2 is over all the particles. It can be shown that the modified Virial Theorem should be written as (Chernin et al. 2009; Chernin et al. 2011):

$$\langle K \rangle = -\frac{1}{2} \langle U_1 \rangle + \langle U_2 \rangle. \tag{15.19}$$

How much does the inclusion of the dark energy influence the usual virial mass estimates? Equation (15.19) may be rewritten in terms of the total mass M, a characteristic velocity V and a characteristic sizes R:

$$M = V^2 \frac{R}{G} + \frac{8\pi}{3} \rho_V R^3 \qquad (15.20)$$

which can be expressed in convenient units as:

$$M/M_\odot = 2.3 \times 10^8 \left(V/\mathrm{kms}^{-1} \right)^2 (R/\mathrm{Mpc}) + 0.9 \times 10^{12} (R/\mathrm{Mpc})^3. \qquad (15.21)$$

Note that the additional positive term is interestingly equal to the absolute value of the effective (anti)gravitating mass of dark energy contained in the spherical volume of radius R. This term is a quantitative measure of the lost-gravity effect.

For instance, if we take the first (classical) term in Equation (15.20) to be $2 \times 10^{12}\, M_\odot$ and if the size R is about 1 Mpc, then the second term contributes about 30% of the total mass. We see that the relative contribution of the dark energy scales as the crossing time squared $(R/V)^2$. In that case the lost-gravity effect is 10–30 times larger in a group like the Local Group than in a rich cluster like the Coma cluster.

These results demonstrate the effect of the lost gravity in gravitationally bound systems embedded in the dark energy background. The relative motion of the bodies is controlled by the gravity (of dark matter and baryons) which is partly counterbalanced by the dark energy anti-gravity. Consequently, both the Kahn–Woltjer and virial mass estimates for dark matter and baryons must be corrected for dark energy.

The zero-gravity radius can serve as a natural boundary of a system. When one makes virial mass estimates for a distant galaxy group, the data are generally restricted to the bright galaxies in the inner part of the group while the dwarf galaxies in the outer parts remain undetected. Thus it is expected that the derived virial mass does not include the whole dark mass contained by the system. Motivated by the results on the intriguingly low average matter density in the local universe, based on virial masses of groups (e.g., Makarov and Karachentsev 2011), Chernin et al. (2012) considered the possibility that the dark matter halos typically extend up to the zero-gravity radius of a system. Using common density distribution functions (such as the isothermal density profile and the Navarro–Frenk–White halo) and the size and the mass of the observed virial core, they could derive the zero-gravity radius R_V for the groups. Integration of the mass distribution up to R_V resulted in the total mass which could be from 2 to 6 times larger than the original virial mass, depending on the group catalogue and the way the virial mass had been calculated. Chernin et al. (2012) concluded that the local universe within about 200 Mpc may not be an underdensity region (by a factor of 2 or 3) as had been suggested, but could be quite typical or perhaps even slightly overdense.

15.4   Towards local measurement of dark energy

When considering dark energy on galaxy group scales, it is convenient to write the dark energy density (in round numbers) as

$$\rho_V / 7 \times 10^{-30}\,\text{g/cm}^3 = (M/2 \times 10^{12}\,M_\odot)/[(8\pi/3)(R_V/1.3\,\text{Mpc})^3]. \qquad (15.22)$$

Thus one could determine the dark energy density around a galaxy group if its mass M and the zero-gravity radius R_V were known from observations. Physically, there is inward acceleration at distances $< R_V$ and outward acceleration at distances $> R_V$ (but within the region where the simple point-mass model is adequate). However, the expected accelerations in the nearby velocity field are very small (of the order of 0.001 cm/s/yr) and it is out of question to measure directly such variations of the radial velocities of dwarf galaxies within a few megaparsecs.

It is interesting to mention here that Sandage (1962) considered how the redshift changes with time, when a **distant** galaxy is observed. A formula for the very slow change expected in an expanding universe was derived by McVittie in the Appendix of Sandage (1962). This effect has been discussed in the context of Friedmann models with dark energy (e.g., Loeb 1998). For a universe with $\Omega = \Omega_M + \Omega_V$ the rate of change of the redshift is

$$\left(\frac{dz}{dt}\right)_0 = -(1+z)H_0\left[(\Omega_M(1+z) + (1-\Omega) + \Omega_V(1+z)^{-2})^{1/2} - 1\right]. \qquad (15.23)$$

For a dust universe ($\Omega_V = 0$), all redshifts are decreasing at the present time. If $\Omega_V > 0$, the redshifts may be either increasing or decreasing, depending on the value of z (nearby objects: z increasing; distant objects: z decreasing). In terms of radial velocity, the predicted change $dv/dt = c(dz/dt)_0$ for distant galaxies ($z \sim 1$) is ~ 1 cm/s/yr.[3] Of course, when we currently use distant supernovae, we do not measure such tiny changes in redshifts, but derive the behavior of the scale factor $R(t)$ from the magnitude-redshift relation using test objects at different cosmic times and distances (redshifts).

We also cannot follow a dwarf galaxy in its trajectory for millions of years in order to see the location of the minimum velocity which defines the zero-gravity distance (Section 15.2). Several test objects with the same energy, but with different

---

3 This fundamental cosmological test may be within the reach of the future 42 m E-ELT telescope, using the Lyman α forest in the redshift range $2 < z < 5$ and other absorption lines in the spectra of high redshift QSOs, though even then requiring some 4000 hours of observing time over a period of 20 years (Liske et al. 2008). A redshift region where the redshifts are increasing with time would be the hallmark of a positive cosmological constant.

moments of expulsion from the center, would trace the needed constant-energy curve for the present observer. However, this would require widely different times of expulsion, while the usual view is that the objects were expelled in the distant past within a relatively narrow time interval (e.g., Chernin et al. 2004). They also had different energies, so what we see now is a locus of points at different distances from the center and lying on different energy curves; they make the observed velocity— distance relation. Such an expected observed distance—velocity relation does not have a clear signature of the zero-gravity distance (e.g., a minimum).

Chernin et al. (2006) noted that the size of the group or the zero-velocity distance is a strict lower limit to R_V, giving an **upper** limit to the local dark energy density ρ_{loc} (for a fixed group mass). It is more difficult to determine a strict upper limit to R_V, which would give the very interesting **lower** limit to the dark energy density ρ_{loc}.

One possibility to estimate an upper limit to R_V is to find the distance where the local outflow reaches the global Hubble rate (Teerikorpi and Chernin 2010). This distance R_{ES} is expected to give the size of the volume from which the mass M has been gathered during the formation of the group (the Einstein–Straus vacuole). As was noted above (Equation (15.7)), one can calculate its present value in terms of the zero-gravity radius assuming the present average cosmic mass density: $R_{ES}(t_0) = (2\rho_V/\rho_{ES})^{1/3} R_V$. This is also the distance where the global Hubble ratio is reached, because at this point the enclosed mass is the same as for the uniform global Friedmann model, hence the expansion rate is the same. For $\Omega_M = 0.3$, $\Omega_V = 0.7 R_{ES} = 1.7 R_{EV}$. This is the same 1.7 as in the global scale factor ratio leading to $1 + z_V = 1.7$. Namely, the requirement that the global acceleration is zero when $\rho_V = 1/2\,\rho_M(z_V)$ leads in terms of the current mass density to the condition $2\rho_V = (1 + z_V)^3 \rho_M$ or $(2\rho_{loc}/\rho_M)^{1/3} = (1 + z_V) = 1.7$. The ratio 1.7 between the distance where H_0 is reached and R_V is approximately valid also if the local dark energy density differs from the global one.

Using this idealized model and $R_{ES} \approx 2.6 \pm 0.3$ Mpc (Karachentsev et al. 2009; also Fig. 5.4), one may estimate that $R_V = 1.5 \pm 0.2$ Mpc and the dark energy density around the Local Group is $0.8 \lesssim \rho_{loc}/\rho_V \lesssim 2$ for the Local Group mass of $4 \times 10^{12}\,M_\odot$. The limits would change directly proportional to the adopted mass value.

A similar range was derived from the Virgo cluster (Chernin et al. 2010). These tentative results, which illustrate the interpretation of short-scale outflows in terms of dark matter and dark energy, encourage more extensive studies of the out-flows around galaxy systems. So, Hartwick (2011) has analyzed a large sample of galaxy groups using a new method based on positional and redshift information; he found that the radial outflow around the groups goes over, as expected, to the global Hubble law beyond the distance of about $2R_V \approx R_{ES}$, where the zero-gravity distance was calculated from the virial mass and the global dark energy density.

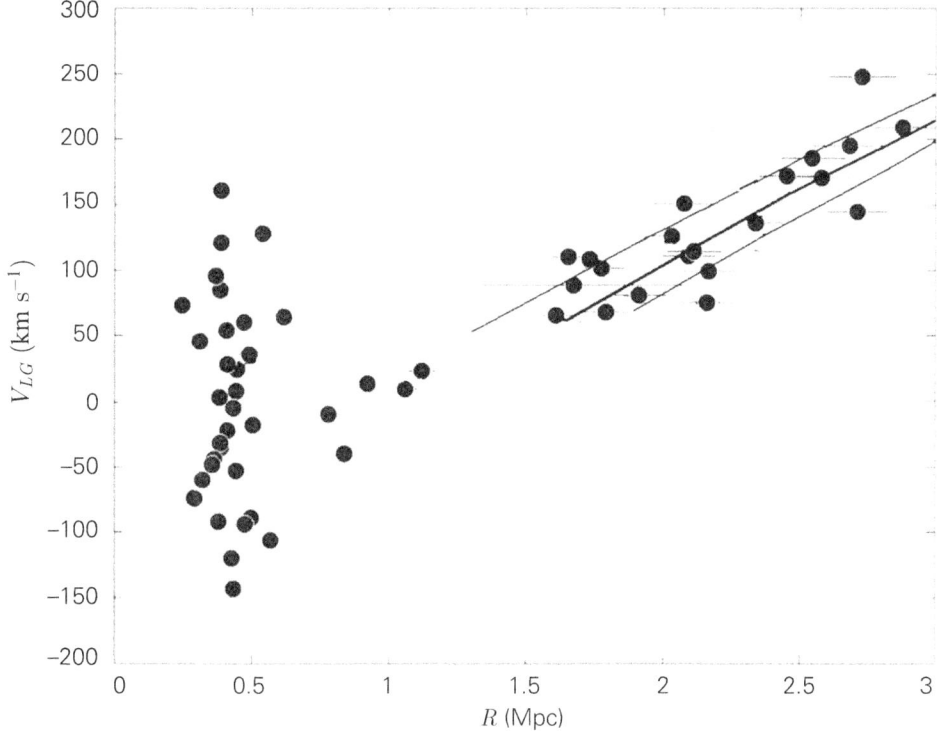

Figure 15.5. Velocity vs. distance relative to the center of mass of the M31/Milky Way Galaxy binary. Note the $\pm$ outward/inward speeds of the central bound core members of the Group, a gap (at $V \sim 0$ km/s and $R \sim 1$ Mpc), and the linearly increasing receding V with R from the gap outward. The heavy central line is a best fit which corresponds to $M = 4 \times 10^{12}\ M_\odot$ from the Kahn–Woltjer method and the Concordance $\rho_v = 7 \times 10^{-30}$ g/cm^3. The lower and upper lines correspond to $\pm 2 \times 10^{12}$ solar masses around the best fit mass. On the other hand, assuming $M = 4 \times 10^{12}\ M_\odot$, the upper and lower curves put the local dark energy density into the range from 5×10^{-30} g/cm^3 to 10×10^{-30} g/cm^3. The diagram is adapted from Chernin et al. 2009.

The present-day distance-velocity relation: Dark energy can modify the velocity distance relation, as seen today. The flight time from the center of the group ($x_0 \ll 1$) to the normalized distance $x = r/R_V$, for a particle with energy $E = -\alpha GM/R_V$ can be parameterized in terms of the vacuum Hubble time $T_V = 1/H_V$:

$$t = T_V \int_{x_0}^{x} \left(x^2 + 2/x - 2\alpha\right)^{-1/2}; \tag{15.24}$$

while the normalized velocity y is related to x as

$$y = \left(x^2 + 2/x - 2\alpha\right)^{1/2}. \tag{15.25}$$

One can calculate for each location x the required energy α for a particle to reach this distance after the flight time t, and thus the normalized velocity y. From the derived relation $y = y(x)$, the locus of present distance-velocity positions is (xR_V, yH_VR_V). So, fixing the mass M and the dark energy density ρ_V (which give the distance R_V and the vacuum constant H_V needed in the normalization), and the normalized flight time t/T_V (say, the age of the universe) we can calculate the observed velocity-distance relation. The present position of the particle (a dwarf galaxy) on its energy curve depends on how long ago it was ejected from the center.

For instance, Chernin et al. (2009) varied the mass and compared the predicted Hubble relation with the outflow around the Local Group. In Figure 15.5 we show three predicted curves, assuming that the local dark energy density is the same as the global one 7×10^{-30} g/cm^3. One sees that the Local Group mass $2 \times 10^{12} M_\odot$ puts the Hubble curve too high and $6 \times 10^{12} M_\odot$ too low, while $4 \times 10^{12} M_\odot$ provides an admissible fit. These curves can be also reproduced using other parameter values. So the lowest curve is also close to the prediction using $M = 2 \times 10^{12} M_\odot$, $\rho_{loc} = 0$ g/cm^3, and $t = 13.6$ Gyr, and for any higher mass this zero-dark energy prediction is still more off the observed relation. Or we may keep the mass $M = 4 \times 10^{12} M_\odot$ and ask what values of dark energy then reproduce the upper and the lower curve. Then the upper curve corresponds to $\rho_{loc} = 10 \times 10^{-30}$ g/cm^3 and the lowest curve to $\rho_{loc} = 5 \times 10^{-30}$ g/cm^3, or the range $0.7 \lesssim \rho_{loc}/\rho_V \lesssim 1.4$. This is consistent with the above result using the edge of the Einstein–Straus vacuole where $R_{ES} = 1.7R_V$. See also Peirani and de Freitas Pacheco (2008).

Though we cannot yet determine very accurately the local dark energy density ρ_{loc}, because of the remaining uncertainty in the Local Group mass, these observations clearly point to $\rho_{loc} > 0$ and are consistent with the assumption that $\rho_{loc} = \rho_V$.

15.5   The Hubble law and dark energy

The question of the regularity of the local Hubble law has its roots in the 1930s, in the works by Howard Robertson and Arthur Geoffrey Walker who realized that there is an intimate connection between the homogeneity and the velocity of expansion. General geometrical reasoning, often illustrated by Eddington's expanding balloon analogy, shows that if a uniform space expands so that it remains uniform, then the linear velocity—distance relation can be seen everywhere: velocity = constant × distance. This theoretical relation is also often called the Hubble law, along with the original, observed redshift—distance law. In fact, understanding the connection between the supposed uniformity of matter distribution and the Hubble law was the first great success of the expanding world model.[4]

---

4   From Einstein's equations it follows that in expanding uniform space there indeed is the exact expansion velocity—distance relation $V_{exp} = H \times r$, while the redshift is a complicated function of the distance $z = z(r)$. Only for small redshifts is the redshift a linear function of the distance: $z \approx (H/c)r$, $z \ll 1$.

Sandage, Tammann and Hardy (1972) recognized a contradiction between the observed linear Hubble law and a possible hierarchical galaxy distribution (or more generally, a very non-uniform distribution). At that time they used the existence of the good local Hubble law as an argument against de Vaucouleurs' hierarchical model. A strong deflection from a linear distance—velocity relation expected within the hierarchical inhomogeneities has been confirmed by theoretical calculations (Haggerty and Wertz 1972; Fang et al. 1991; Gromov et al. 2001). In general, such a good linear Hubble expansion flow seems to contradict the standard picture where the Hubble law is the strict consequence of homogeneity of matter distribution.

"In fact, we would not expect any neat relation of proportionality between velocity and distance for these [closeby] ... galaxies",

Weinberg (1977) remarked about the same thing.

Baryshev et al. (1998) emphasized that the observed linear redshift-distance relation inside the inhomogeneous power-law (fractal) density distribution creates the "Hubble – deVaucouleurs paradox" (also called the "Hubble enigma" by Chernin et al. 2000).

Using asymptotically homogeneous Lemaître–Tolman–Bondi models (i.e., asymptotically Friedmann-like on large scales), Gromov et al. (2001) found the necessary conditions for the linear Hubble law existing within inhomogeneities modeled as a fractal structure with dimension $D = 2$. The larger the scale of homogeneity R_{hom} the smaller should be the density parameter Ω_M; e.g., for $R_{hom} = 100$ Mpc the linear Hubble law could exists at distances $r > 1$ Mpc only if the density parameter $\Omega_M < 0.01$ (confirming one possible explanation given by Sandage et al. (1972)—a very low cosmic density). Later Sandage (1999) again pointed out that the linear Hubble flow is surprisingly quiet ("cold") in the local volume where Hubble found his law: the velocity dispersion about the mean Hubble relation is ≤ 60 km/s (in fact, even less than 40 km/s; e.g., Ekholm et al. 2001). Sandage said that

"explanation of why the local expansion field is so noiseless remains a mystery".

Moreover, the local rate of expansion is similar, if not identical, to the global rate H_0. Whiting (2003) added to the list of oddities that the dispersion does not depend on the mass of a galaxy, but is the same for giants and dwarfs (Teerikorpi et al. (2005) suggested that, along with the other features, also this might be due to the effect of dark energy).

Soon after the discovery of the acceleration of the universe from very distant supernovae, Chernin et al. (2000) and Chernin (2001) suggested that the cold local Hubble flow could be a signature of the dominance of the cosmological vacuum (the Λ-term) also on local scales. As we saw above, the dominance begins quite close the Local Group and other similar systems. Baryshev et al. (2001) extended this explanation to include dark energy which may vary in time. If the equation of state parameter w is in the range $-1 < w < -1/3$, the dark energy density was higher in the past and the cooling down of peculiar velocities—see below—could start earlier and be more effective. Recall that the relation between density and scale factor for a substance with the equation of state $P/c^2 = w\rho$ is $\rho \sim R^{-3(1+w)}$.

The main point was that in such regions of the universe where dark energy dominates new structures do not condense and linear perturbations of density and peculiar velocities decay. This effect was considered for vacuum-dominated regions by Chernin et al. (2000) using Zeldovich's (1965) analysis of gravitational instability. The analysis shows that when the antigravitating vacuum dominates on certain scales, there the gravitational instability is terminated and the nonlinear perturbations (which were earlier growing), are then freezing out or even being adiabatically cooled (the peculiar velocity $\delta V \sim R^{-1}$; $R(t)$ is the scale factor).

Lahav et al. (1991) derived a formula for the growth rate of the density perturbation δ, $f = d(\ln \delta)/d(\ln R(t))$. At the present epoch,

$$f(z = 0) \approx \Omega_M^{0.6} + (1/70)\Omega_V\left(1 + \frac{1}{2}\Omega_M\right) \qquad (15.26)$$

It has been often thought that dark energy has insignificant local effects. The formula shows that for a **fixed** matter density parameter Ω_M, adding the cosmological vacuum into the model has practically no effect for the present growth rate, which also determines the peculiar velocities around the growing density fluctuations and hence the regularity of the Hubble flow. However, comparing flat ($k = 0$) models, the present epoch growth rate in the $\Omega_V = 0$ model is about a factor of two greater that in the $\Omega_V = 0.7$ model (see below).

Lahav et al. (1991) saw this immunity to Λ or dark energy as reflecting the cosmic vacuum as a perfectly uniform background which does not have local force effects. A galaxy does not "feel" the presence of dark energy. However, dark energy influences the behavior of the global scale factor and in this way enters the differential equations for a growing individual matter density contrast δ. The small (average) influence derives from the relatively short time interval when dark energy dominated during the late epoch of structure formation. The global balance redshift $z_V = (2\rho_V/\rho_M)^{1/3} - 1 \approx 0.7$ corresponds to the look-back time of about 45% of the universe's total age.

However, in our lumpy universe there are situations where local effects of dark energy are expected. The fact alone that one has to modify the Virial Theorem for a gravitationally bound system in order to account for the antigravity of dark energy shows that dark energy can have local effects. Also, because of the lumpy distribution of matter, dark energy may have dominated at quite different time intervals in different regions of the universe, depending on the local matter contrast. Within large voids which represent "pacific oceans" of dark energy, test particles would freely trace the repulsive effect of dark energy for a long time, leading to a cold Hubble flow with the asymptotic "vacuum Hubble constant" $H_V = (8\pi G\rho_V/3)^{1/2}$, in comparison with the global average expansion rate which is $H_0 = (8\pi G(\rho_M + \rho_V)/3)^{1/2}$

for the flat $\Omega = 1$ universe. For instance, with $(\Omega_M, \Omega_V) = (0.25, 0.75)$, $H_V = 61$ km/s/Mpc for $H_0 = 70$ km/s/Mpc.

It is also interesting to note that typical scales of gravity-dominated regions and large peculiar velocities (which means a poor Hubble law) are connected. It was shown in Teerikorpi et al. (2005) that the correlation length r_0 in the correlation function describing the general distribution of galaxies (see Section 14.6) is related to the zero-gravity distance for a typical mass concentration as $R_V = 0.8 r_0 \approx 4 h^{-1}$ Mpc, or the "equal energy" sphere discussed above has the radius $R_{EE} = r_0$. There are indications from the large galaxy redshift surveys (SDSS, 2dF) that while the pair-wise peculiar velocities of galaxies are high (500–600 km/s) within scales $\lesssim r_0$, beyond $2 r_0$ the measured velocity dispersion is $\ll Hr$ which implies a smooth Hubble flow in such dark energy dominated regions with accelerating expansion and no further structure formation.

When we discuss the influence of various amounts of dark energy, we have to specify a class of Friedmann models. If we just add dark energy into a model, its global parameters, including curvature, change. We have the strong condition, from the fluctuations of the CMB as we discussed in the previous chapter, that the universe has a flat spatial geometry. So we are constrained to consider the situation $\Omega = \Omega_M + \Omega_V = 1$. In that case the present growth rate comes to depend significantly on the fraction of the vacuum Ω_V in the model. This is well seen from Equation (15.26) and it was also demonstrated by Axenides and Perivolaropoulos (2002) that for a small Ω_M (i.e., large Ω_V) the growth gets suppressed. If $w > -1$, this effect is still stronger because then dark energy density was larger in the past.

The corresponding behavior of peculiar velocities is seen in fig. 2 of Peebles (1984), and also in fig. 2 of Axenides and Perivolaropoulos (2002). For example, compared with the case $\Omega_V = 0.0$, peculiar velocities as calculated from the growth rate are a factor of 2 smaller when $\Omega_V = 0.7$ (in both cases $\Omega = 1$). In this sense, in a flat universe the expected peculiar velocities around mass concentrations representing similar relative density fluctuations do depend on the fraction of dark energy it contains.

As we discussed above, the local effect of dark energy was first considered in the expansion field around our Local Group and other nearby groups of galaxies, where the low scatter Hubble outflow having about the global Hubble rate was explained as due to the dominating dark energy starting just outside of the group. Axenides and Perivolaropoulos (2002) concluded from their analytic calculations that the cooling mechanism, even though it works in the sense of suppressing peculiar velocities, is not enough (using $w = -1$) to explain the local low scatter of the Hubble law. Such simplified models of isolated mass concentrations may not be adequate and one would like to see the effect of dark energy or absence of it in realistic cosmological N-body simulations.

In their pioneering CDM (i.e., cold dark matter, no dark energy) N-body simulations of this problem Governato et al. (1997) identified many "Local Groups" to match the observed dynamics and environments of our Local Group. Importantly, calculation of velocity dispersions in 5 Mpc volumes around the groups led to the following results: For the $\Omega = 1$ CDM model the peculiar velocities had dispersions in the range from 300 to 700 km/s, while for the $\Omega = 0.3$ CDM model the range decreased to 150–300 km/s, still too high as compared with the observed < 50 km/s.

Then the simulations by Macciò et al. (2004) confirmed that inclusion of the cosmological constant in the calculations leads to significantly lower velocity dispersions in Local Volume-like regions than what happens without dark energy. They also showed that replacing the cosmological constant ($w = -1$) by a dark energy model with $w = -0.6$, peculiar velocities are still reduced by about 15%. This is roughly as expected from the simple analytic calculation in Baryshev et al. (2001) where it was argued that the longer adiabatic cooling in the dark energy model with $w = -2/3$ leads to lower peculiar velocities.[5]

Macciò et al. (2004) concluded that two things are essential for producing the low local velocity dispersion: 1) a correct cosmology, requiring the Λ or dark energy component, and 2) a rather low environmental density contrast, such as may be around the Local Group. If the density contrast is large, then large peculiar velocities are expected even in dark energy cosmologies. This is understandable because then the gravity-dominated regions are larger. Analysis of the Millenium simulations by Niemi and Valtonen (2009) has confirmed Macciò et al. (2004).

Simulations by Hoffman et al. (2008) and Martinez-Vaquero et al. (2009) failed to find the dark energy flow, but this may be due to a resolution problem in their work. With improved resolution, Nurmi et al. (2010) show clearly that in many instances dark energy flows are observed near simulated galaxy groups (see Figure 15.6)).

A different kind of local Hubble flow calculation was made by Chernin et al. (2004). They traced the trajectories of local galaxies back to the epoch of the formation of the LG and found initial conditions which were very different from those that would directly lead to the linear Hubble flow. With simulations they identified the dark energy as the agent which introduces the subsequent regularity in the nearby flow.

15.6 Redshift asymmetry as signature of dark energy

In the Friedmann model the homogeneous distribution of matter and the regular linear Hubble law are tightly connected. In reality, we know that the distribution

---

5 One may suppose that a still lower velocity dispersion might have been found for a third example, the coherently evolving (i.e., dark energy density $\propto$ dark matter density) model with $w = -2/3$, where the time allowed for cooling is still longer. This model was interesting, because it produces a Hubble relation which is close to the standard Λ-model fitting of the SNIa observations for $z < 1.5$ (Teerikorpi et al. 2003).

of gravitating matter is very lumpy and this causes deviations from the good Hubble law in the form of peculiar velocities on small and even larger scales (bulk flows). Such distortions are indeed observed, e.g., around the nearby Virgo cluster. However, in addition to such rather well understood deviations, astronomers have reported on more striking exceptions to the rule (the Hubble law of redshifts), which if true, could tell something new about the universe.

In science such anomalies

"may be either genuine or (more often) wrong observations or true observations having conventional but not yet known explanations"

(Aurela 1973). It is important to search for ordinary explanations, before concluding that a fundamentally new phenomenon is present in some anomaly. Even conventional explanations may have interesting new implications, in this case for dark matter and dark energy, as we will discuss below.

The suggested redshift anomalies come in a wide variety of sources. Most famously, in 1966 Halton Arp, at the Palomar Observatory, noticed that radio sources, among them quasars, tended to be close to, or aligned across, some of the galaxies in his *Atlas of Peculiar Galaxies*. If true, such associations between high-redshift quasars and low-redshift galaxies violate the Hubble law: the quasars would have a large extra component in their redshift, in addition to the cosmological one.

The current idea is that quasars are active galactic nuclei having essentially the same redshift as their host galaxies. This mainstream view regards Arp's configurations as chance associations: when looking at the space populated by objects at different distances one necessarily finds "optical double stars". One can calculate their number expected by chance, from the number of galaxies and quasars in the sky. Burbidge (2001) listed investigations which use well-defined samples of quasars and compare their positions with bright low-z galaxies. These suggested statistically significant correlations and have inspired Arp himself and others (e.g., Hoyle, Burbidge and Narlikar 2000) to study the idea that quasars are ejected from the nuclei of galaxies and are composed of freshly created matter. This clearly would require non-standard physics.

With large numbers of galaxies and quasars now available from the Sloan Digital Sky Survey and other surveys, studies have generally indicated correlations in the mentioned sense, usually interpreted in the framework of weak gravitational lensing by foreground large-scale structures, with amplification increasing the number of detected background objects (e.g., Gaztañaga 2003; Scranton et al. 2005). However, Arp's associations are compact ($\sim$100 kpc) systems around individual galaxies. Thus if the higher-z companions are actually distant background objects, their line-of-sights intersect the halos and halo objects could act as lenses (Barnothy 1974; Canizares 1981). Baryshev and Ezova (1997) took the lenses to be intermediate (meso) mass objects such as globular clusters, dwarf galaxies, and clusters of dark matter with masses of 10^3 to 10^9 solar masses. The meso-lensing effect by such sub-halo objects could magnify background point sources by about five magnitudes and would make initially star-like images split or spread to 10–100 milliarcsecond sized forms. Yonehara et al. (2003) arrived at similar conclusions in their work based on the predicted cold dark matter lumps. Hierarchic gravitational clustering leads to the picture where the massive haloes contain thousands of CDM lumps with masses in the range from 10^6 to 10^9 $M_\odot$. The expected mesolensing may allow one to detect the substructures observationally (Surdej et al. 1993; Bukhmastova and Baryshev 2008). Large interferometers like VLTI and Keck I-II telescopes have achieved an optical resolution of about 5 milliarcseconds, though for relatively bright objects. A systematic mapping of the dark halo substructure using gravitational lensing has not yet been done, an interesting task also for future still larger telescopes. As a byproduct, such studies might illuminate the question of Arp's associations—rather than radically new physics, they may tell something important about dark matter!

Another suggested redshift anomaly may be indirectly due to dark energy. Here we have in mind ordinary galaxies with possible small extra redshifts. In particular, this has been studied from the differential velocities (redshifts) of companion galaxies with respect to the main galaxy in a group, since Arp (1970) noted that primary galaxies appear to have mainly smaller redshifts than their companions. Jaakkola (1971) and Jaakkola and Moles (1976) pointed out that the effect seems to depend on the morphological type of the companion, excess redshifts (of the order of 100 km/s) appearing especially for Sbc and Sc galaxies. These results were sometimes interpreted as evidence for anomalous redshifts, though much smaller than in quasars. Jaakkola and Moles considered also a possibility of selection effect: the group data might contain non-members (projected field galaxies taking part in the Hubble flow). Because the background volume is larger than the foreground volume, one might thus ascribe to the group more galaxies from the background (larger redshifts) than from the foreground (smaller redshifts). However, Jaakkola and Moles (1976) did not investigate such a selection quantitatively, and were inclined to conclude that it does not work. Let us consider this question in some detail.

Groups and larger systems of galaxies are seen as clumps on the sky close together in angle. For nearby groups, e.g., the M81 group, they may be spread through many degrees on the sky. The members of a given group will have similar cosmological redshifts but with confusing additional Doppler red and blue shifts due to internal motions.

Typically, these groups have a massive central spiral galaxy (or a binary for our own Local Group). In early group catalogs, assignment of membership was subjective and nonsystematic. Huchra and Geller (HG, 1982) devised a computer program to objectively and consistently compile galaxy groups and their members. Galaxies nearby on the sky with similar cosmological redshift zs are presumed to be close together in 3D space. However, redshift distances, $D \approx cz/H_0$, are uncertain in part due to inclusion of additional internal motion Doppler shifts.

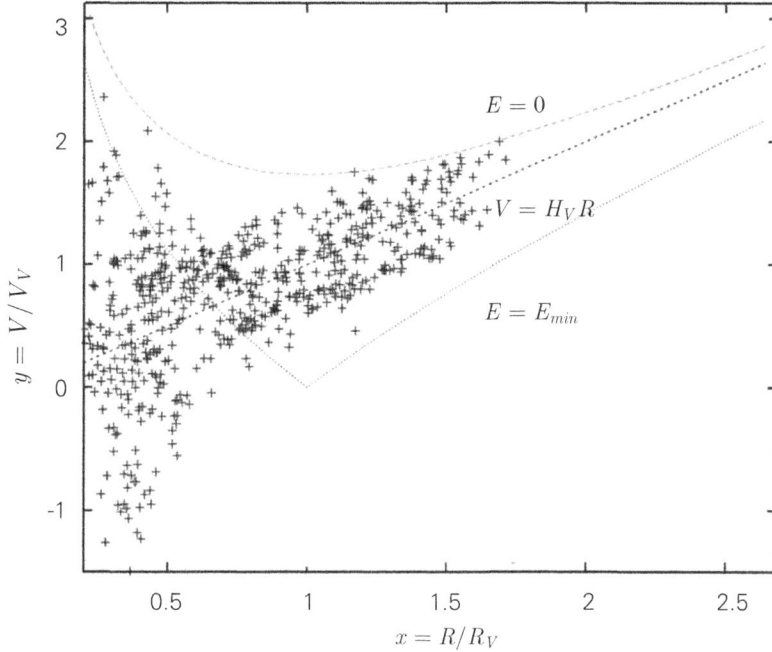

Figure 15.6. The normalized Hubble diagram for dwarf galaxies of about equal to or more than 10^8 solar masses each, around a typical small group of galaxies of 10^{12} solar masses. The figure is based on high resolution cosmological simulations by Nurmi et al. (2010). Symbols are defined in Section 15.5 (Courtesy of Pasi Nurmi).

One would expect equal numbers of red and blue shifts relative to brightest HG members due to internal motions. However, Sulentic (1984), in nearby HG spiral dominated groups found 119 red vs 77 blue. $(119 - 77)/(119 + 77) = 0.21 \pm 0.1$. This is a considerable asymmetry; there is a less than 1% probability that it would occur by chance in a truly symmetric population. As an explanation, Sulentic proposed that this is a moderate example of Arp's quasar non-Doppler red-shifts which are somehow created inside individual group members.

Motivated by Sulentic's work, a simple explanation of the excess was proposed by Byrd and Valtonen (1985), an outward motion of outer members relative to the primary galaxy. In their quantitative study they showed that if most of the group population is composed of unbound expanding members then the way the groups are selected will lead to an artificial redshift excess for the companions. It is useful to discuss this explanation in some detail.

Sulentic's finding represented a paradox, if the groups are viewed as gravitationally bound systems where a galaxy is equally likely to move one way or another. What if the groups are not gravitationally bound, but they are groupings of galaxies in a vacuum-dominated smooth Hubble flow? We may study this situation by taking a slice from the cone that defines the angle of the group in the sky (Figure 15.7).

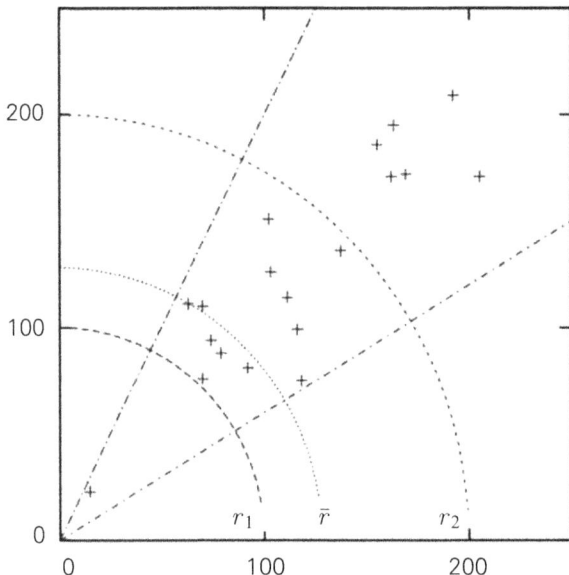

Figure 15.7. Identification of a group of galaxies from the Hubble flow. The sector (cone in three dimensions) defines the group area in the sky, while the inner and outer radii at r_1 and at r_2, respectively, limit the group in depth. The most likely distance of the dominant galaxy $\bar{r}$ is also indicated. The three spherical surfaces divide the cone into a front volume and to a back volume. The latter is clearly bigger, and is likely to contain more galaxies. Even though galaxies of the back volume are harder to see, more of them are anyhow observed, and there is an excess of redshifted galaxies (in the back) in comparison with the blueshifted galaxies (in the front of the dominant galaxy).

The inner edge of the slice is at the distance r_1 from us and the outer edge of the slice at the distance r_2. Somewhere in this volume lies the apparently brightest galaxy. We first demonstrate that the brightest galaxy is most likely in the front part of the volume.

We need the well-known probability distribution of true luminosities L of galaxies which is called the **Abell luminosity function**:

$$n(L)dL = n_0 L^{-a} dL \tag{15.27}$$

where $n(L)$ is the number of galaxies in the luminosity interval $[L, L + dL]$ in a unit volume of space. The power-law index $a = 1.75$, for $L < L_*$ and $a = 3$, for $L \geq L_*$, which means that the distribution has a "knee" at L_*. The value of L_* is not very far from the luminosity of our Galaxy (Abell 1975).

The true (absolute) luminosity L and the apparent luminosity l of a galaxy at the distance r from us are related by $L = 4\pi r^2 l$. Let this galaxy be the dominant galaxy of the group. Then the probability distribution of its distance from us is

$$f(r)\,dr = n(L)(dL/dr)r^2\,dr. \tag{15.28}$$

Here we have assumed that the group volume is uniformly distributed by galaxies, and have taken account of the increase (as r^2) of the number of galaxies per shell of unit thickness dr when r increases. Using Equation (15.27) and the inverse square law, this becomes

$$f(r)\,dr = Br^{-2a+3}\,dr \tag{15.29}$$

where B is a constant when n_0, a and l are given. By integrating this distribution from r_1 to r and from r to r_2, and putting the two integrals equal to each other, we obtain the median position of r:

$$r = \bar{r} = \left(\frac{2}{1+\beta^{-2a+4}}\right)^{1/(2a-4)} r_2 \tag{15.30}$$

where $\beta = r_1/r_2$. Typically $\beta = 0.5$ in the Huchra–Geller catalog. If we take $a = 3$ which is justified for all except the nearest groups, we get

$$\bar{r} = \sqrt{2/5}\,r_2. \tag{15.31}$$

Thus instead of being at the midpoint between r_1 and r_2, at $0.75\,r_2$, the brightest galaxy is likely to be well forward of it at $\simeq 0.63\,r_2$.

It is now clear that the front part of the group volume is much smaller than the back volume, with respect to the dominant galaxy. In the Hubble flow, the galaxies in the front volume are blueshifted relative to the brightest galaxy, and the galaxies of the back volume are redshifted. Thus the number of redshifted galaxies N_R should be greater than the number of blueshifted galaxies N_B, and there should be a redshift asymmetry Z defined as

$$Z = \frac{N_R - N_B}{N_R + N_B}. \tag{15.32}$$

But the matter is not this simple. It is easier to see the galaxies of the front volume than of the back volume, and with the fixed apparent magnitude limit, the actually observed number of front volume galaxies is greatly enhanced by the luminosity function which rises steeply toward fainter galaxies. Thus we have to combine the volume effect and the luminosity function effect, which work in opposite directions, to calculate the overall observed redshift asymmetry Z.

Let the absolute luminosity of the faintest observable galaxy at distance r be L, and the corresponding quantities at the outer edge of the group volume be r_2 and L_2, respectively. They are related by

$$L = L_2(r/r_2)^2.$$

The number of observed galaxies at the distance r is obtained by integrating the luminosity function of Equation (15.27) from L to ∞:

$$N = n_0 \int_L^\infty L^{-a}\,dL = \frac{n_0}{a-1}\left[L_2(r/r_2)^2\right]^{1-a}. \tag{15.33}$$

By an integration over r we get the number for the front volume

$$N_B = C \int_{r_1}^{\bar{r}} r^{2(1-a)} r^2 \, dr = \frac{C}{5-2a} \left(\bar{r}^{5-2a} - r_1^{5-2a} \right)$$

where C is a constant. Similarly we calculate the number of galaxies in the back volume:

$$N_R = \frac{C}{5-2a} \left(r_2^{5-2a} - \bar{r}^{5-2a} \right).$$

Therefore when $a = 3$

$$Z = \frac{r_2^{-1} + r_1^{-1} - 2\bar{r}^{-1}}{r_2^{-1} - r_1^{-1}} = \frac{2\beta r_2 / \bar{r} - 1 - \beta}{1 - \beta}. \tag{15.34}$$

If $\bar{r} = 0.63 \, r_2$ and $\beta = 0.5$

$$Z = 0.17.$$

The observations give $Z = 0.21 \pm 0.1$, fully consistent with the assumption of the expansion Hubble flow in the outer part of the group (Byrd and Valtonen 1985, Valtonen and Byrd 1986), now naturally seen as due to the dark energy domination.

It is important to note that the explanation given by Byrd and Valtonen was based not only on a selection effect, but also on an idea on the dynamics of the galaxy groups (a pure field galaxy population, around a small bound group, would not offer an adequate explanation of the redshift excess).

About 25 years later, this explanation could be essentially verified by Niemi and Valtonen (2009) using synthetic galaxy groups extracted from the cosmological N-body simulation called the Millenium Simulation performed by the Virgo Consortium.

Niemi and Valtonen found that about one half of all groups in their mock catalogues are actually gravitationally unbound, and a large and statistically highly significant redshift excess appears only for such unbound groups, when in addition their first-ranked galaxies have been misidentified. On the contrary, gravitationally bound groups did not show any significant redshift excess.

Essentially, Niemi and Valtonen showed from numerical simulations of Λ-dominated cosmological models that groups of galaxies should possess expanding populations which reveal redshift asymmetries when viewed from outside at a reasonably close distance. Thus the observed redshift asymmetries may also be viewed as indirect evidence for non-zero local dark energy (as implied by the ΛCDM model).

As we discussed above, independent evidence for expanding populations around the Local Group and some other nearby groups, behaving as expected from the repulsive "antigravity" of dark energy, have been also found from direct distance-velocity observations. Hartwick's (2011) analysis of more distant groups points at the same direction.

As Valtonen et al. (2008) noted,

the observation of "anomalous redshifts" in 1970 proved highly significant in the case of groups of galaxies, and it pointed to the existence of dark energy even though it took many decades to fully understand the implications.

Chapter 16
Cosmological inflation

Why is the universe so symmetric? The universe of galaxies appears more or less same wherever we look, as long as we neglect small details. Also the cosmic expansion measured from galaxy redshifts takes place at the same rate in all directions. And the last scattering surface of the cosmic microwave photons is uniform to an amazing degree once the effects of local motions have been removed from the data. What is the cause of such uniformity? In the standard models of the universe such as the Einstein–de Sitter model the horizon becomes smaller and smaller the closer to the origin we go, and the horizon size at the recombination period now corresponds to an angular size of a only couple of degrees in the sky (as we calculated in Chapter 14.7). How were the universal temperature and other physical parameters coordinated over regions which have always been outside the range of each other's influence? There must be a flaw in this horizon argument. The answer given these days is **cosmological inflation** which we describe in this chapter. Before we get that far, we start by a discussion of the nature of vacuum.

16.1 Physics of the vacuum

The vacuum of non-zero energy has long been known in quantum physics. It is the ground state (the state of the lowest possible energy) of quantum fields. The physical vacuum was introduced in early relativistic quantum theory by the English physicist from Cambridge University Paul A.M. Dirac (a Nobel Prize winner in 1933). In 1927, he introduced the notion of a vacuum with infinite and negative energy as a possible realization of the lowest energy state for electrons. Positrons are treated as "holes" in this infinitely deep "sea" of negative energy. Soon afterward, G.A. Gamow mentioned that Dirac's vacuum sea must drastically affect cosmology via its gravity. We now know that the physical vacuum is definitely present in the experiments of atomic and particle physics.

Going into more detail, a quantum field may be represented by a Fourier transform (Box 13). Each mode of a fixed wavelength λ behaves like a simple harmonic oscillator. The energy of the quantum oscillator is

$$E = \hbar\omega\left(n + \frac{1}{2}\right), \tag{16.1}$$

where $\omega = 2\pi c/\lambda$ is the frequency of an oscillator, and n is an integer. This energy may have only discrete values corresponding to the number n. The energy difference between two energy states of the oscillator differing in n by 1 is $\hbar\omega$. When n is very big, the difference is small compared to the energies, E. Under these conditions, the discrete nature of the oscillator is not noticeable, so that the field can be described by classical field theory with continuous changes of energy. However, if the number n is small, the quantum effects are strong. Most important is the non-zero energy of the state with $n = 0$, the ground state of the field, the state of its lowest energy. No particles of the field exist in this state, so this is the vacuum state. The energy of the vacuum state is $\hbar\omega/2$. This non-zero minimal energy is referred to as the **zero-point energy**. Each mode of all the quantum fields contributes its zero-point energy to the total vacuum energy.

The zero-point energy is manifested in a number of effects which are directly measured in laboratory experiments. The most famous is the effect that is due to the ground state of the electromagnetic field which was theoretically predicted in 1948 by the Dutch physicist Hendrik B.G. Casimir while working at Philips Research Laboratories in Eindhoven. The effect arises due to a dependence of the state of the field on boundary conditions. Mathematically, not only the field equations but also the boundary conditions determine the representative set of quantum oscillators. Casimir considered the boundary conditions for two parallel conducting plates ("mirrors") in empty space, i.e., in the absence of external electromagnetic fields. Casimir calculated that the set of zero-point oscillators between the plates would be altered so that a macroscopic attractive force would appear between the plates. This force per unit area on the plate is

$$F_C = -\frac{\hbar c \pi^2}{240 d^4} \tag{16.2}$$

The force per unit area is proportional to the inverse fourth power of the distance d between the plates, (not more than a few microns in real experiments much less than the sizes of the plates).

Qualitatively, the effect assumes that the set of quantum oscillators in the space between the plates is different from that outside them. The space between the plates is finite in the direction normal to the plates, and thus oscillations with the wavelengths $\lambda = 2\pi c/\omega > d$ cannot be supported. Thus while the external energy density is produced by all the possible zero-point oscillations, the energy density between the plates is produced by an incomplete set of oscillations. The vacuum energy density is considered negative (like in Dirac's sea), and thus the vacuum pressure is positive. As a pressure gradient rises it makes the plates move towards each other i.e., attract each other.

Experiments since 1957 have confirmed the Casimir effect with increasingly better accuracy. They measure differences in the energy state of the physical vacuum but not the absolute value of the density of the physical vacuum. The major physical cause is that in any physical interactions, only differences of energy (at, say, two space-time events or two states of the system) are measured. The entire amount of energy does not affect the interactions. The result of the measurements of the observable quantum quantities does not depend on it. Because of this, in physical theory, the zero level for energy may be arbitrarily chosen, and does not change the expected values of observable quantities in any experiment. In particular, the lowest energy may be adopted as zero level, and then the vacuum energy density may be taken to be zero without loss of generality.

The only exception among forces to the above effect is gravitational interaction. The total amount of energy (not energy difference) produces gravity. As a result, total energy has a real physical sense only in gravitational physics, and total energy can be measured only in gravity experiments or systems. The supernova distance vs. redshift observations described earlier in this book are just of this kind of measurement. The universe itself serves as an experimental set-up.

In quantum theory (which is non-gravitational), the total energy of the physical vacuum may be calculated as a sum of all zero-point states with wavelengths from zero to infinity, or frequencies from infinity to zero. It is obvious that the sum diverges at infinite frequencies, so the vacuum energy becomes infinite, a straightforward result of quantum field theory.

The infinite total energy gives also an infinite energy density. A mode with a wavelength λ contributes the energy $\hbar\omega/2$ in a volume of size $\approx \lambda^3$. The zero-point energy density

$$\rho(0)_\lambda \approx \hbar\omega\lambda^{-3} \propto \lambda^{-4} \tag{16.3}$$

The sum over all wavelengths is

$$\rho(0) = \sum_\lambda \rho(0)_\lambda \tag{16.4}$$

which diverges at the shortest wavelengths, or highest frequencies.

Note that the dependence on the wavelength λ in Equation (16.3) is the same as the dependence of the Casimir force on the distance d between the conducting plates. The Casimir force F_C per unit area for distance d has the dimension of an energy density.

But despite these divergences, quantum theory is quite successful except for those related to gravity. Infinities of this origin do not produce observable effects in strong, weak, and electromagnetic interactions. However, infinities and the divergence in energy are catastrophic for gravitational interactions such as, cosmology. Infinite energy means infinitely strong gravity which "curves" space-time infinitely, so that space-time cannot actually exist at all. This is what Gamow had probably in mind when he commented upon the cosmological effect of Dirac's vacuum. But despite Dirac's vacuum, the universe does exist, the real curvature of space-time is finite, and in fact, is rather moderate.

Thus, the infinities of quantum theory make no physical sense. Instead, they reflect the incompleteness of our current knowledge about the phenomena in which quantum physics and also gravitational physics (i.e., General Relativity) play equally essential parts.

One way to avoid infinity in $\rho(0)$ is to limit the wavelength band of quantum oscillators and to assume that for all oscillators, λ is larger than some minimum $\lambda(0)$. Physics suggests the Planck length $L_{Pl} = G\hbar/c^3 \approx 10^{-33}$ cm (see Section 17.7) as a "natural" cut-off wavelength $\lambda(0)$. This is indeed a special length scale. The wavelength $\lambda(0)$ is the only combination of the three fundamental physical constants $\hbar, c, G$ that has the dimension of length. Planck quantities look like "natural" units of the physical conditions in which quantum physics (represented by $\hbar$), Relativity (c) and gravity (G) are involved at an equal degree. The Planck length is actually the absolute minimal length because the standard concepts of space-time are valid only for space scales > the Planck length. It is believed that at smaller scales, space (or actually space-time) is quantized—even though no one knows what this may exactly mean.

The "natural" vacuum density from the wavelength cut-off at the Planck length is

$$\rho(0) \approx \hbar c \lambda^{-4} \approx \hbar c L_{Pl}^{-4} = \rho_{Pl} \qquad (16.5)$$

where ρ_{Pl} is the Planck density $\rho_{Pl} \approx 10^{94}$ g/cm^3. The Planck mass is defined similarly, $M_{Pl} = \left(\frac{\hbar c}{G}\right) \sim 2 \times 10^{-5} g$, i.e. mass of light everyday bodies >> elementary particles. Note that the Planck density is enormously high as compared with the "observed" cosmic vacuum density ρ_V. The ratio

$$\rho_V / \rho_{Pl} \approx 10^{-123} \qquad (16.6)$$

is a quantitative measure of the miserably poor current knowledge in the field!

Besides the Planck density, which is clearly too large, the only other "natural" value for the physical vacuum density by fundamental theory is zero. In 1968, a Russian physicist at Moscow State University, Yakov B. Zeldovich, argued that the zero density might come from a kind of symmetry between bosons and fermions. This symmetry was suggested in a general way a few years later and is called supersymmetry (Box 10). Supersymmetry implies that the vacuum of fermion fields has the same energy as the vacuum of boson fields—in absolute value. But the signs of the energies are opposite. The energy of physical vacuum of bosons is positive while that of fermions is negative (like the previously mentioned Dirac's vacuum of electrons). For exact supersymmetry, the net vacuum energy sums up to zero.

Supersymmetry looks like a kind of theory where the absolute zero-point of energy may be meaningful. It suggests that gravity and space-time geometry must be involved in supersymmetry, in this or in some other way.

The observed universe is clearly not in a supersymmetric state. If supersymmetry exists at all, it must be broken. Or it may be exact "from the very beginning" but was slightly violated at some moment in the early universe—for instance, it may happen near the 1 TeV era, as some considerations suggest. At this moment and later on, the two vacuum contributions did not exactly cancel each other. Thus a tiny excess of the positive vacuum energy of the bosons over the negative vacuum energy of the fermions may appear. Then the sum of the two energies may be non-zero, small and positive.

As Zeldovich demonstrated in 1968, the net effect of all zero-point contributions with all possible wavelengths can be described in terms of hydrodynamics. The density $\rho(0)$ and the pressure $P(0)$ may be introduced and ascribed to the physical vacuum. Zeldovich calculated these quantities under some special additional conditions and found that the "zero-point fluid" has the equation of state $P(0) = -\rho(0) c^2$. This is the same equation of state that was recognized earlier for cosmic vacuum as represented by the cosmological constant. Many attempts to prove that the resulting vacuum energy density and pressure must be at the cosmological level have thus far failed (see Dolgov 2003 for a review).

Studies of the nature of physical vacuum have become especially active since the discovery of the dark energy cosmic vacuum. How, the origin of the cosmic vacuum, its microscopic structure and the nature of its constant density are unknown. Is the cosmic vacuum identical to the vacuum of the quantum fields? The concept of supersymmetry suggests the answer "Yes", qualitatively. "No" is suggested by a recent idea concerning macroscopic extra dimensions of space. This idea treats the observed vacuum as a three-dimensional "shadow" of a multidimensional vacuum which is assumed to be truly fundamental. It is unknown if extra dimensions really exist (see Chapter 17).

The observational discovery of the cosmic dark energy vacuum has significantly and importantly contributed to fundamental physics. Even if the cosmic vacuum has nothing in common with the vacuum of quantum fields, cosmology gives the strongest observational upper limit to the density of the physical vacuum. Now, if the cosmic vacuum is identical to the physical vacuum, cosmology provides the first direct observation and fairly precise measurement of a basic constant and invariant quantity in quantum-gravity physics.

16.2   Why does the universe expand?

A most attractive and reasonable answer to why the universe expands was suggested by Erast B. Gliner (1965) of the Joffe Institute (St. Petersburg, Russia). He assumed that the initial state of the universe was a vacuum. This is not the same cosmic dark energy vacuum discovered in supernova observations decades later, but a sort of initial, primordial vacuum with a density ρ_I near the Planck density $\rho_{Pl} \approx 10^{94}$ g/cm^3. Despite this extremely high density, all other physical properties of the primordial vacuum are the same as that of the observed much lower density cosmic dark energy vacuum. In particular, the primordial vacuum's equation of state is $P_I = -\rho_I c^2$ with a constant positive density ρ_I. Its effective gravitating density, $-2\rho_I$, is thus negative, so the primordial vacuum produced strong antigravity. It is also assumed that there were no particles or fields of matter initially, so the primordial vacuum alone determined the space-time. The primordial vacuum space-time was static and described by the de Sitter static metric (see Box 14):

$$ds^2 = S(r)c^2 d\tau^2 - S(r)^{-1} dr^2 - r^2 d\Omega^2 \qquad (16.7)$$

where $d\Omega^2 = d\theta^2 + \sin^2\theta d\phi^2$ and $S(r) = 1 - (r/A_V)^2$, r is the radial coordinate, τ is a time coordinate which is different from the proper time t of the Friedmann metric element and A_V is a length constant. In the present day universe the value of A_V is

$$A_V = (\kappa \rho_V)^{-1/2} = 1.6 \times 10^{28} \, \text{cm} \qquad (16.8)$$

where $\kappa = \frac{8\pi G}{3c^2}$ is an appropriate combination of natural constants so that a length results. The inflation period value of $A_V = A_I$ was very much smaller. Note that Equation (15.2) gives the corresponding time scale $A_V/c = 5 \times 10^{17}$ s.

Box 14. De Sitter static model.

The Schwarzschild metric makes no inclusion of the cosmological constant Λ which can be corrected by adopting the following metric (Rindler 1977)

$$ds^2 = c^2 \left(1 - \frac{2GM}{c^2 r} - \frac{1}{3}\Lambda r^2\right) dt^2 - \frac{dr^2}{1 - \frac{2GM}{c^2 r} - \frac{1}{3}\Lambda r^2} - r^2 d\theta^2 - r^2 \sin^2\theta d\phi^2.$$

$$\text{(B14.1)}$$

For slow motion this leads to orbits in a central potential

$$\Phi = -\frac{GM}{r} - \frac{1}{6}\Lambda c^2 r^2. \qquad (\text{B14.2})$$

The second term above represents a repulsive central force whose magnitude is $\frac{1}{3}\Lambda c^2 r$ (Equation (8.5)). It causes, among other things, an additional precession of orbits. Now the magnitude of the precession must be too small to be measurable in the Solar System.

If we let $M \to 0$, the static metric is that of empty space, except for the vacuum:

$$ds^2 = c^2 \left(1 - \frac{1}{3}\Lambda r^2\right) dt^2 - \frac{dr^2}{1 - \frac{1}{3}\Lambda r^2} - r^2 d\theta^2 - r^2 \sin^2\theta d\phi^2 \qquad (\text{B14.3})$$

as was discovered by de Sitter in 1917. An apparent singularity occurs at $r = (3/\Lambda)^{1/2}$ but this is purely a coordinate problem, not a real singularity. This de Sitter static space is a generalization of the Minkowski space (Equation (1.11)) for the case of the vacuum. The metric represents a hypersphere of curvature $-\frac{1}{3}\Lambda$, with a curvature radius $A_V = |\frac{1}{3}\Lambda|^{-1/2}$.

The first term of the metric (the 00-component) gives rise to the redshift ("de Sitter's effect") in the light arriving from distant objects in this static model.

When t goes to infinity and the scale factor of the expansion becomes exponential, $R(t) \propto \exp(ct/A_V)$, the Friedmann metric is asymptotically reduced to the de Sitter metric by means of a simple coordinate transformation. This is also true for the models with positive and negative curvature of the co-moving space for which the scale factor $R(t) = A_V \cosh(ct/A_V)$ and $R(t) = A_V \sinh(ct/A_V)$, respectively; the scale factor is normalized here on the curvature radius.

In this new reference frame and coordinate system in which the de Sitter solution is given, the spatial section is different from the co-moving space of the Friedmann metric. However, the cosmic vacuum fills any of these spaces uniformly, and its density is the same in all of them. This is because the vacuum density is a covariant quantity, and if we have measured it in the co-moving space (as in the real supernova observations), we may be sure that it is the same in any other space section.

The space-time of the present universe can approximately be described by the de Sitter solution, and in this sense the accelerating universe is transforming to a nearly static space-time. The faster the universe expands, the closer it is to the de Sitter steady state metric, so that this approximation works better.

However, the static vacuum in the static space-time can cause motion of matter. If two particles appear in the primordial vacuum, they cannot be at rest. Instead, they tend to move towards each other under the action of their masses' mutual gravity. But the primordial vacuum acts on them by its own antigravity which might be much stronger than the mutual attraction of the particles, because of the vacuum's enormous density. Antigravity makes the particles move apart from each other. The distance R between them thus grows exponentially as a function of time: $R \propto \exp(ct/A_I)$, where $A_I = (\kappa\rho_I)^{-1/2}$.

As is well known in modern physics, virtual pairs of particles and antiparticles can spontaneously appear from physical vacuum. This quantum process of the pair birth can straightforwardly be assumed in the initial vacuum state because quantum processes are as significant as gravity or antigravity at the Planck density we computed. With no antigravity, the two particles would be attracted and move towards each other to annihilate very soon. However, the very strong antigravity of the primeval vacuum prevents the annihilation by making the particles move away from each other. By this process, the virtual particles might become real particles.

Thus, according to Gliner, the primordial vacuum produces particles and makes them move apart each other. Thus, the general cosmological expansion might be created.

This ingenious elegant idea was developed by Erast B. Gliner and Irina G. Dymnikova (1975) as well as Lev E. Gurevich (1975) of Joffe Institute in the 1970s. Since the early 1980s, Alan Guth, Andrei Linde, Alexei Starobinsky and many other cosmologists have attempted to elaborate the idea and develop it into a concise and productive theory. Hundreds of specific models have been suggested to describe quantitatively the physics of the primordial vacuum, matter generation in it and the process of matter acceleration. This process of the generation of the matter in the universe and its acceleration is commonly referred to as **inflation**. Inflation occurs at an initial pre-Friedmann stage of the cosmic evolution which lasts a very tiny fraction of a second.

As yet there is no single reliable model of inflation. Many suggested models suggested during the last 30 years have some common features which follow. One feature, an initial finite volume of space is considered as the initial state of the universe. The initial volume's size is of $\sim 1\,000$ times the Planck length, $L_{Pl} \approx 10^{-33}$ cm. It is assumed that the volume is a de Sitter universe (or something which is very similar to it) with a vacuum density near the Planck density, or that order. It is also typically assumed that the domain is in the state of exponential (or nearly exponential) expansion. This exponential expansion proceeds for millions of Planck times, $t_{Pl} \approx 10^{-43}$ sec, which ends with the decay of the vacuum and the generation of matter. The expanding volume is near 1 cm at the end of the inflation. After that the Friedmann cosmic evolution starts. After inflation and the consequent Friedmann expansion, the domain has at present the size about 10–100 times the observable horizon radius of today.

Interesting descriptions and many details of inflation models may be found in the excellent books by the major authorities in inflation: A. Linde, *Particle Physics and Inflationary Cosmology* (1990); A. H. Guth, *The Inflationary Universe* (1997).

This scenario described above raises some questions. The simplest is: What exactly does expand when there is no matter at inflation? No clear answer to this and other simple questions about inflation has as yet been found.

From the theoretical point of view, an unlucky flaw of inflation models is that they are all non-covariant being based on $3 + 1$ split of space-time. From a 4-dimensional point of view, there is only the static de Sitter space with no expansion, in what is treated as exponential inflation.

The main obstacle for development of inflation models is that current fundamental physical theory cannot handle the Planck densities, lengths and times—such physics is as yet unknown. For example, the law of gravity is reliably known only down to the sub-millimeter spatial scales while the Planck length is 30 orders of magnitude smaller. Using General Relativity and current particle physics for inflation requires an extrapolation of these standard theories to conditions very far from the scope of current experimentally verified theory.

In view of these difficulties, it is hardly accidental that the inflation models failed to predict the existence of the real cosmic dark energy vacuum prior its observational discovery. After the discovery, the models now cannot explain the measured value of the vacuum density. Below we will describe some models of inflation in greater detail. But first we discuss the main reasons why such a model is considered necessary.

16.3   Why is the universe uniform?

The observed universe is well-organized as a whole. If it is really finite with the present size near the present horizon radius, practically the whole volume of the universe is seen. An observer finds it uniform and isotropic everywhere. The isotropy manifests itself most obviously in the isotropy of the cosmic microwave background (CMB) radiation. As we have discussed, the CMB looks perfectly isotropic (after extraction of the observer's motion and the imprints of the initial perturbations) up to the largest angular scale of about 180°. The CMB was emitted at the recombination epoch, which was the last chance for the CMB photons to come into "agreement" about their angular distribution over the sky. Before recombination happened, photons interacted with baryonic plasma, and in this way they were able to "communicate" with one other. After recombination, the relic photons propagated freely, practically without any communication among themselves.

However, the possibility of interaction is limited at recombination. Any interaction agent propagates with a finite velocity. The speed of light is the maximal velocity of the propagation. The horizon radius, ct_{rec}, is thus the principal distance limit for the interaction at recombination. In the CMB, the horizon radius $r = ct_{rec}$ is seen at an angle $\theta \cong (t_{rec}/t_0)(1 + z_{rec})$. Putting $t_{rec} \cong 10^{13}$ s appropriate to our earlier discussion, $t_0 \cong 4 \times 10^{17}$ s for the present day, $z_{rec} \cong 1\,100$ as observed, one gets $\theta \cong 2°$. Thus at the "wall of the last scattering", there are very approximately $(180/2)^2 \cong 8\,000$ areas of this angular size that have never been in interaction and simply never "saw" each other. More formally, one may say that all these areas were causally disconnected (for a more formal calculation, see Section 14.7).

Thus, the CMB we now observe comes from areas separated by angles which may be significantly larger than the maximal interaction angle θ. Considering the observed uniformity, how does radiation coming from one part of the sky "know" how strong it must be to match so precisely the radiation coming from, say, the opposite part of the sky and points in between? The same question about causality connection concerns also the spatial cosmic matter distribution which is uniform on large scales—up to the CMB horizon. We see the parts of the universe, which are separated by such distances as the horizon radius for the first time today. Why do they look so similar?

And, finally, from the supernova observations and other observations billions of light years away down to a few million light years in our own Local Group of galaxies, why is the cosmic vacuum apparently uniform in all the universe over its history?

This is the **uniformity** or **causality problem** in cosmology. This problem is due to the fact that the universe had a starting point some 13.7 Gyr ago, and so its age is finite. Therefore the light can thus cover only a finite distance in this finite amount of time, and so the causally connected areas are limited in size.

Gliner's primordial vacuum idea provides an interesting approach to the causality problem. If the initial state of the universe was a vacuum, what is today an almost infinitely large volume could have been, in principle, within one causally connected area early in the history of the universe. Imagine light beam going from a source to an observer. From the equation of light propagation, $ds^2 = 0$ (Equation (16.7)), written for the de Sitter solution in the Friedmann metric. The form of solution of the equation is

$$\chi = c \int_0^{t_0} d\tau/a(\tau), \tag{16.9}$$

where τ is the proper time of the metric, and $a(\tau) \propto \exp(c\tau/A_I)$ is the exponential scale factor of the metric. The integral diverges at the lower limit, so that the radial coordinate χ is infinite. This is because $a(\tau)$ rises with time faster than linearly, which is only possible for the accelerated vacuum-driven solutions. (At the later Friedmann stage of the expansion, there is no such acceleration or divergence). Identifying the radial coordinate χ with the coordinates of the particles generated by the primeval vacuum, all the particles turn out to be within the causally connected region.

This was discussed by A. Guth (1981) for inflation models which assume that the initial size of the cosmological domain has a size which is the size of the causally connected area at the initial moment. The size is given by the Hubble radius, $c/H_I = A_I$, a constant determined by the initial vacuum density. For example, if the density of the primordial vacuum $\approx 10^{88}$ g/cm^3, the initial size of the volume $\approx 10^{-30}$ cm. Within this causally connected area, there were interactions that were able to make the volume uniform. After that, the volume expanded exponentially, as assumed, for the time $\cong 60(G\rho_I)^{-1/2}$, and preserved its uniformity. At the end of inflation, its size was about 1 cm. The uniformity is preserved also during the subsequent Friedmann evolution—up to the present epoch. Thus, according to inflation models, the universe looks uniform because it was causally connected and uniform at the very beginning.

Figure 16.1 illustrates the evolution of horizons (the Hubble length) in the expanding universe. The physical scale is representative of the scales where we are able to make observations of uniformity, i.e., in the currently observable universe and in

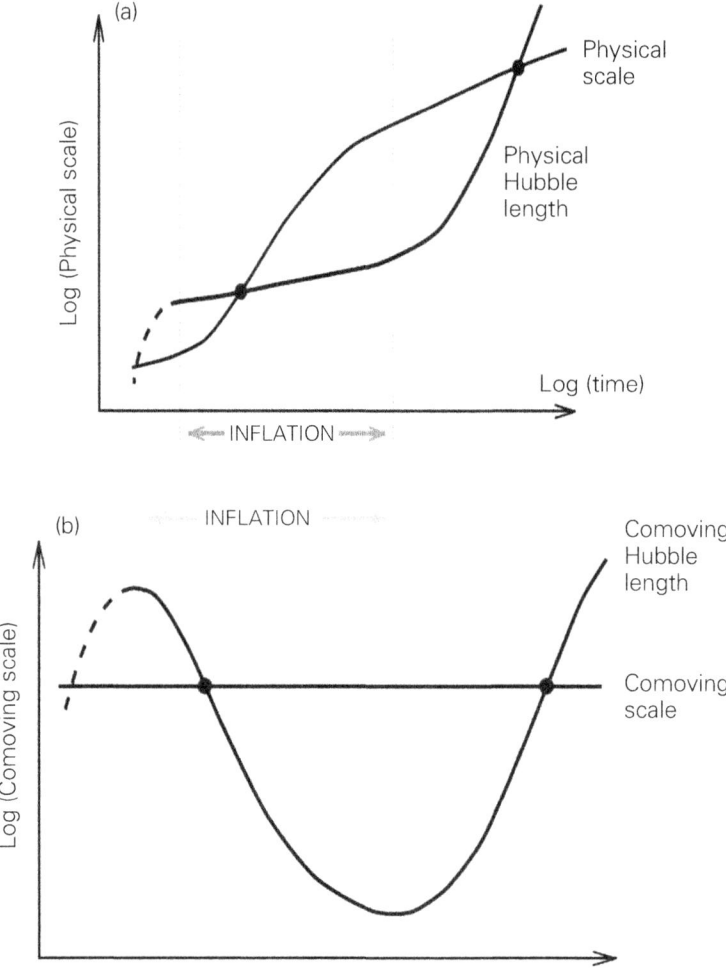

Figure 16.1. Upper diagram (a): A comparison of the horizon scale (Physical Hubble length) with the expansion scale (Physical scale) in the early universe according to the inflation model. Lower diagram (b): The physical Hubble length is scaled to the physical expansion scale (the Co-moving scale), to make it the Co-moving Hubble length. The horizontal axis is logarithmic in time, in arbitrary units. The period of inflation is marked on the time axis. Drawing after Liddle 1999.

scales a few orders of magnitude below it. All these scales are covered practically within the thickness of a single line in the figure.

At the beginning of the inflation, our scale of interest (the physical scale) is smaller than the scale of physical interactions (the physical Hubble length), and thus all deviations from uniformity are ironed out. Perturbations may survive in small scale fluctuations only. However, at some point during the inflationary expansion (marked by the first circle on the time axis) the lines cross each other, and our scale of interest becomes larger than the horizon. At this point any perturbations become imprinted or "frozen" in the universe. Any causal physics in our scale of interest becomes ineffective.

Long after the inflation is over, the lines cross again (the second circle on the time axis), and then physical interaction may again resume in our scale of interest. In the standard model without inflation, the Hubble scale curve would be always below the physical scale curve to the left of this circle, thus creating what is called the **horizon problem**. With the inflation period, the Hubble scale curve pops up again above the physical scale left of the first circle, at very early times, making uniformity possible even in the largest observable scales. The inflation period also solves a second problem called the **flatness problem**. The huge expansion during the inflation period by a factor of 10^{30} or more "irons" out any strong curvature that there may have been initially in the universe and produces a spatially flat universe with $k = 0$, as we observe. This may be seen from Friedman's Equation (8.14) which can be rather simply transformed into the form

$$c^2 k = (\Omega - 1) R^2 H^2.$$

If in the beginning the curvature was non-zero ($k = -1$ or $+1$), then the density parameter Ω has to approach unity (i.e., the flatness), if RH increases with time. On the other hand $RH = \mathrm{d}R/\mathrm{d}t$ increases if $d(dR/dt)dt$ is positive, which just means accelerated expansion!

These solutions to cosmological problems are attractive; but they raise some questions. What was the physical agent or process which provided the causal connection at the initial vacuum state? There must be processes leading to the very small amplitude $>$ the horizon non-uniformities observed in the CMB. How might the radiation isotropy be established at the epoch when radiation did not exist? These and other simple questions are open. The causality problem needs more study.

16.4   Scalar fields

Scalar fields are the most popular way to describe inflation. It is assumed that the Universe is filled by an unknown field which was extremely strong in the early universe and which has long since decayed to very low levels, if not to zero exactly. In particle physics, a scalar field is used to represent particles of zero spin. It transforms as a scalar under coordinate transformations. It is usually assumed that the scalar field is a function of time alone.

In condensed matter physics scalar fields are used in connection of phase transitions where they are called order parameters. Correspondingly in particle physics theories scalar fields are a crucial ingredient for spontaneous symmetry breaking. Scalar fields have appeared also elsewhere in physics, for example in Yukawa's theory of nuclear force, in higher dimensional Nordström–Kaluza–Klein theories, in supergravity, and in superstring theories. Supersymmetry theories predict the existence of spin 0 (scalar) partners of spin 1/2 leptons.

The most famous example of spontaneous symmetry breaking is the Higgs field which is thought to break the electro-weak symmetry. The existence of the Higgs field is still to be verified experimentally. Scalar fields are also expected to be associated with the breaking of other symmetries, such as those of Grand Unified Theories. Scalar fields are also postulated in scalar-tensor theories of gravitation which are connected with string theory. The cosmological constant Λ is an example of a scalar field in cosmology. Thus it is quite natural to suggest that a scalar field or field of unknown kind is responsible for the inflationary expansion of the universe. There is no particular scalar field theory of wide acceptance; however, there are some general properties of the fields that are likely to be satisfied. Here we follow the discussion by Hobson, Efstathiou and Lasenby: *General Relativity. An Introduction for Physicists (2006)*.

The scalar field that drives inflation is usually called the inflaton field. It cannot replace General Relativity since we know that gravitation cannot be a scalar theory, since spin 0 (scalar) theories show no coupling to light—no deflection of starlight by the Sun, for example. But there is no evidence ruling out a general scalar addition to General Relativity. The scalar field can be added to the general relativistic Lagrangian, and the theory can be then derived by the usual methods. The Friedmann equations simply obtain extra terms from the scalar part of the action, and one can do the usual cosmological calculations.

Here we will describe a much used model of a scalar field ϕ where the diagonal elements of the energy-momentum tensor, density ρ and pressure P, have simple expressions. We take the density to be

$$\rho c^2 = \left[\tfrac{1}{2}\dot{\phi}^2 + V(\phi)\right]. \tag{16.10}$$

The first term on the right hand side is equivalent to kinetic energy and $V(\phi)$ corresponds to potential energy. The pressure P is given by

$$P = \left[\tfrac{1}{2}\dot{\phi}^2 - V(\phi)\right]. \tag{16.11}$$

In general the kinetic and potential energies evolve in time and their ratio has no special value. But it is interesting to look at three special cases.

A **free field** has no potential energy and hence its equation of state is $\rho c^2 = P$ or $w = P/\rho c^2 = 1$. In the **slow roll** case, to be described in more detail below, the kinetic energy is negligible and the equation of state is $\rho c^2 = -P$ or $w = -1$. When the field undergoes **coherent oscillations**, the time average of the kinetic and potential energies are equal, and the equation of state is $P = 0$ or $w = 0$. Then the scalar field acts like ordinary non-relativistic matter.

Rather than being an unchanging field like the cosmological constant, the scalar field can vary with time. One idea is that the potential leads to a scalar field which **tracks** the equation of state, evolving from $w = 1/3$ in the radiation dominated epoch, to $w = 0$ about the time of matter-radiation equality, to $w < 0$ today and approaching $w = -1$ in the future. This makes a very attractive form of dark energy (called **quintessence** or the fifth element, borrowing its name from the belief in antiquity that the celestial spheres are made of the fifth element, quintessence, in contrast to the ordinary elements of earth, fire, air, and water). This idea has been introduced primarily to explain the present day relatively modest acceleration of expansion.

In a given theory, we will specify the potential $V(\phi)$, at least up to some parameters which one could hope to measure. There is no well established fundamental theory that one could use as guidance for the choice of the potential. People tend to regard $V(\phi)$ as a function to be chosen arbitrarily, with different choices corresponding to different models of inflation. An example of such potentials is a massive scalar field

$$V(\phi) = \tfrac{1}{2} m\phi^2. \tag{16.12}$$

The parameter m is related to the mass of the particle which carries the interaction of the scalar field. We will consider a simple potential displayed in Figure 16.2.

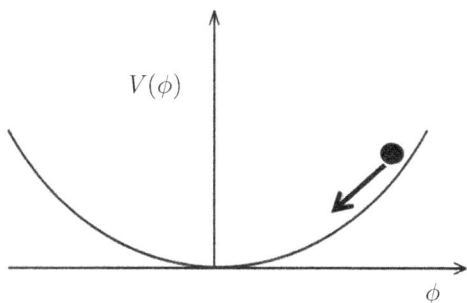

Figure 16.2. A generic inflationary potential where the scalar field is displaced from the minimum. It rolls down trying to reach the bottom of the potential. After a drawing by Liddle 1999).

16.5   Equation of motion

The equations for an expanding universe containing a homogeneous scalar field are easily obtained by substituting Equations (16.10) and (16.11) into the Friedmann thermodynamic Equation (10.12)

$$\dot{\rho} + \left(\rho + \frac{P}{c^2} \right) \left(\frac{3\dot{R}}{R} \right) = 0. \tag{16.13}$$

Taking the time derivative of Equation (16.10),

$$\dot{\rho} c^2 = \ddot{\phi}\dot{\phi} + \frac{dV}{d\phi}\dot{\phi} \tag{16.14}$$

and using we get

$$\ddot{\phi} + 3H\dot{\phi} + V' = 0. \tag{16.15}$$

We may recognize this equation from a simple dynamical problem of determining the motion of a rolling ball on a curved surface, under the force of gravity, and including a friction term (the middle term on the left hand side). In this equivalent problem ϕ is the horizontal coordinate and V the vertical coordinate. See Figure 16.2. Equation (16.15) is the equation of motion of the scalar field.

In the very early universe we may assume that the scalar field dominates the energy density. We may also assume that in the Friedmann equation the curvature term kc^2 can be neglected. Then we have

$$\dot{R}^2 = \frac{8\pi G}{3} \rho R^2. \tag{16.16}$$

Substitute the scalar field density from Equation (16.10):

$$H^2 = \frac{8\pi G}{3} \left[\frac{1}{2} \dot{\phi}^2 + V(\phi) \right]. \tag{16.17}$$

From Equations (16.15) and (16.17) we may solve ϕ and H for any given potential $V(\phi)$. From Equation (13.137), the acceleration is given by:

$$\ddot{R} = -\frac{4\pi G}{3} \left(\rho + \frac{3P}{c^2} \right) R, \tag{16.18}$$

which after substitution of Equations (16.10) and (16.11) becomes

$$\ddot{R} = -\frac{8\pi G}{3c^2} \left[\dot{\phi}^2 - V(\phi) \right]. \tag{16.19}$$

Therefore the condition for acceleration in the early universe is

$$\dot{\phi}^2 < V(\phi). \tag{16.20}$$

We can make this condition even stronger by saying that

$$\dot{\phi}^2 \ll V(\phi). \tag{16.21}$$

In our rolling ball example this condition corresponds to a slow roll on a gentle slope. Thus in cosmology we call this condition the **slow roll approximation**. In this approximation $\ddot{\phi}$ can be neglected in Equation (16.15) and we have simply

$$3H\dot{\phi} + V' = 0. \tag{16.22}$$

The cosmological field equation (Equation (16.17)) is simply

$$H^2 = \frac{8\pi G}{3} V(\phi). \tag{16.23}$$

Differentiate this:

$$2H\dot{H} = \frac{8\pi G}{3} V'\dot{\phi}, \tag{16.24}$$

and substitute V' from Equation (16.22). We get

$$\dot{H} = -4\pi G\dot{\phi}^2. \tag{16.25}$$

We may rewrite the slow roll condition (Equation (16.21)) in a different way. Equations (16.22) and (16.23) give

$$\dot{\phi}^2 = (V')^2/9H^2 = (V')^2/24\pi G V \tag{16.26}$$

whereby

$$\varepsilon \equiv \frac{3}{2}\frac{\dot{\phi}^2}{V} = \frac{1}{16\pi G}\frac{(V')^2}{V^2} \ll 1. \tag{16.27}$$

Thus ε measures the slope of the potential. The curvature is measured by

$$\eta \equiv \frac{1}{8\pi G}\frac{V''}{V}.$$

By differentiating Equation (16.27) with respect to ϕ we find that $\eta \ll 1$. The conditions

$$\varepsilon \ll 1 \quad \text{and} \quad \eta \ll 1 \tag{16.28}$$

say that the potential has to be sufficiently flat for the inflation to occur. A special case when the potential is exactly flat, i.e., $V(\phi)$ is constant, is easily solved from Equation (16.23); we have already encountered this solution in Chapter 8 where it was called the de Sitter universe. Inequalities (16.28) are called the **slow roll conditions**. Inflation will occur when the slow-roll conditions are satisfied.

16.6   Reheating after inflation

Inflation dilutes all fields that there may have been initially to extremely low densities. The only exception is the inflaton field. Inflation ends in **reheating** where the inflaton's energy density is converted into conventional matter. After reheating we may follow the standard Big Bang theory.

If the potential has a local minimum such as is illustrated in Figure 16.2, the slow roll will end at some point and the field will start oscillating around the minimum. The oscillation is gradually damped by the $3H\dot{\phi}$ friction. Finally the field becomes stationary at the bottom of the potential at some value V_{min}. If this value is greater than zero, the conditions for inflation are again satisfied, but now with a constant value of ϕ. It corresponds to a late inflation and could be a physical explanation of the Λ term in Einstein equations, assuming that the value of V_{min} is just right and corresponds to the measured accelerations in today's universe.

When the scalar field begins to oscillate, the process transfers energy from the coherent oscillations of the inflaton field to creation of particles. The reheating epoch may be very short and efficient, and hence most of the energy density in the inflaton field at the end of inflation may be available for conversion into particles.

Since the time evolution of the cosmic scale is

$$R(t) \propto \exp(Ht)$$

the number N of e-foldings within the time range from t_1 to t_2 is

$$N = \int_{t_1}^{t_2} H dt = \int_{\phi_1}^{\phi_2} \frac{H}{\dot{\phi}} \, d\phi = -8\pi G \int_{\phi_1}^{\phi_2} \frac{V}{V'} \, d\phi,$$

where we have used Equation (16.22) and (16.27) in the last step. We may estimate $V' \sim \frac{V}{\phi}$, and if $\Delta\phi = |\phi_2 - \phi_1|$, $N \sim 4\pi G(\Delta\phi)^2$. The solution of the horizon and flatness problems requires $N > 60$ which means $\sqrt{4\pi G}\Delta\phi \gg 1$. This is an additional requirement for plausible scalar fields.

16.7   Density perturbations and gravitational waves

One of the most important properties of inflationary cosmology is that it explains the origin of the spectrum of density perturbations, i.e., the Harrison–Zeldovich spectrum, or a scale-free power-law spectrum. It also predicts the existence gravitational waves which have not yet been observed. The density perturbations evolve at lower scales to provide seeds for the formation and clustering of galaxies. At larger scales they explain the anisotropies in the microwave background radiation. The initial perturbation spectrum is observed most directly at the low l numbers of the CMB spectrum where it provides the observed plateau.

The gravitational waves do not affect the formation of galaxies, but may contribute extra microwave anisotropies on the large angular scales. The density perturbations are often called scalar perturbations and the gravitational waves tensor perturbations; this terminology refers to their transformation properties. As we said previously, the perturbations that existed initially at the beginning of inflation may have survived "frozen" throughout the inflation period and may have provided the seeds for further evolution of density structure.

The mechanism for producing density perturbations in the inflation cosmology is quantum fluctuations. Inflation is trying to make the universe perfectly homogeneous, but it cannot defeat the Uncertainty Principle which requires that always some irregularities remain. By this process, it is possible for inflation to solve the homogeneity problem and at the same time leave enough irregularities behind to explain why the present universe is not completely homogeneous.

At present, observations are only quite weakly constraining concerning different inflation models. The spectral index of initial density perturbations n is known to lie near one, with the plausible range, depending on what parameters one allows to vary, stretching from perhaps 0.8 to 1.2. As it happens, that is more or less the range which current inflation models tend to cover, and so most models survive. It is important for inflation model building to get an accurate measurement of n, say with an error bar of around 0.01 or better. Such a measurement would exclude the vast majority of the models currently under discussion. The currently operational Planck CMB observatory should reach around this accuracy level.

At the moment there is no evidence for a gravitational wave contribution to the CMB background. If such a contribution can be identified, it will be very strong support for inflation. Even a good upper limit would be able to rule out some models. At present, inflation is the most promising candidate theory for the origin of perturbations in the universe. Different models lead to clearly different predictions for these perturbations, and hence high-accuracy measurements are able to distinguish between models, excluding either all or the vast majority of them.

Since its inception, inflationary cosmology has been a gallery of different models. More and more models have been proposed and very few models have been discarded. However, the near future holds promise of finally being able to throw out inferior models. If inflationary cosmology survives as our model for the origin of structure, we may hope that only a narrow range of models remains to choose between.

Chapter 17
Cosmic internal symmetry

Empedocles (490–430 BC) proposed that everything in the material world is made of four ultimate substances which are fire, air, water, and earth. Plato used later the term "elements" for these. The concept of the four basic elements of Nature was developed by Aristotle, and became the standard dogma for two thousand years. Nowadays, the standard cosmological model that emerged in 1998–1999 as a result of the discovery of cosmic vacuum, or dark energy, assumes that everything in the universe is also made of four different basic ingredients: dark energy, dark matter, baryons, and radiation (Chapter 8). In this final chapter of this book, we review current results related mainly to the energy composition of the universe. We will show that a simple time-independent correspondence—a kind of internal symmetry—exists that unifies the four cosmic energies despite obvious differences in their physical properties. This symmetry suggests a new approach to some old and modern cosmological problems that otherwise seem unrelated.

17.1   Time-independent parameters of the Metagalaxy

The space volume available for cosmological observations has a radius $R_0 \sim 10^{28}$ cm ~ 10 Gpc. This radius is comparable, in order of magnitude, to the Hubble radius $c/H_0 \sim 10^{28}$ cm, where H_0 is the present-day Hubble constant, and to the horizon radius ct_0, where t_0 is the present age of the universe. The largest distances which can be reached with the best modern instruments have approached these limits since the mid-1990s. We will refer to the largest observable part of the universe as the Metagalaxy. Theoretically, the Metagalaxy may be or may not be a part of a larger or even infinite universe (such as described by the zero-curvature Friedmann model); but this cannot be proved or disproved in direct astronomical observations. In this sense, questions about anything "outside" or "beyond" the Metagalaxy have no physical meaning. The Metagalaxy is a causally connected and practically isolated physical system. While initially the notion of the Metagalaxy is related to its present-day state, it may be extended also to the past and future of the system as well. Taking the Metagalaxy to be the 3D expanding co-moving volume with the present-day radius R_0, one may use the standard cosmological model (Chapter 8) to study its evolution. The size of the system grows with the cosmological expansion as $R = R_0(1 + z)^{-1}$. In this chapter, we will deal mostly with empirical studies in cosmology, and because of this we will preferentially use the term "Metagalaxy" rather than "universe".

The energy content of the Metagalaxy will be the focus of our discussion here. The recipe of the cosmic mix is given by the observed densities of: cosmic dark energy or vacuum (V), dark (D) matter, baryons (B) and, finally, radiation (R) which is mostly the ultra-relativistic gas of the CMB photons (Chapter 8). Recall that

$$\Omega_V = 0.73; \quad \Omega_D = 0.23; \quad \Omega_B = 0.04; \quad \Omega_R = 9 \times 10^{-5}. \tag{17.1}$$

These are measured in the units of the present-day critical density:

$$\rho_c(t_0) = (3/8\pi G)H_0^2 = 0.94 \times 10^{-29}\,\mathrm{g/cm^3}, \tag{17.2}$$

According to the standard cosmological model, dark energy treated as cosmic vacuum has a constant density, ρ_V, which is associated with the Einstein cosmological constant.

As we have discussed earlier in this text, the three non-vacuum energies are time-dependent whose densities decrease with cosmological expansion as follows:

$$\rho_D \propto \rho_B \propto R(t)^{-3}, \quad \rho_R \propto R(t)^{-4}.$$

The two "dark" ingredients, the dark energy vacuum and dark matter, have comparable densities that differ at present by not more than half an order of magnitude:

$$\Omega_V/\Omega_D \simeq 3. \tag{17.3}$$

In fact, all four energy densities are "astrophysically" comparable at present. Their values are within four orders of magnitude. This is a clear indication of the special character of the present epoch of the cosmological evolution. Indeed, at other moments of cosmic time, the densities may differ from the present-day values and from each other by many more orders of magnitude. For instance, in the era of Big Bang Nucleosynthesis (BBN) when the age of the universe was measured in a few minutes, radiation dominated with a density a billion times the baryon density and 10^{32} times the vacuum density. In the remote future, the non-vacuum densities will tend to zero and the vacuum density becomes completely dominant.

Note especially the following close coincidence seen at the present time which is time-independent:

$$\Omega_D/\Omega_B = 5.8. \tag{17.4}$$

Since dark matter and baryons are non-relativistic, the dimensionless parameter of Equation (17.4) is conserved over the entire history of the universe starting from the BBN era.

Another cosmic dimensionless parameter that is conserved is the ratio of the number density of the CMB photons to the number density of baryons. Using the present-day values, $n_R(t_0) = 410$ cm^{-3}, and $n_D(t_0) = \rho_B/m \simeq 10^{-7}$ cm^{-3} (where m is the proton mass), we calculate

$$n_R/n_B \simeq 2 \times 10^9. \tag{17.5}$$

The cosmic density parameter, which is the ratio of the total energy density to the critical density, $\Omega = 1$, is also a conserved cosmic dimensionless parameter in standard cosmology with a flat 3D co-moving space. This parameter, as well as the vacuum density and the ratios of Equations (17.4) and (17.5), are strongly time-independent physical parameters of the Metagalaxy.

There are also three "extensive" time-independent physical parameters of the Metagalaxy using current local values and the Metagalaxy radius. The total mass of dark matter in it,

$$M_D = \frac{4\pi}{3} \rho_D(t_0) R_0^3 = 9 \times 10^{54}\,\text{g} \simeq 5 \times 10^{21} M_\odot, \tag{17.6}$$

the total baryonic mass,

$$M_B = \frac{4\pi}{3} \rho_B(t_0) R_0^3 = 2 \times 10^{54}\,\text{g} \simeq 1 \times 10^{21} M_\odot, \tag{17.7}$$

and the total number of the CMB photons,

$$N_R = \frac{4\pi}{3} n_R(t_0) R_0^3 \simeq 2 \times 10^{87}. \tag{17.8}$$

What is the physical nature of the key time independent parameters of the Metagalaxy? Are there hidden relations among them or are there possible links to fundamental constants of nature? These questions are the major subjects of our discussion below.

17.2   The Friedmann integrals

We introduce now the Friedmann integrals which will be used in the empirical analysis of the data of the previous section. The expansion dynamics of the Metagalaxy whose total radius is $R(t)$ is controlled by the Friedmann "dynamical" equation (Equation (8.14) with $k = 0$, with the density components of Equation (8.10), including radiation density, shown explicitly):

$$\dot{R}(t)/c^2 = (A_V/R)^{-2} + A_D/R + A_B/R + (A_R/R)^2. \tag{17.9}$$

The Equation (17.9) describes a parabolic expansion in a flat 3D co-moving space, as adopted in standard cosmology. The four constants A_V, A_D, A_B, A_R

represent the four cosmic energies. These are the Friedmann integrals (hereafter FINTs) which are time-independent genuine physical characteristics of each of the four cosmic energies. Historically, the length-dimension constant A was introduced by Friedmann (1922) to represent non-relativistic matter in his dynamical equation.

The FINTs come from the Friedmann "thermodynamic" equation (Equation (10.12) and Equation (16.13)) which is applied to each of the energies individually:

$$\frac{\dot{\rho}}{(\rho + w)} = -3\,\frac{\dot{R}}{R}. \tag{17.10}$$

Here the constant pressure-to-density ratio $w = P/\rho c^2 = -1,\ 0,\ 0,\ 1/3$ for the dark energy vacuum, dark matter, baryons, and radiation, respectively. The integral of the equation is:

$$A = \left[\kappa \rho R^{3(1+w)}\right]^{\frac{1}{1+3w}}, \tag{17.11}$$

where $\kappa = \frac{8\pi G}{3c^2}$ is the Einstein gravitational constant and G is the Newton gravitational constant.

Because of their origin from Equation (17.10) as constants of integration, the FINTs are *a priori* completely independent of one another and not restricted by any non-trivial theoretical constraints. Each of them expresses a conservation law for the corresponding energy ingredient.

We can see from Equation (17.11) that the FINT for vacuum does not depend on the Metagalaxy radius $R(t)$:

$$A_V = (\kappa \rho_V)^{-1/2} = 15 \times 10^{27}\ \text{cm}. \tag{17.12}$$

Substituting the relation found earlier, Equation (17.1) FINT is directly related to the cosmological constant: $A_V = (3/\Lambda)^{1/2}$. Thus, Equation (17.12) is the "law of density conservation" for the dark energy vacuum. Contrary to ordinary fluids for which only extensive parameters are usually conserved, we now see an "intensive" parameter conservation for this very special medium.

Note that a new remarkable cosmic coincidence now appears. The universal constant length A_V proves to be very near the present-day radius of the Metagalaxy, R_0. This coincidence is not too mysterious, as we will see below. It reflects the special character of the current epoch of the cosmic evolution as the epoch of the transition from gravity domination to antigravity domination.

As for the non-vacuum energies, Equation (17.11) gives the conservation laws for the three extensive cosmic quantities of Equations (17.6)–(17.8). Indeed, the values A_D and A_B are directly related to the total masses of dark matter, M_D, and the total mass of baryons, M_B, respectively:

$$A_D = \kappa \rho_D R^3 = \frac{3\kappa}{4\pi} M_D = 5 \times 10^{27} \text{ cm}, \tag{17.13}$$

$$A_B = \kappa \rho_B R^3 = \frac{3\kappa}{4\pi} M_B = 0.9 \times 10^{27} \text{ cm}. \tag{17.14}$$

The FINT for radiation is related to the total number of the CMB photons N_R:

$$A_R = (\kappa \rho_R)^{1/2} R^2 \simeq (\kappa \hbar / c)^{1/2} N_R^{2/3} = 0.1 \times 10^{27} \text{cm}, \tag{17.15}$$

where $\hbar = h_P / 2\pi$, and h_P is the Planck constant.

Equations (17.12)–(17.15) show that the four FINTs are approximately comparable within two orders of magnitude:

$$A_V \sim A_D \sim A_B \sim A_R \sim 10^{27 \pm 1} \text{ cm}. \tag{17.16}$$

This approximate magnitude equality of the four FINTs is hardly a mere arithmetical accident. It is instead a significant physical feature of the Metagalaxy. The recipe of the cosmic energy composition may look strange or overcomplicated, when given in terms of the observed densities. It proves however to be quite simple when rewritten in terms of the Friedmann integrals.

The identity in Equation (17.16) is obviously more significant than the less exact coincidence of the four observed cosmic densities. More importantly, the density coincidence only takes place for the present-day epoch, i.e., only once in cosmic history. Contrary to density, the coincidence of the above four FINTs is a time-independent feature of the Metagalaxy. Although found with the data of the present epoch of cosmic evolution, Equation (17.16) is valid for all the epochs whenever the four energies exist.

For instance, baryons have existed at least since the earliest cosmic history era documented with current observational data—the BBN era. They are non-relativistic after the era of ~ 1 GeV temperatures, redshift $\sim 10^{12}$ and cosmic age $\sim 10^{-6}$ s. In the future, baryons will exist at least until the epoch of the hypothetical proton decay, i.e., to the cosmic age of $\geq 10^{32}$ years. It means that the FINT for baryons remains the same over 45 decades of the cosmic time. The integrals for vacuum and radiation are practically eternal. The integral for dark matter exists for all the cosmic past or future time when dark matter particles are non-relativistic. This existence may begin at the cosmic age of a few picoseconds (1 ps $= 10^{-12}$ s), if the dark matter particles are "weakly interacting massive particles" (WIMPs) with a mass ~ 1 TeV.

The identity of the four FINTs suggests that there exists a special correspondence among the cosmic energies which may be treated as a kind of symmetry (Chernin 2001–2003). Any symmetry describes a similarity of objects in a set. This is the most general definition of symmetry suggested by Herman Weyl (1951). If symmetry does not concern space-time relations, it is referred to as internal symmetry—in contrast to geometrical symmetries (like spatial isotropy and uniformity). A typical example of internal symmetry is provided by particle physics—the symmetry between the proton and the neutron. These two particles differ in mass, electric charge, life-time, etc., but they constitute a set which is a hadron doublet with a common constant value (1/2) of the so-called isotopic spin.

Equation (17.16) describes the similarity of the four cosmic energies. This similarity may be referred to as "cosmic internal symmetry" (which we hereafter shorten to "COINS"). These energies are obviously different in many respects. It is most noteworthy that one of them is a vacuum, while the three others are non-vacuum energies. Among the non-vacuum energies, baryons and electrons are fermions, while the CMB photons (and gravitons) are bosons. But despite all this, the four energies constitute a regular set—a quartet—with the Friedmann integral, $A \sim 10^{27}$ cm, as its common constant physical parameter which is comparable for all the members.

Note briefly some major features of the new COINS:

1) It unifies the four basic elements of the cosmic energy composition.
2) It is a time-independent symmetry in the evolving Metagalaxy.
3) This is a robust symmetry: it is weakly sensitive to details of the observational data.
4) It is a covariant symmetry, since it is formulated in terms of the Friedmann integrals which are the scalar (invariant) quantities. Identified in the co-moving 3D space, COINS exists in any other spatial sections and in the 4D space-time as a whole.
5) It is not exact, but an approximate or slightly broken symmetry. The symmetry violation is measured by the accuracy of the symmetry relation of Equation (17.16).
6) It implies that there is a correspondence between the total dark matter mass, the total baryonic mass and the total number N_R of relativistic particles (the CMB photons):

$$M_D \sim M_B \sim \left(\frac{\hbar}{c\kappa}\right)^{1/2} N_R^{2/3}. \tag{17.17}$$

7) Its existence implies also that there is a consistency of extensive cosmic quantities M_D, M_B, N_R and the intensive quantity ρ_V:

$$\rho_V \sim \kappa^{-3} M_D^{-2} \sim \kappa^{-3} M_B^{-2} \sim \left(\kappa^2 \hbar / c\right)^{-1} N_R^{-4/3}. \tag{17.18}$$

8) Finally, it may be an aspect of (or an implication from) supersymmetry, since it puts in correspondence cosmic bosons (CMB photons) and fermions (baryons), together with the dark energy vacuum.

Historically, the FINT identity for radiation and non-relativistic matter was first recognized (Chernin 1968) soon after the discovery of the CMB radiation. With the discovery of dark matter and cosmic vacuum, the identity was extended to all the four energies (Chernin 2001).

17.3   The physics behind COINS

It may be supposed that COINS has its roots in the physics of the early universe when the four cosmic energies came into existence. Studies of this process meet considerable obstacles because the microscopic structure of dark energy is completely unknown and the identity of the hypothetical dark matter particles is still obscure. It is also mysterious that there are no antiprotons and antineutrons in the "ordinary" baryonic fraction of cosmic energy—a drastic contradiction with the particle-antiparticle symmetry of fundamental physics (Chapter 11). Nevertheless, we will try to approach the problem, adopting the view (as in the standard cosmological model) of the dark energy vacuum with a time-independent and perfectly uniform density. We also assume that dark matter is a collection of weakly interacting massive particles (WIMPs) which are stable or long-lived. Under these assumptions, a model can be developed which may show how, in principle, COINS might originate in the early universe.

We assume dark matter particles are WIMPs and assume that their mass is near the characteristic mass/energy of the electroweak interaction:

$$m_D \simeq M_{EW} \simeq 10^{-21}\,\mathrm{g} \simeq 1\,\mathrm{TeV}/c^2. \tag{17.19}$$

The WIMPs must interact with each other and all the other matter only via the weak interaction and gravity. They exist together with their antiparticles, but, in practice, their annihilation does not occur because of weakness of the particle-antiparticle interaction and the low number density at the present epoch. The particle-antiparticle annihilation stopped long ago, when the cosmic temperature fell during expansion to the level of

$$T(t_D) \simeq m_D c^2/k_B \sim 10^{16}\,K, \tag{17.20}$$

(where k_B is the Boltzmann constant) and the expansion rate, $1/t$, wins out over the annihilation rate. The annihilation rate

$$\Gamma = n_D(t_D)\sigma_D c. \tag{17.21}$$

where σ_D is the WIMP annihilation cross-section, and $n_D(t_D)$ is the WIMP particle density at $t = t_D$ when the particles become non-relativistic. Equation (17.21) is obtained in an ordinary way in physical kinetics by counting how many particles there are in the "collision tube" of cross section σ_D per unit time. Since the particles moved with a speed practically equal to the speed of light c, the length of the tube is c (per unit time) and the volume of the tube is $\sigma_D c$. With a number density of n_D the number of particles in the tube was $n_D \sigma_D c$. For every particle in the tube, there would be one collision, and thus the annihilation rate, i.e., collisions per unit time, is also the same number. The expansion became faster than annihilation since the moment when

$$\Gamma \simeq 1/t_D, \tag{17.22}$$

where (Equation (10.21))

$$t_D = \left(\frac{3}{32\pi G \rho_R(t_D)}\right)^{1/2}. \tag{17.23}$$

This standard cosmological relation is valid at the epoch of radiation domination, and the radiation density

$$\rho_R(t_D) \simeq \frac{k_B T(t_D)}{c^2} \left(\frac{k_B T(t_D)}{\hbar c}\right)^3. \tag{17.24}$$

With this density,

$$t_D \sim 10^{-12}\,\mathrm{s}. \tag{17.25}$$

This is the earliest moment in the history of the Metagalaxy to which standard physics can be applied without additional assumptions and extrapolations. The corresponding redshift (Equations (10.34), (10.39) and (17.20))

$$z(t_D) = z_D = \left(\frac{k_B T_0}{m_D c^2}\right)^{-1} \sim 10^{15}, \tag{17.26}$$

where $T_0 \sim 2.75$ K is the present-day temperature of the CMB photons. The radius of the Metagalaxy at the epoch $t = t_D$

$$R(t_D) = R_0 (1 + z_D)^{-1} \sim 10^{-15} R_0. \tag{17.27}$$

At $t > t_D$, the WIMP dark matter survived against annihilation and existed since then as a thermal relic of the early universe.

A kinetic process of this type is called "freeze-out". The kinetics of cosmological freeze-out has been long studied—see, for instance, the books by Zeldovich and Novikov (1983), where the freeze-out of cosmic neutrinos was described, and Kolb and Turner (1990), and also Dolgov *et al.* (1988). We follow here mostly the work by Arkani-Hamed *et al.* 2000.

Turning back to the key kinetic equations of the process (Equations (17.21) and (17.22)), we may now specify the annihilation cross-section σ_D with the use of a simple approximate relation valid for massive ($mc^2 > 50$–100 GeV) particles (borrowed from particle physics),

$$\sigma_D \sim \left(\frac{\hbar}{m_D c}\right)^2. \tag{17.28}$$

With this σ_D and the cosmic time t_D, we find the dark matter density at the freeze-out time:

$$\rho_D(t_D) = m_D n_D(t_D) \sim \left(\frac{Gc^5}{\hbar^7}\right)^{1/2} m_D^5. \tag{17.29}$$

After that epoch, the gas of WIMPs expanded like non-relativistic fluid, and its density decreased with redshift as $(1+z)^3$ (Equations (17.20)–(17.24)). Note the high sensitivity of the result of Equation (17.29) to the value of the WIMP mass at $t > t_D$ since $\rho_D \propto m_D^5$.

Let us now introduce the Friedmann integrals for the freeze-out model. The FINTs for dark matter, A_D, and for radiation, A_R, may be expressed (with Equations (17.13), (17.14)) via the corresponding densities and the radius of the Metagalaxy at the freeze-out epoch:

$$A_D = \kappa \rho_D(t_D) R(t_D)^3 \simeq \kappa \left(\frac{c^5 G}{\hbar^7}\right)^{1/2} m_D^5 R(t_D)^3, \tag{17.30}$$

$$A_R = [\kappa \rho_R(t_D)]^{1/2} R(t_D)^2 \simeq \left[\kappa \frac{k_B T(t_D)}{c^2} \left(\frac{k_B T(t_D)}{\hbar c}\right)^3\right]^{1/2} R(t_D)^2. \tag{17.31}$$

The ratio of the FINTs

$$\frac{A_R}{A_D} \simeq (1+z_D) \frac{\hbar^2}{Gm_D^3 R_0} \sim 10^3, \tag{17.32}$$

if $m_D \sim M_{EW}$ (Equation 17.19). Considering a factor $>$ ten has been ignored in deriving Equation (17.32), the FINTS are within two orders of magnitude of one another—in agreement with the empirical estimates of the section above. The identity of the FINTs emerges now as an outcome of the WIMP annihilation freeze-out in the first picoseconds of the cosmological expansion. Specifically, it is the kinetics described by Equations (17.21) and (17.22) that provides a link between radiation and dark matter. In terms of the FINTs, this link is expressed in the time-independent covariant equality of Equation (17.32) which is valid for all the further life-time of the two energies.

There is a critical observational test for the freeze-out model. Indeed, the result depends sensitively (as already mentioned) on the assumed mass of the WIMP. In particular, the ratio of Equation (17.32) has the correct value, provided that the WIMP mass

$$m_D \simeq \frac{\hbar c}{(G k_B T_0 R_0)^{1/2}} \left(\frac{A_R}{A_D}\right)^{-1/2}.$$ (17.33)

is near the value of the electroweak mass/energy, M_{EW}. In fact, this is a verifiable prediction of the model. If current and future experiments on direct detection of the dark matter particles find a significantly different mass, this model must be ruled out.

But if the WIMP mass $\sim 1\,\mathrm{TeV}/c^2$ is confirmed by the experiments, it would seemingly mean that the physical nature of COINS—with respect to the two energies of the four—is due to the electroweak-scale physics at the first picoseconds of the cosmological expansion.

Vacuum and baryons remain out of the model; obviously, the physics of their origin cannot be reduced to freeze-out kinetics only. As to baryons, a concept of baryogenesis has been suggested which is based on the Sakharov–Kuzmin mechanism. The mechanism includes three physical conditions: (1) baryon number violation, (2) C- and CP-symmetries violation, and (3) a departure from thermal equilibrium in the cosmic medium (Sakharov 1966, Kuzmin 1970). Baryogenesis at the electroweak temperatures is especially interesting because the three criteria can seemingly be satisfied at that epoch (Kuzmin et al. 1985, Riotto and Trodden 1999, Dolgov 1992, Rubakov and Shaposhnikov 1996). As to the dark energy vacuum, it has been proposed (see Chapter 16) that electroweak physics might be responsible for the origin of the cosmic vacuum via supersymmetry violation at TeV temperatures (see Zeldovich 1968, Dolgov 2004 and references therein). These two concepts in combination with the freeze-out model show that the epoch of electroweak energies might look as a real start of the cosmic evolution, the "Electroweak Big Bang", in which the cosmic energies and their symmetry relation come into existence due to common physics.

New prospects in this direction are discussed by Arkani-Hamed et al. (2000—see also references therein) who suggest a simple formula for the vacuum density in terms of only two fundamental physical constants, the electroweak mass $M_{EW} \sim 1\,\mathrm{TeV}$ and the Planck ("reduced") mass:

$$M_{Pl} = \left[\frac{\hbar c}{K}\right]^{1/2} \sim 10^{-5} g \sim 10^{15}\,\mathrm{TeV}/c^2.$$ (17.34)

where $K = 8\pi G/3$. The formula does not come from any basic theory, it is simply a result of numerological exercises; but it does give an acceptable, on the order of magnitude, value of the vacuum density:

$$\rho_V \sim (M_{EW}/M_{Pl})^8 \rho_{Pl} \simeq 7 \times 10^{-29}\,\mathrm{gcm}^{-3},$$ (17.35)

where

$$\rho_{Pl} = \frac{c^5}{K^2 \hbar} \simeq 7 \times 10^{91} \, \mathrm{gcm}^{-3} \sim 10^{15} \, \mathrm{TeV}/c^2/\mathrm{cm}^3. \tag{17.36}$$

In the system of "natural units" which is widely used in particle physics, the Planck constant $\hbar$, the speed of light c and the Boltzmann constant k_B are set to be equal to one. Then the dimensions of mass, energy and momentum are identical. The dimensions of time and length are both reciprocal mass. The Einstein gravitational constant $\kappa = 8\pi G/(3c^2)$ has the dimension of mass^{-2}. In these units, Equation (17.35) takes the form

$$\rho_V \sim (M_{EW}/M_{Pl})^8 M_{Pl}^4 \sim X^{-8} M_{Pl}^4. \tag{17.37}$$

The big dimensionless number here,

$$X \sim M_{Pl}/M_{EW} \sim 10^{15}, \tag{17.38}$$

is known as the hierarchy number (a simple approximate relation, borrowed from particle physics). It gives a measure of the huge gap between the two fundamental energy scales. The nature of the gap is not yet well understood and is one of basic problems of fundamental physics (see also below). An attempt to justify the relation of Equation (17.37) was made by Klinkhamer and Volovik (2009) with the use of so-called "q-theory"; see also Marochnik et al. (2010).

In these units and with Equation (17.37), the FINT for the vacuum

$$A_V \sim (M_{Pl}/M_{EW})^4 M_{Pl}^{-1} \sim X^4 M_{Pl}^{-1}. \tag{17.39}$$

The hierarchy number enters not only Equations (17.37) and (17.39) for vacuum, but also the relations of the freeze-out physics above:

$$t_D \sim X^2 M_{Pl}^{-1}, \quad \sigma_D \sim X^2 M_{Pl}^{-2}, \quad \rho_D(t_D) \sim X^{-5} M_{Pl}^4. \tag{17.40}$$

In addition, the redshift at t_D may be expressed (coincidentally?) via the hierarchy number as well: $z_D \sim X$.

17.4   Coincidence problem

It was already mentioned that the densities of the two energies in the dark sector of the cosmic mix, ρ_V and ρ_D, are approximately identical (see Section 17.1). This raises a question: Why should we observe them to be so nearly equal now? While the vacuum density (or the cosmological constant) is time independent, the dark matter density dilutes as R^{-3} as the Metagalaxy expands. Despite the evolution of $R(t)$ over many orders of magnitude, we appear to live at an epoch at which the two densities are roughly the same. This is the "cosmic density coincidence" problem which is commonly considered to be a severe challenge to the current cosmological concepts.

In fact, all the four observed energy densities, the two dark ones and the two others, are of nearly the same order of magnitude (Section 17.1). Why are they nearly coincident now? This is a more general question, and COINS may seemingly suggest a possible answer to it without any additional assumptions. We will show now that the time-independent identity of the four FINTs is actually behind the temporary (and accidental, in this sense) coincidence of the four densities.

Let us rewrite the definition of the FINTs with the use of Equations (17.12)–(17.15) and the symmetry relation of Equation (17.16) and the Planck mass M_{Pl} in the form:

$$\rho_V = \kappa^{-1} A_V^{-2} \sim \frac{M_{Pl}^2}{A^2}, \tag{17.41}$$

$$\rho_D = \kappa^{-1} A_D R^{-3} \sim \frac{A M_{Pl}^2}{R^3}, \tag{17.42}$$

$$\rho_B = \kappa^{-1} A_B R^{-3} \sim \frac{A M_{Pl}^2}{R^3}, \tag{17.43}$$

$$\rho_R = \kappa^{-1} A_R^2 R^{-4} \sim \frac{A^2 M_{Pl}^2}{R^4}. \tag{17.44}$$

These relations indicate that the four densities must become identical and all have the value $\sim \frac{M_{Pl}^2}{A^2}$ on the order of magnitude, at the epoch when $R(t_0) \sim A_V \sim A$. Thus, the four cosmic densities are observed to be nearly coincident because of 1) COINS as basic time-independent symmetry, and 2) the special character of the moment of observation at which the size of the Metagalaxy is approximately equal to the Friedmann integral.

The approximate equality of the constant FINT A and the radius of the Metagalaxy $R(t)$ is the label of the current epoch of the cosmic evolution (see Section 17.1). The present epoch is special indeed because of a clear physical reason: this is the epoch of transition from matter domination to vacuum domination in cosmic dynamics. At the exact moment of transition when $\ddot{R} = 0$ the two densities are related by the equation $2\rho_V = \rho_D$. The current Metagalaxy is not far from this moment, and therefore the two densities must anyway be close to each other.

At the transition epoch, the solution for matter dominated expansion, $R(t) \propto t^{2/3}$, and the solution for vacuum dominated expansion, $R(t) \propto \exp(ct/A)$, are both valid in a rough approximation. Because of that, the rate of expansion given by the first solution, $H = \dot{R}/R \simeq (\frac{2}{3})(1/t_0)$ is near the same value from the other solution, $H \simeq c/A$. As a result, we have

$$R_0 \simeq c/H_0 \simeq ct_0 \simeq A. \tag{17.45}$$

In fact, any coincidence that takes place at the present epoch is due—in this way or another—to the only approximate equality $R_0 \simeq A$. Another question is why do we happen to live at such a special epoch? This is among the matters that are effectively discussed on the basis of the Anthropic Principle (see Weinberg 1977). The Anthropic Principle states that the evolution of mankind has required certain steps, and that the cosmic evolution could not have proceeded through these steps in a time frame which is considerably shorter than 10 billion years. There is the time required to produce enough heavy elements in the universe for the formation of planets and for the building blocks of life. Then the biological evolution through natural selection of species has required its own time. The Earth came into being after about 9 billion years of cosmic evolution, while the emergence of mankind on the Earth required another nearly 5 billion year span of evolution. One can argue that these time scale are "typical" in the sense that, within a rather wide margin, the emergence of intelligent beings capable of studying the universe could not have happened much earlier, and that it was perhaps inevitable to happen around our cosmic time.

In terms of the standard cosmology, the Anthropic Principle requirements may perhaps all be reduced to practically one simple condition: the present epoch is the epoch when the size of the Metagalaxy reaches the universal value of the Friedmann integral.

17.5   Dicke's flatness problem and solutions

According to observations, primarily the WMAP data on the anisotropy of the CMB (Spergel *et al.* 2007), the observed total energy density in the Metagalaxy, $\rho = \rho_V + \rho_D + \rho_B + \rho_R$, is equal to the current critical cosmic density, ρ_c, with a high accuracy:

$$\Omega(t_0) \equiv \rho/\rho_c = 1 \pm 0.02. \tag{17.46}$$

This is perhaps the most important result of observational cosmology obtained after the discovery of the cosmic acceleration and dark energy. It implies that the 3D co-moving space of the Metagalaxy is flat or at least nearly flat and the cosmological expansion proceeds in a parabolic, or near parabolic, regime (see Chapter 8).

But why is $\Omega(t_0)$ so close to unity? The question was first discussed by the American experimental physicist Robert H. Dicke well-known for his fundamental contribution to many areas of modern science, including cosmology. He mentioned that the universe must initially be extremely finely tuned to yield the observed balance between the total energy density of the universe, $\rho(t)$, and the critical density ρ_c (Dicke 1970). In the 1970's, the observational constraints on $\Omega(t_0)$ were much weaker than now, and it was known only that the parameter is between 0.1 and 10. Nevertheless, Dicke recognized clearly the principal significance of the question.

Dicke's argument (slightly modified) is as follows. The ratio of the total density to the critical density called the density parameter, $\Omega(t)$, is exactly unity in the flat 3D co-moving space at any time, but it is a function of time in a curved space. Suppose the space is not flat and $\Omega(t_0)$ deviates from unity at present. The observed value of $\Omega(t_0)$ is then a result of the dynamical evolution started probably near the beginning of the cosmological expansion. The "initial conditions" for the expansion were fixed at that early era, and one may restore them with the use of the Friedmann expansion theory. The Friedmann dynamical Equation (8.14) for the curvature radius $a(t)$ of a non-flat 3D space is this:

$$\dot{a}^2 = \kappa c^2 \rho a^2 - c^2 k. \tag{17.47}$$

where we have substituted $\kappa = \frac{8\pi G}{3c^2}$ and replaced $R(t)$ by $a(t)$. Here k is the curvature parameter (which should not be confused with the Boltzmann constant). Recall that k is zero for the flat 3D co-moving space, and ± 1 for the space of positive or negative curvature, respectively (Chapter 8). We may write the density parameter as

$$\Omega(t) = \kappa c^2 \rho(t) \left(\frac{a}{\dot{a}}\right)^2,$$

which is derived from Equations (8.28) and (17.47). When $k = \pm 1$, this parameter deviates from unity. Using Equation (17.47), the relative deviation from unity is:

$$\frac{|\Omega - 1|}{\Omega} = \frac{1}{\kappa \rho a^2}. \tag{17.48}$$

The Equation (17.48) value increases with cosmic time from the origin as $t^{2/3} \propto (1+z)^{-1}$ during the matter domination epoch and as $t \propto (1+z)^{-2}$ during the earlier epoch of radiation domination. The value is

$$\frac{|\Omega_* - 1|}{\Omega_*} \sim 10^{-6} \frac{|\Omega_0 - 1|}{\Omega_0} \tag{17.49}$$

at the start of the matter domination epoch when $z = z_* \approx 1000$. The value is

$$\frac{|\Omega_{BBN} - 1|}{\Omega_{BBN}} \sim 10^{-15} \frac{|\Omega_0 - 1|}{\Omega_0}. \tag{17.50}$$

at the epoch of the Big Bang Nucleosynthesis (BBN), when $z = z_{BBN} \sim 10^9$.

We see from Eqn. (17.50) that if the initial conditions for the cosmological expansion are fixed at the BBN epoch, the initial value of the density parameter should be tuned with an accuracy of better than $\sim 10^{-15}$ to meet even the observational restrictions of the 1970s. The accuracy of $\sim 10^{-27}$ is needed, if the initial conditions are set at the electroweak era when $z \sim 10^{15}$. Such a fine tuning was reasonably considered by Dicke as strange and unacceptable. But no solution to the problem was suggested at that time.

The discovery of dark energy enables us to see the Dicke's paradox in quite a different light. Keeping the same statement of the problem as above, one may use again the Friedmann equation of Equation (17.47), but with the vacuum item. Written now for the radius $R(t)$ of the Metagalaxy (i.e. the co-moving radius of the current horizon radius), the equation takes the form:

$$\dot{R}^2 = \kappa c^2 \rho R^2 - c^2 k \alpha^2. \tag{17.51}$$

Here $\alpha = R/a$ is a constant dimensionless parameter of 3D space geometry, and $\kappa \rho R^2 = (R/A_V)^2 + A_D/R + A_B/R + (A_R/R)^2$.

It is seen from Equation (17.51) that at earlier times when R goes to zero, the constant term $c^2 |k| \alpha^2$ (which represents the effect of the 3D curvature) is much less than the non-vacuum items, $\kappa \rho R^2 = A_D/R + A_B/R + (A_R/R)^2$, on the right-hand side of this equation. But the term $c^2 |k| \alpha^2$ is also much less than the vacuum item, $(A_V/R)^{-2}$, at the opposite limit of late times when R goes to infinity.

Thus, curvature effect vanishes at the limit of both zero and infinite time. Therefore the effect is maximal at an intermediate moment of time between these two limits. The curvature effect may be quantified by the same relation as Equation (17.48):

$$\frac{|\Omega(t) - 1|}{\Omega(t)} = \frac{\alpha^2}{\kappa \rho R^2}. \tag{17.52}$$

In accordance with what is said above, this value goes to zero in the limits of both small and large time. It has its maximum when $\ddot{R} = 0$ and the gravity of matter and antigravity of vacuum balance each other. At that moment t_V of "zero-gravity",

$$R(t_V) = \left(\frac{1}{2} A_D A_V^2\right)^{1/3}, \quad \rho = 3\rho_V. \tag{17.53}$$

(In these equations, we neglect a small contribution of baryons and radiation to the total density around the moment $t = t_V$.) Then the maximal deviation of the density parameter from unity

$$\left|\frac{\Omega - 1}{\Omega}\right|_{\max} \simeq \frac{1}{2} \alpha^2 (A_V/A_D)^{2/3}. \tag{17.54}$$

The modern strongest observational restriction $\frac{|\Omega - 1|}{\Omega_0} \leq 0.02$ is met if

$$\frac{1}{2} \alpha^2 (A_V/A_D)^{2/3} \leq 0.02. \tag{17.55}$$

With $A_V \simeq 1.5 \times 10^{28}$ cm and $A_D \simeq 0.5 \times 10^{28}$ (see Section 17.2), one finds finally for the constant parameter α:

$$\alpha = R/a \leq 0.1. \tag{17.56}$$

Thus, any model with a curved 3D space fits the strongest modern observational restrictions, if its curvature radius is about 10 times the radius of the Metagalaxy. The criterion is marginally met in the Luminet model of the topologically-compact space (Chapter 14). The observational restrictions of the 1970s (see above) are met, if $\alpha \simeq 1$. To see the contrast with the fine-tuning argument, one may compare numbers like 1 or 0.1 with such extremely small numbers like 10^{-16} or 10^{-27}.

The criterion of Equation (17.56) is time-independent; treated as "initial conditions", it may be fixed at any moment of time in the past—for instance, at the BBN or electroweak eras without any fine tuning. In this way, Dicke's paradox is clarified and eliminated (Chernin 2003, Adler and Overduin 2005, Lake 2005).

In fact, not fine-tuning, but rather a robust symmetry is actually behind the observed phenomena of the nearly flat 3D co-moving space. COINS which controls the balance between vacuum antigravity and dark matter gravity rules out any significant deviations from flatness at present, in the past and in the future of the Metagalaxy.

Note that a solution of Dicke's paradox is suggested also by the inflation theory (Section 17.16). This theory deals with giant vacuum densities near the Planck density and so it needs a far-going arbitrary extrapolation of standard physics to an unknown domain by some 30 orders of magnitude in space and time—from the electroweak scale to the Planck scale. As we see now, the paradox may be resolved without such assumptions via dark energy.

But what if the co-moving space is not approximately but exactly flat? This possibility does not contradict any observational data. Moreover, the standard cosmological model uses the version of flat space (not as a rigorous result but as a simple and very good approximation to reality). Dicke's statement of the problem does not embrace the case of exactly flat space; in this case, our considerations above are not valid either; nor does the inflationary model. If the "problem of exactly flat space" arises someday (it is, however, not quite clear how it could actually occur), we will have to search for entirely different approaches to it.

17.6   Big numbers

The ratio of the number density of the CMB photons to the number density of baryons, $n_R/n_B \sim 10^{10}$, (the Big Baryonic Number of Section 17.1) is a major conserved dimensionless physical parameter of the Metagalaxy; it is responsible, in particular, for the BBN outcome. In terms of the Friedmann integrals, one has

$$B \sim \left(\frac{c}{\hbar}\right)^{1/2} A_R^{3/2} A_B^{-1} m_B M_{Pl}^{-1/2} \sim 10^9, \tag{17.57}$$

where $m_B = 1$ GeV is the baryon mass.

According to COINS, $A_R = A_B = A_V = X^4 M_{Pl}^{-1}$ (Sections 17.2, 17.3); then the Big Baryonic Number turns out to be

$$B \sim X^2 (m_B / M_{Pl}). \tag{17.58}$$

This gives numerically $B \sim 10^{11}$ which is not bad as a rough order-of-magnitude estimate.

The freeze-out physics of Section 17.3 suggests that the "Big Dark Number" may also be of interest:

$$D \equiv n_R / n_D \sim 10^{12}, \tag{17.59}$$

where n_D is the number density of dark matter particles, and it is again assumed that the WIMP mass $\sim M_{EW}$. In terms of the Friedmann integrals, one has with $A_R \sim A_D$

$$D \sim \left(\frac{c}{\hbar}\right)^{1/2} A_R^{3/2} A_D^{-1} M_{EW} M_{Pl}^{-1/2} \sim X \sim 10^{15}. \tag{17.60}$$

Numerically, this estimate is not too far from the real figure above.

Let us address the gross extensive conserving parameters of the Metagalaxy (Equations (17.39) and (17.42)). With the COINS and using the relation $R_0 \sim A_V$, one finds for the total dark mass in the Metagalaxy:

$$M_D \sim A M_{Pl}^2 \sim X^4 M_{Pl} \sim 10^{60} M_{Pl}. \tag{17.61}$$

Then the total number of dark matter WIMPs (with ~ 1 TeV mass) particles:

$$N_D \sim X^5 \sim 10^{75}. \tag{17.62}$$

In a similar way, one finds for the total number of baryons,

$$N_B \sim X^4 M_{Pl} / m_B \sim 10^{78}, \tag{17.63}$$

and the total number of the CMB photons (Equation (17.17))

$$N_R \sim X^6 \sim 10^{90}. \tag{17.64}$$

Thus, the big cosmic numbers, both dimensional and dimensionless, are closely linked with each other via COINS and the hierarchy phenomenon of fundamental physics.

17.7 Extra dimensions?

The hierarchy number has long been in the focus of discussions in particle physics without any relation to cosmology. Now it has come to cosmology, and reveals its key significance for the vacuum density, Friedmann integrals, COINS and the COINS-associated figures (see above). The number X gives a measure of the intrinsic strength of gravity in comparison with electroweak interaction. In terms used commonly in particle physics, the strength of gravity is given by the gravitational constant, G, which is the inverse square of the Plank mass M_{Pl} (in natural units with $\hbar = c = 1$). The appropriate strength for electroweak interaction is given by the inverse square of the mass $M_{EW} \sim 1\,\mathrm{TeV}$. So gravity is X^2 times weaker than the electroweak interaction. One may remember that the Compton wavelength of the particle with the Planck mass is equal to the Schwarzschild radius for this mass, and this is the realm where gravity with black holes, gravitational waves and gravitons, etc. is an essentially quantum phenomenon. In cosmology, the relations involving X give an obvious example for a very close interrelation between the physics of the smallest and the physics of the largest in nature. They also show definitely which sector of fundamental theory is especially significant for cosmology.

The physical nature of the hierarchy phenomenon is still unknown in particle physics, obviously because matters like quantum gravity remain almost entirely obscure.

Nevertheless an interesting idea recently suggested and yet completely hypothetical seems to suggest an entirely new prospect in the field. This is the idea of extra spatial dimensions. Nima Arkani-Hamed, Savas Dimopoulos and Georgi Dvali, working at Harvard, Stanford and New York universities, respectively, proposed in 1988 that the energy hierarchy might be eliminated from fundamental theory, if there exist spatial dimensions additional to the ordinary three dimensions. It is assumed that the extra dimensions have a macroscopic size which is much larger than the characteristic lengths of particle physics and much less than the cosmological lengths such as the size of the Metagalaxy or the horizon radius. In this sense, this macroscopic phenomenon is able to mediate between the smallest and the largest, which is most interesting for cosmology.

The hypothesis assumes that the extension of space is finite along the extra directions. In this case, such dimensions are called compactified. The space of extra dimensions is finite and compact. It means that there are upper limits for any distance that can be covered by light or any other "traveler" in the motion along the extra dimensions. This may be illustrated, for instance, by the motion on the surface of the cylinder in the directions perpendicular to its axis. One may say that the extra dimensions are "curled" into rings of a finite length. The extra dimensions might constitute, together with the real 3D space, a multidimensional space which is treated as the "true space of nature", in this new framework.

This is not the first attempt to incorporate extra dimension into cosmology. Historically, multidimensional spaces with the number of dimensions larger than 3 have long been discussed in geometry and physics. It was in particular recognized that the laws of nature may be different in spaces of different dimensions. The German philosopher Immanuel Kant was the first to discuss the problem two centuries ago and recognized a relation between Newton's law of gravity and the dimensionality of space. He argued that as we measure the gravity force at increasing distance r from a gravitating body, its strength is proportional to $1/r^2$, or the inverse of the surface area of a sphere of the radius r. This is because all lines of force penetrate the surface of the sphere. But this is only true in our ordinary three-dimensional space. If the space would be N-dimensional, the surface area penetrated by the force lines would be proportional to r^{-N+1}, so that there is no Newtonian inverse square law of gravity in a space with $N \neq 3$ dimensions. If for example, the space is 4-dimensional, the law of gravity will be not a law of inverse square, but a law of inverse cube. In a 5-dimensional space, gravity force will decrease with distance as the inverse fourth power of the distance, and so on.

After Kant, extra dimensions were studied in 1914 by the Finnish physicist Gunnar Nordström in an attempt to reconcile electromagnetism and gravity. The theory was developed independently by the Swedish physicist Oskar Klein and the German physicist Theodor Kaluza in the 1920s, but it proved to be unsuccessful. Nowadays the idea of extra dimensions is exploited in the string theory which offers to be a "theory of everything". This theory uses 6 or 7 extra dimensions of the Planck length $\sim 10^{-33}$ cm and promises to explain all the basic characteristics of particle physics including the spectrum of the masses of the particles (but this is still a remote goal). The concept by Arkani-Hamed *et al.* is declared to be string-motivated; but actually it may be developed with no reference to the string theory.

It is most interesting that a central part is ascribed to the electroweak energy scale M_{EW}, in this concept. It is assumed that there is one and only one "truly fundamental" energy scale M_* in nature, and this scale is close to, if not exactly coincident with, the electroweak scale M_{EW}. Because of this, it was and is still expected that a new arena is open for particle physics: with the energy $M_* \sim 1$ TeV, new dimensions are believed to be within reach of the experiments of this decade particularly at the Large Hadron Collider and the Tevatron at which experiments at such an energy will be feasible.

The Planck scale M_{Pl} loses now its fundamentality and is reduced to a combination of the truly fundamental energy scale M_* and the size R_* of the compact macroscopic extra dimensions:

$$M_{Pl} = (M_* R_*)^{n/2} M_*, \tag{17.65}$$

where n is the number of compact extra dimensions which are proposed to be of the same size. The "natural units" are used hereafter with $c = \hbar = 1$.

The gravitational constant in three-dimensional space becomes (together with the Planck mass) of secondary significance. It is given by the relation

$$G = M_{Pl}^{-2} = (M_* R_*)^{-n} M_*^{-2}. \tag{17.66}$$

But gravity remains a special field, the only field which exists in the whole multi-dimensional true space. All the other fields of matter are confined to the 3D hyper-surface which is the "ordinary" space. In this case, such a hypersurface is called a membrane, or in short brane.

With the equations above, one may find the size of extra dimensions:

$$R_* = \left[\frac{M_{Pl}}{M_*}\right]^{2/n+1} M_{Pl}^{-1} \sim 10^{\frac{30}{n}-18} \, \text{cm}. \tag{17.67}$$

Strictly speaking, this is not an elimination of the hierarchy, but its reformulation in the new terms of M_* and R_*: instead of energy separation from M_{Pl} to M_{EW}, a spatial separation R_* is introduced into the theory.

As to the number of extra dimensions n, it is argued that the case $n = 1$ is hardly reasonable. In this case, the size of the one extra dimension would be of the order of 10^7 km, as is seen from the above equation. An extra dimension that large would long ago have made itself obvious in the dynamics of the Solar System. As the number of the extra dimensions increases, their size gets smaller, according to the equation, so that for, say, six equal extra dimensions, the size is $\sim 10^{-13}$ cm. Most probably, we should focus on two extra dimensions; in this case, they would be sub-millimeter in length:

$$R_* \simeq 10^{-3} \, \text{cm}, \quad n = 2. \tag{17.68}$$

Not quite accidentally, this is the largest extra dimension length which is permitted by the current experiments with gravity at small scales. Indeed, the Newtonian inverse square law for gravity has so far been checked and confirmed down to distances near 0.1 mm (Adelberger et al. 2003, 2009). Deviations from this law at larger distances are ruled out, but they are not excluded at shorter distances. As is seen from this, the idea of sub-millimeter size extra dimensions has probably good experimental prospects: it will be proved or disproved when the accuracy of the measurements in sub-millimeter gravity experiments becomes a bit higher than now, and this is feasible in the perhaps next few years, as the experimentalists believe. If gravity at shorter scales, $r < R_*$, will decrease with distance r faster than $1/r^2$, it would mean that extra dimensions do exist.

Specifically, with two extra dimensions, the strength of gravity between two masses m_1, m_2 decreases with distance as

$$\frac{G_5 m_1 m_2}{r^4}, \quad r < R_*, \tag{17.69}$$

where G_5 is the 5D gravity constant related to the "ordinary" gravity constant by

$$G_5 = G R_*^2. \tag{17.70}$$

The last equality follows from the condition that the law G_5/r^4 has to give the same answer as the Newtonian law, G/r^2, at the distance $r = R_*$. Obviously, gravity at larger scales, $r > R_*$, cannot feel finite (compact) extra dimensions. At the distances $r < R_*$, no deviations are expected from the inverse square law for the electrostatic Coulomb force, since this field is supposed to be confined into the brane which is the ordinary 3D space.

Let us now turn to the cosmological implications of the macroscopic extra dimensions. It is clear that the Friedmann standard cosmology is valid only at the spatial scales where gravity is not affected by the extra dimensions. The criterion is simple: the horizon radius is larger than the size of the extra dimensions. This criterion is obviously met in the expanding Metagalaxy since the cosmic time, $t > R_*/c \simeq 3 \times 10^{-14}$ s, which is near the electroweak era. This means that the standard cosmology does not "remember" its multidimensional pre-Friedmann history, if all the processes on our brane proceed in thermodynamic equilibrium at the temperatures ~ 1 TeV. But it is possible that, in principle, some features of the real world might be due to the non-equilibrium kinetics of the earlier multidimensional epoch of the cosmic evolution.

At the fundamental level, extra-dimensions alternate the basic constants that enter the COINS and the quantities associated with it. First of all, the hierarchy number, $X = M_{Pl}/M_{EW}$, is replaced now with the product

$$X = (M_* R_*)^{n/2}. \tag{17.71}$$

Accordingly, the Friedmann integral gets the form:

$$A \sim (M_* R_*)^{3/2n} M_*^{-1}. \tag{17.72}$$

Then in the case of two extra dimensions

$$A \sim (M_* R_*)^2 R_* \sim 10^{27} \text{ cm}, \quad n = 2. \tag{17.73}$$

The vacuum density expressed in terms of the new basic constants looks like

$$\rho_V \sim (M_* R_*)^{-2n} M_*^4, \tag{17.74}$$

and in the case of two extra dimensions, we have:

$$\rho_V \sim R_*^{-4}, \quad n = 2. \tag{17.75}$$

This is a surprising result: the vacuum density can be expressed via the size of the extra dimensions alone, if the number of extra dimensions $n = 2$. For this most important cosmological quantity, the hierarchy is indeed eliminated (Chernin 2002a).

In the case $n = 2$, the cosmological constant $\Lambda = 3\kappa\rho_V$ becomes

$$\Lambda \sim (M_* R_*)^{-4} R_*^{-2}, \quad n = 2. \tag{17.76}$$

Together with the Friedmann integral, the gross parameters of the Metagalaxy are given in terms of the fundamental energy and length; so that we have (for $n = 2$):

*the total dark mass

$$M_D \sim (M_* R_*)^5 M_*, \tag{17.77}$$

*the total number of dark matter (WIMP) particles

$$N_D \sim (M_* R_*)^5, \tag{17.78}$$

*and the total number of the CMB photons

$$N_R \sim (M_* R_*)^6. \tag{17.79}$$

In the spirit of extra dimensions, all we observe in three-dimensional space are not more than shadows of the true multidimensional entities. In particular, true vacuum exists in the multidimensional space, and the observed cosmic vacuum is just its 3D "projection" to the cosmological brane. The true vacuum is defined in the multidimensional space, and its density

$$\rho_{V5} \sim R_*^{-6}, \tag{17.80}$$

for two extra dimensions. This true vacuum density is also free from the hierarchy effect.

Thus, taking the relations above at their face value, one may conclude that the basic cosmological parameters, and the vacuum density among them, come all from the extra-dimensional physics. The extra-dimension physics is imprinted in the cosmological constant and the vacuum density, in the gross cosmological figures and the big dimensionless constants, *etc.*; that is if extra dimensions really exist.

Equation (17.80) needs a special comment. It means that the physical nature of the cosmic vacuum is entirely due to gravity alone since gravity is the only field which lives in all $3 + 2 = 5$ spatial dimensions. If this is so, vacuum is not related in any way to other fields which "live" only on brane. On the contrary, if the observed vacuum is due to fermionic and bosonic fields of matter and processes like supersymmetry violation (see Section 17.3), the vacuum must be confined in the brane—together with the fields of matter which generate it. In this later case, the physical sense of the relation between the vacuum density and the extra-dimension size is not obvious, while numerically, the size $R_* \sim 10^{-3}$ cm is well suitable to the role of the cut-off wavelength $\lambda(0)$ (see Chapter 16) in the spectrum of zero-point quantum oscillations.

17.8   New naturalness

COINS enables one to look from a new point of view at the two problems of
cosmology that were already mentioned in Chapter 16. As is said there, it has been
widely believed that such a fundamental quantity as the density of the physical
vacuum must be very special, and there only two possibilities—the Planck density
or zero—and nothing else. The real vacuum density has proven to be neither
Planckian nor zero which has led to the "naturalness" problem: Why is the vacuum
density ρ_V 120 orders of magnitude smaller than its "natural" non-zero value $\sim M_{Pl}^4$
COINS provides a new framework for naturalness judgments. Indeed, due to the
COINS symmetry relation $A_V \sim A_D \sim A_B \sim A_R$, vacuum with its observed density
is now not an isolated and very special type of cosmic energy. It appears now to be
a regular member of the quartet into which all the cosmic energy ingredients are
unified. The real vacuum density looks quite natural in this set, since the Friedmann
integral for vacuum is almost exactly equal to the FINTs for the three non-vacuum
energies. We may actually say that vacuum with the observed density has found its
proper natural place among other cosmic energies. On the contrary, the vacuum
with the Planck or zero density would look embarrassingly strange in this quartet:
if with the Planck density, its FINT would be different from the three other FINTs
in at least 60 powers of ten and the FINT for zero-density vacuum would be infinite
in value.

The idea of extra dimensions (see the section above) provides additional evidence
in favor of the naturalness of the real vacuum. This is simplicity: vacuum density is
just $\rho_V \sim R_*^{-4}$, and this relation is as simple as the "natural" (actually inadequate)
formula $\rho_{Pl} \sim M_{Pl}^4$.

The other problem mentioned in Chapter 16 concerns the observed uniformity of
the Metagalaxy. It is treated in the inflation theory, and it was found that a necessary condition for the set-up of uniformity may be met at inflation: the cosmic
space might be causally connected on the largest observed scales. However some
questions remain without answers. One is about the physical agent which would
execute the causal connection in the initial vacuum state. As a colleague of the
Pulkovo Observatory said once at a seminar in St. Petersburg, there was the speed
of light and the light cone, but there was no light at inflation.

It seems "natural" that vacuum—not hypothetical primordial one, but the really
observed vacuum of the Metagalaxy—is perfectly uniform and the origin of its
uniformity has seemingly raised no questions. But why vacuum is uniform? An
assumed answer may be like this: the vacuum must be uniform because its density
is given by the cosmological constant Λ which is universal, i.e., the same in all
areas of the whole space-time, whether the regions are causally connected or
disconnected. Accepting this explanation, let us try to look how it may work in the
COINS framework.

COINS suggests that this explanation may be extended to the non-vacuum ingredients of cosmic energy as well. Indeed, all the four energies are unified in a quartet, and each of them is determined by the Friedmann integral. The symmetry relation, $A_V \sim A_D \sim A_B \sim A_R \sim \Lambda^{-1/2}$, indicates that the FINT has the same status in theory as the cosmological constant itself has: it is the same in all space-time, including both causally connected and disconnected areas. In addition, there is an explicit relation between the FINT and the two fundamental physical constants M_{Pl} and M_{EW} which are also universal (see above): $A \sim \Lambda^{-1/2} \sim X^4 M_{Pl}^{-1}$.

The non-vacuum densities are not constant, contrary to the vacuum density. Therefore some "initial conditions" must be formulated for them. For example, the initial conditions for the Friedmann stage of cosmic evolution may be fixed at the electroweak era. At that time, the radiation density (see Section 17.3) is given by the constant energy M_{EW} only:

$$\rho_R(t_{EW}) \sim M_{EW}^4. \tag{17.81}$$

Using the argument above, one may say that this density is uniform in all the space initially at $t = t_{EW}$, since the constant M_{EW} is universal. In a similar way, the initial, at $t = t_{EW}$, density of dark matter (see again Section 17.3) is given by a combination of the two fundamental energies:

$$\rho_D(t_{EW}) \sim X^{-5} M_{Pl}^4, \tag{17.82}$$

so it is also uniform, since it is given by the universal constants. The same is true for the baryon density: if it is an outcome of the electroweak baryogenesis, the density must also be given by the same universal constants (plus the baryon mass?).

This line of argument needs more stimulating suggestions from the current and future developments in electroweak physics, both theoretical and experimental.

17.9   Protogalactic perturbations

Let us cross from the Metagalaxy as a whole to its astronomical populations. Current concepts relate the origin of galaxies, galaxy clusters, superclusters and the whole Cosmic Web, to the process of gravitational instability. Gravitational instability amplifies seed protogalactic perturbations in the initially homogeneous and uniform cosmic medium, and gives rise eventually to the formation of the observed cosmic structures (as demonstrated in Chapter 13).

The cosmic vacuum is one of the major dynamical factors which control the process. Indeed, the vacuum antigravity terminates the development of gravitational instability in the linear regime at the zero-gravity moment ($t = t_V, z = z_V$). After that, at the vacuum domination epoch, the perturbations can grow only in the nonlinear regime when the relative amplitude of density perturbation is near or larger than unity, and gravity is stronger than antigravity in the over-density volume.

This result implies that the initial cosmological perturbations must be finely tuned in amplitude to come to the nonlinear regime between the redshifts of, say, $z \approx 3 - 10$, when the oldest galaxies are observed, and $z = z_V \approx 1$, when vacuum antigravity stops the linear perturbation growth. Curiously enough, we face here a fine-tuning problem which is very similar to Dicke's paradox of Section 17.5.

We may discuss the problem using again the method suggested (but with no regard to dark energy vacuum) by Zeldovich (1965). See Section 13. Let us assume that the "unperturbed" state of the medium is described by the standard model with flat 3D co-moving space and parabolic expansion. A "perturbed" area embedded in this background is an over-density, which is assumed to be an expanding spherical volume that is described as a part of a Friedmann model with the 3D space of positive curvature and the elliptic regime of expansion. The elliptic expansion proceeds more slowly than the parabolic one does. Because of this, the density in the perturbed volume decreases more slowly than the background density does. As a result, the density of the perturbation area, ρ, becomes larger than the background density, ρ_c (which is the critical density), and the density contrast, $\delta \equiv \frac{\rho - \rho_c}{\rho_c}$ between the over-density and the background grows with time. That is the gravitational instability in action. (Note that the assumed spherical shape of the over-density does not introduce any restrictions to the results, since the dynamics of linear perturbations does not depend on their shape.)

The equation for the radius $r(t)$ of the over-density is a Newtonian analogue of the Friedmann dynamical equation:

$$\dot{r}^2 = \kappa c^2 \rho r^2 - 2|E|, \tag{17.83}$$

where ρ is the total (vacuum, dark matter, baryons, radiation) cosmic density and $E = \mathrm{Const}(t) < 0$ is the total (per unit mass) mechanical energy of the particle at the radial distance $r(t)$ from the center of the over-density; the motion is non-relativistic, so that $|E| \ll c^2$. In the equation for the background expansion, the second (curvature) item on the right-hand-side of this equation is zero, and the density is equal the critical density as was said earlier.

Proceeding as in the derivation of Equation (13.105) earlier then from Equation (17.83) the density contrast is

$$\delta \equiv \frac{\rho - \rho_c}{\rho_c} = \frac{|E|/c^2}{\kappa \rho_c r^2}. \tag{17.84}$$

From Equation (17.84) the density contrast increases as r^2 proportional to t and also proportional to $(1 + z)^{-2}$ at the epoch of radiation domination, and varies as r, i.e., proportional to $t^{2/3}$ which is proportional to $(1 + z)^{-1}$ at the matter domination epoch. This is in complete agreement with the law of gravitational instability in the expanding universe discovered by Lifshitz (1946)—see Chapter 13. At vacuum

domination, the density contrast $\propto r^{-2} \propto \exp(-2ct/A_V) \propto (1+z)^2$, which re-
produces correctly the asymptotic of the exact solution (Chernin, Nagirner and
Starikova (2003) and references therein) for the transition regime from the growth
to decline of the perturbations.

Let us follow the evolution of the largest (on the spatial scale) over-densities
which grow up to $z \sim 1$. If perturbations are generated at the BBN or electroweak
eras, their initial amplitudes must be tuned with the accuracy better than 10^{-15} or
10^{-27}, respectively, to guarantee nonlinearity ($\delta \sim 1$) within the appropriate redshift
range. This estimate follows directly from the relations above. The numerical similarity with Dicke's paradox (Section 17.5) is not accidental: comparing the considerations of this section with the treatment of Section 17.5, one can see that the
deviations from the flatness and the density perturbations here evolve in the same
way with time.

The fine-tuning problem in gravitational instability may be eliminated in a
way which is similar to that of Section 17.5. Indeed, it can be seen from Equation
(17.84) that the density contrast δ goes to zero in the limits of both small and large
times. Therefore it has a maximum in an intermediate era, and this happens at the
moment t_V of zero-gravity when $d^2 R/dt^2 = 0$.

At t_V, the density $\rho \approx 3\rho_V$ and the radius of the Metagalaxy (Equation 17.53)

$$R(t_V) = \left(\frac{1}{2} A_V^2 A_D\right)^{1/3}.$$

Indeed, it is easy to see from Eqn. (17.84) that

$$\delta_{\max} \simeq \frac{1}{(3\kappa\rho_V r(t_V)^2/(|E|/c^2))} = \frac{1}{3\left(\frac{1}{2}\right)^{2/3}\left(\frac{A_V}{A_D}\right)^{2/3}\beta^2}, \qquad (17.85)$$

where $\beta^2 = (r/R)^2 c^2/(|E|) = \text{Const}(t)$ is the constant parameter of the over-density.
This is identical to Equation (14.28) taking $R = ct$, $r = R$ and $|E| = \delta\Phi$, except for
a small coefficient. The explanation of this spectrum by inflation is sometimes taken
as a proof that inflation theory is correct. Equation 17.85 is perhaps a more fundamental way of getting the spectrum, with no reference to inflation.

Equation (17.85) shows that $\delta_{\max} \sim 1$, if the parameter $\beta \approx 1$. It is with this (and
larger) value of the constant parameter β that the density perturbations may survive
against the vacuum antigravity and develop further to form eventually the entire
variety of the large-scale cosmic structures.

No fine tuning of the amplitude is needed, as we see. The gravity-antigravity
interplay controlled by COINS is behind the dynamics of the structure formation
in the Metagalaxy. An order-of-unity constant parameter guarantees the quantitatively correct perturbation evolution, in the linear regime. This parameter may be
fixed as the "initial conditions" for the perturbation growth at any epoch in the

past, because the parameter is time independent (for more details see Chernin 2005).

Finally, we may use the results of this section to specify the "initial" perturbation amplitude at the electroweak era. At that time, the matter density is practically equal to the radiation density $\rho_R(t_{EW}) \sim M_{EW}^4$ (in natural units—see Section 17.3) and the redshift $z \sim X \sim 10^{15}$. With the relation of Equation (17.85) above and $\beta \sim 1$, one finds:

$$\delta_{EW} \sim \frac{1}{\kappa \rho R^2} \sim X^{-2} \sim 10^{-30}, \tag{17.86}$$

where the COINS relation, $A_D \sim A_R \sim A_V \sim X^4 M_D^{-1}$, was used.

Thus, the density contrast at the redshift $z = z_{EW}$ is given in terms of the universal hierarchy number X alone. In this way, COINS together with the freeze-out physics provides the protogalactic perturbations with a "natural" initial amplitude.

It seems challenging that the key questions of twenty-first century cosmology may be formulated in terms of standard physics, and they may perhaps be resolved on the standard grounds. The basic physics is known—fresh ideas are most wanted . . .

Bibliography

Aarseth, S. J. and Saslaw, W. C. (1972) "Virial mass determinations of bound and unstable groups of galaxies", *Astrophys. J.*, **172**, 17.

Abell, G. O. (1975) "Clusters of Galaxies" in *Stars and Stellar Systems*, **9**, *Galaxies and the Universe*, Ed. A. Sandage, M. Sandage, and J. Kristian (Chicago: Univ. Chicago Press), 601.

Adler, R., Bazin, M., and Schiffer, M. (1975) *Introduction to General Relativity*, 2nd ed. New York, McGraw-Hill.

Adler, R., Overduin (2005) "The nearly flat universe" *Gen. Rel. Gravitation*, **37**, 1491.

Aguirre, A. N. (1999) "Dust versus cosmic acceleration". *Astrophys. J.*, **512**, L19.

Allen, S. W., Schmidt, R. W., Ebeling, H., Fabian, A. C., and van Speybroeck, L. (2004) "Constraints on dark energy from Chandra observations of the largest relaxed galaxy clusters", *Mon. Not. Roy. Astron. Soc.*, **353**, 457.

Alpher, R. A., Bethe, H., and Gamow, G. (1948b) "The origin of chemical elements", *Phys. Rev.*, **73**, 803.

Alpher, R. A., Follin, J. W., and Herman, R. (1953) "Physical Conditions in the Initial Stages of the Expanding Universe" *Phys. Rev.* **92**, 1347.

Alpher, R. A. and Herman, R. C. (1949) "Remarks on the evolution of the expanding universe", *Phys. Rev.*, **75**, 1089.

Alpher, R. A., Herman, R. C., and Gamow, G. (1948a) "Thermonuclear Reactions in the Expanding Universe" *Phys. Rev.* **74**, 1198.

Amanullah, R. C., Lidman, D., Rubin, G. et al. (Supernova Cosmology Project) (2010) "Spectra and Hubble Space Telescope Light Curves of Six Type Ia Supernovave at $0.511 < z < 1.12$ and the Union2 Compilation." *Astrophys. J.*, **716**, 712. arXiv:1004.1711

Annila A. (2011) "Least-time paths of light". *Mon. Not. Roy. Astron. Soc.*, **416**, 2944.

Arfken, G. B. and Weber, H. J. (1995) *Mathematical Methods for Physicists*, Fourth edition, Academic Press, San Diego.

Arkani-Hamed, N., Dimopoulos, S., and Dvali, G. (1998) "The hierarchy problem and new dimensions at a millimeter" *Phys. Lett. B*, **429**, 263.

Arkani-Hamed, N., Hall, L. J., Kolda, C., and Murayama, H. (2000) "New Perspective on Cosmic Coincidence Problems," *Phys. Rev. Lett.* **85**, 4434. http://link.aps.org/doi/10.1103/PhysRevLett.85.4434

Arp, H. (1966) "Atlas of Peculiar Galaxies" *Astrophys. J. Supp.*, **14**, 1.

Arp, H. C. (1970) "Redshifts of companion galaxies" *Nature*, **225**, 1034.

Aurela, A. M. (1973) "Potentially fundamental anomalous observations" *Ann. Univ. Turkuensis A*, **162**, 85.

Axenides, M. and Perivolaropoulos, L. (2002) "Dark energy and the quietness of the local Hubble flow" *Phys. Rev. D*, **65**, 127301.

Bahcall, J. N. and Salpeter, E. E. (1965) "On the interaction of radiation from distant sources with the intervening medium", *Astrophys. J.*, **142**, 1677.

Bahcall, J. N. (1984a) "The distribution of stars perpendicular to galactic disk", *Astrophys. J.*, **276**, 156.

Bahcall, J. N. (1984b) "Self-consistent determinations of the total amount of matter near the sun", *Astrophys. J.*, **276**, 169.

Bahcall, J. N., Greenstein, J. L., and Sargent, W. L. W. (1968) "The absorption-line spectrum of the quasi-stellar radio source PKS 0237-23", *Astrophys. J.*, **153**, 689.

Bahcall, N. A. and Fan, X. (1998) "The most massive distant clusters: determining Ω and σ_8", *Astrophys. J.*, **504**, 1.

Bahcall, N. A., Fan, X., and Cen, R. (1997) "Constraining Ω with cluster evolution", *Astrophys. J.*, **485**, L53.

Balland, C. and Blanchard, A. (1997) "On the uncertainty in x-ray cluster mass estimates from the equation of hydrostatic equilibrium", *Astrophys. J.*, **487**, 33.

Barnes, J. E. (1992) "Transformation of galaxies. I. Mergers of equal-mass stellar disks", *Astrophys. J.*, **393**, 484.

Barnes, J. E. and Hernquist, L. (1996) "Transformations of galaxies. II. Gas dynamics in merging disk galaxies", *Astrophys. J.*, **471**, 115.

Barnothy, J. (1974) "Galaxy-Quasar Associations and Globular Clusters" *BAAS*, **6**, 212.

Baryshev Yu. and Ezova Yu. (1997) "Gravitational Mesolensing by King Objects and Quasar–Galaxy Associations" *Astronomy Reports*, **41**, 436.

Baryshev, Yu., Sylos-Labini, F., Montuori, M., Pietronero, L., and Teerikorpi, P. (1998) "On the fractal structure of galaxy distribution and its implications for cosmology" *Fractals*, **6**, 231.

Baryshev, Yu. V., Chernin, A. D., and Teerikorpi, P. (2001) "The cold local Hubble flow as a signature of dark energy" *Astron. Astrophys.*, **378**, 729.

Baryshev Yu. and Teerikorpi P. (2012) *Fundamental Questions of Practical Cosmology*, Astrophysics and Space Science Library 383, Springer Verlag.

Becker, R. H., Fan, X., White, R. L. et al. (2001) "Evidence for reionization at $z \approx 6$: Detection of a Gunn-Peterson trough in a $z = 6.28$ quasar", *Astron. J.*, **122**, 2850.

Benetti, S., Cappellaro, E., Mazzali, P. A. et al. (2005) "The diversity of Type Ia supernovae: evidence for systematics?" *Astrophys. J.*, **623**, 1011.

Bennett, C. L., Halpern, M., Hinshaw, G. et al. (2003) "First-year Wilkinson Microwave Anisotropy Probe (WMAP) observations: Preliminary maps and basic results", *Astrophys. J. Suppl.*, **148**, 1.

Bethe, H. (1939) "Energy production in stars", *Phys. Rev.*, **55**, 434.

Beutler, F., Blake, C., Colless, M. et al. (2011) "The 6dF Galaxy Survey: baryon acoustic oscillations and the local Hubble constant". *Mon. Not. Roy. Astron. Soc.*, **416**, 3017.

Binney, J. and Tremaine, S. (1987, 2008) *Galactic Dynamics*, Princeton University Press, Princeton, NJ.

Bisnovatyi-Kogan, G. S. and Zeldovich, Ya. B. (1970) "Growth of Perturbations in an Expanding Universe of Free Particles" *Astronomicheskii Zhurnal.*, **47**, 942.

Biviano, A., Durret, F., and Gerbal, D. (1996) "Unveiling hidden structures in the Coma cluster", *Astron. Astrophys.*, **311**, 95.

Blitz, L. (1979) "The rotation curve of the Galaxy to $R = 16$ kiloparsecs". *Astrophys. J.*, **231**, L115.

Blumenthal, G. R., Faber, S. M., Primack, J. R., and Rees, M. J. (1984) "Formation of galaxies and large-scale structure with cold dark matter" *Nature*, **311**, 517.

Blumenthal, G. R., Pagels, H., and Primack, J. R. (1982) "Galaxy formation by dissipationless particles heavier than neutrinos", *Nature*, **299**, 37.

Bonamente, M., Joy, M. K., and Lieu, R. (2003) "A massive warm baryonic halo in the Coma cluster", *Astrophys. J.*, **585**, 722.

Bond, J. R., Szalay, A. S., and Turner, M. S. (1982) "Formation of galaxies in a gravitino-dominated universe", *Phys. Rev. Lett.*, **48**, 1636.

Bond, J. R. and Efstathiou, G. (1984) "Cosmic background radiation anisotropies in universes dominated by nonbaryonic dark matter", *Astrophys. J.*, **285**, L45.

Bondi, H. and Gold T. (1948) "The Steady—State theory of the expanding Universe". *Mon. Not. Roy. Astron. Soc.*, **108**, 252.

Bottema, R. (1999) "The kinematics of the bulge and the disc of NGC 7331", *Astron. Astrophys.*, **348**, 77.

Bottinelli, L., Gouguenheim, L., Paturel, G., and Teerikorpi, P. (1986) "The Malmquist bias and the value of H_0 from the Tully-Fisher relation", *Astron. Astrophys.*, **156**, 157.

Branch, D. (1987) "High-velocity matter in a classical Type I supernova: the demise of Type Ia homogeneity." *Astrophys. J.*, **316**, L81.

Bukhmastova, Yu. L. and Baryshev, Yu. V. (2008) "Constraining the nature of galaxy haloes with gravitational mesolensing of QSOs by halo substructure objects" In *Problems of Practical Cosmology*, Procs. of the International Conference at Russian Geographical Society, 23–27 June, 2008, Eds. Yu. V. Baryshev, I.N. Taganov, P. Teerikorpi, Vol. 1. TIN, St. Petersburg, p. 18.

Burbidge, E. M., Burbidge, G. R., Fowler, W. A., and Hoyle, F. (1957) "Synthesis of the elements in stars", *Rev. Mod. Phys.* **29**, 547.

Burbidge, G. (2001) "Noncosmological Redshifts" *Pub. Astron. Soc. Pac.*, **113**, 899.

Burke, W. L. (1981) "Multiple gravitational imaging by distributed masses", *Astrophys. J.*, **244**, L1.

Butcher, H. R. (1987) "Thorium in G-dwarf stars as a chronometer for the Galaxy" *Nature*, **328**. 127.

Byrd, G. G., Chernin, A., and Valtonen, M. J. (2007) *Cosmology: Foundations and Frontiers*, URSS Publishers, Moscow.

Byrd, G. G. and Valtonen, M. J. (1978) "Orbital dynamics of the radio galaxy 3C 129" *Astrophys. J.*, **221**. 481.

Byrd, G. G. and Valtonen, M. J. (1985) "Origin of redshift differentials in galaxy groups" *Astrophys. J.*, **289**, 535.

Byrd, G. G., Valtonen, M. J., McCall, M. L., and Innanen, K. A. (1994) "Orbits of the Magellanic Clouds and Leo I in Local Group history" *Astron. J.*, **107**, 2055.

Cameron, A. G. W. (1957) "Nuclear reactions in stars and nucleogenesis", *Pub. Astron. Soc. Pacific*, **69**, 201.

Canizares C. R. (1981) "Gravitational focusing and the association of distant quasars with foreground galaxies" *Nature*, **291**, 620.

Carroll, S. M., Press, W. H., and Turner, E. L. (1992) "The cosmological constant" *Ann. Rev. Astron. Astrophys.*, **30**, 499.

Casimir, H. B. G. (1948) "On the attraction between two perfectly conducting plates" *Proc. Kgl. Ned. Akad. Wet.* **60**, 793.

Cavaliere, A. and Fusco-Femiano, R. (1976) "X-rays from hot plasma in clusters of galaxies", *Astron. Astrophys.*, **49**, 137.

Cayrel, R., Hill, V., Beers, T. C. et al. (2001) "Measurement of stellar age from Uranium decay", *Nature*, **409**, 691.

Célérier, M.-N., Bolejko, K., and Krasiński, A. (2010) "A (giant) void is not mandatory to explain away dark energy with a Lemaître-Tolman model." *Astron. Astrophys.*, **518**, A21.

Cen, R. and Ostriker, J. P. (1999) "Where are the baryons?", *Astrophys. J.*, **514**, 1.

Chernin A. D. (1965) "A Model of a Universe Filled by Radiation and Dustlike Matter" *Astron. Zh.*, **42**, 1124 (*Sov. Astron.* 1966).

Chernin A. D. (1968) *Nature* **220**, 250.

Chernin, A. D. (1981) "The Rest Mass of Primordial Neutrinos and Gravitational Instability in the Hot Universe", *Soviet Astronomy*, **25**, 14.

Chernin, A. (2001) "Cosmic vacuum" *Physics-Uspekhi*, **44**, 1153.

Chernin, A. D. (2002a) "Physical vacuum and cosmic coincidence problem" *New Astron.*, **7**, 113.

Chernin, A. D. (2002b) "Cosmic vacuum and large-scale perturbations in the concordant model" astro-ph//0211489.

Chernin, A. D. (2002c) "The Friedmann integrals and physical vacuum in the framework of macroscopic extra dimensions" astro-ph//0206179.

Chernin, A. (2003) "Cosmic vacuum and the 'flatness problem' in the concordance model," *New Astronomy*, **8**, 79.

Chernin A. D. (2005) "Energy composition of the Universe: time-independent internal Symmetry" *Ap. and Sp. Sci.*, **305**, 143.

Chernin, A. D., Karachentsev, I. D., Kashibadze, O. G. et al. (2007a) "Local dark energy: HST evidence from the vicinity of the M81/M82 galaxy group" *Astrophysics*, **50**, 405.

Chernin, A. D., Karachentsev, I. D., and Makarov, D. I. (2007b) "Local dark energy: HST evidence from the expansion flow around Cen A/M83 galaxy group" *Astron. Astrophys. Transactions*, **26**, 275.

Chernin, A. D., Karachentsev, I. D., Nasonova, O. G. et al. (2010) "Dark energy domination in the Virgocentric flow" *Astron. Astrophys.*, **520**, A104.

Chernin, A. D., Karachentsev, I. D., Valtonen, M. J., Dolgachev, V. P., Domozhilova, L. M., and Makarov, D. I. (2004) "The very local Hubble flow: computer simulations of dynamical history", *Astron. Astrophys.*, **415**, 19.

Chernin, A. D., Nagirner, D. I., and Starikova, S. V. (2003) "Growth rate of cosmological perturbations in standard model: Explicit analytical solution" *Astron. Astrophys.* **399**, 19.

Chernin, A., Teerikorpi, P., and Baryshev, Yu. (2000) "Why is the Hubble flow quiet?" astro-ph/0012021, *Adv. Space Res.*, **31**, 459.

Chernin, A. D., Teerikorpi, P., and Baryshev, Yu. V. (2006) "Non-Friedmann cosmology for the Local Universe, significance of the universal Hubble constant, and short-distance indicators of dark energy" *Astron. Astrophys.*, **456**, 13.

Chernin, A. D., Teerikorpi, P., Valtonen, M. J., Dolgachev, V. P., Domozhilova, L. M., and Byrd, G. G. (2009) "Local dark matter and dark energy as estimated on a scale of ~ 1 Mpc in a self-consistent way" *Astron. Astrophys.*, **507**, 1271.

Chernin, A. D., Teerikorpi, P., Valtonen, M. J., Dolgachev, V. P., Domozhilova L. M., and Byrd, G. G. (2011) "Virial estimator for dark energy" arXiv:1109.1215v1.

Chernin, A. D., Teerikorpi, P., Valtonen, M. J., Dolgachev, V. P., Domozhilova, L. M., and Byrd, G. G. (2012) "Dark energy and extended dark matter halos" *Astron. Astrophys.* (in press). **539**, A4.

Chincarini, G. (1978) "Clumpy structure of the universe and general field" *Nature*, **272**, 515.

Chwolson, O. (1924) "Über eine mögliche Form fiktiver Doppelsterne", *Ast. Nachr.*, **221**, 329.

Clowe, D., Bradač, M., and Gonzalez, A. H. (2006) "A Direct Empirical Proof of the Existence of Dark Matter" *Astrophys. J.*, **648**, L109.

Coc, A., Vangioni-Flam, E., Descouvemont, P., Adahchour, A., and Angulo, C. (2004) "Updated Big Bang Nucleosynthesis compared with Wilkinson Microwave Anisotropy Probe observations and the abundance of light elements", *Astrophys. J.*, **600**, 544.

Colless, M. and Dunn, A. M. (1996) "Structure and dynamics of the Coma cluster", *Astrophys. J.*, **458**, 435.

Colley, W. N., Tyson, J. A., and Turner, E. L. (1996) "Unlensing Multiple Arcs in 0024 + 1654: Reconstruction of the Source Image" *Astrophys J.*, **461**, L83.

Conley, A., Guy, J., and Sullivan, M. (2011) "Supernova constraints and systematic uncertainties from the first three years of the supernova legacy survey," *Astrophys. J. Suppl. Ser.*, **192**, 1.

Cornish, N. J., Spergel, D. N., Starkman, G. D., and Komatsu, E. (2004) "Constraining the topology of the universe," *Phys. Rev. Lett.* **92**, 201302.

Cowie, L. L., Henriksen, M., and Mushotzky, R. (1987) "Are the virial masses of clusters smaller than we think?" *Astrophys. J.*, **317**, 593.

Crone, M. M., Evrard, A. E., and Richstone, D. O. (1996) "Substructure in clusters as a cosmological test", *Astrophys. J.*, **467**, 489.

Dahle, H., Kaiser, N., Irgens, R. J., Lilje, P. B., and Maddox, S. J. (2002) "Weak gravitational lensing by a sample of x-ray luminous clusters of galaxies. I. The data set", *Astrophys. J. Supp.*, **139**, 313.

Davis, M. and Peebles, P. J. E. (1983) "A survey of galaxy redshifts. V. The two-point position and velocity correlations" *Astrophys. J.*, **267**, 465.

Dekel, A. and Aarseth, S. J. (1984) "The spatial correlation function of galaxies confronted with theoretical scenarios", *Astrophys. J.*, **283**, 1.

de Lapparent, V., Geller, M. J., and Huchra, J. P. (1968) "A slice of the universe" *Astrophys. J.*, **301**, L1.

de Lapparent, Valerie, Geller, Margaret J., and Huchra, John P. (1988) "The mean density and two-point correlation function for the CfA redshift survey slices" *Astrophys. J.*, **332**, 44.

De Vaucouleurs, G. and Peters, W. L. (1981) "Hubble ratio and solar motion from 200 spiral galaxies having distances derived from the luminosity index", *Astrophys. J.*, **248**, 395.

Demianski, M., de Ritis, R., Marino, A. A., and Piedipalumbo, E. (2003) "Approximate angular diameter distance in a locally inhomogeneous universe with nonzero cosmological constant", *Astron. Astrophys.*, **411**, 33.

Dicke, R. H. (1970) "Gravitation and the universe" *Memoirs of the American Philosophical Society*, Jayne Lectures for 1969, Philadelphia: American Philosophical Society, 1970.

Dolgov, A. D. (1992) "Non-GUT baryogenesis" *Phys. Rep.* **222**, 309.

Dolgov, A. D. (2003) "Cosmology at the Turn of Centuries" I YA POMERANCHUK AND PHYSICS AT THE TURN OF THE CENTURY. Proceedings of the International Conference. Held 24–28 January 2003 in Moscow, Russia. Edited by A Berkov (Moscow Engineering Physics Institute, Russia), N. Narozhny (Moscow Engineering Physics Institute, Russia), & L. Okun (Institute of Theoretical and Experimental Physics, Russia). Published by World Scientific Publishing Co. Pte. Ltd., 2003. ISBN #9789812702883, pp. 103–123.

Dolgov, A. D. (2004) "Problems of vacuum energy and dark energy". *Results and Perspectives of particle Physics, Proc. 18th Recontre de Physique de la Vallee d'Aosta.*

Dolgov, A. D., Zeldovich Ya. B., and Sazhin M. V. (1988) *Cosmology of the Early Universe.* (In Russian); Moscow Univ. Press, Moscow (English Ed. *Basics of Modern Cosmology*, Editions Frontiers, 1990).

Donahue, M., Voit, G. M., Gioia, I., Luppino, G., Hughes, J. P., and Stocke, J. T. (1998) "A very hot high-redshift cluster of galaxies: more trouble for $\Omega_0 = 1$", *Astrophys. J.*, **502**, 550.

Dyson, F. W., Eddington, A. S., and Davidson, C. R. (1920) "A determination of the deflection of light by the Sun's gravitational field from observations made at the total eclipse of May 29, 1919", *Mon. Roy. Astron. Soc.*, **62**, 291.

Eddington, A. S. (1919) "The total eclipse of May 29 and the influence of gravitation on light", *Observatory*, **42**, 119.

Efstathiou G., Ellis R., and Peterson B. A. (1988) "Analysis of a complete galaxy redshift survey. II—The field-galaxy luminosity function", *Mon. Roy. Astron. Soc.*, **232**, 431.

Einasto, J. and Lynden-Bell, D. (1982) "On the mass of the Local Group and the motion of its barycenter", *Mon. Not. Roy. Astron. Soc.*, **199**, 67.

Einasto J., Kaasik A., Saar E. (1974a) "Dynamical evidence on massive coronas of galaxies", *Nature*, **250**, 309.

Einasto, J., Saar, E., Kaasik, A., and Chernin, A. D. (1974b) "Missing mass around galaxies—Morphological evidence", *Nature*, **252**, 111.

Einstein, A. (1907) "Über das Relativitätsprinzip und die aus demselben gezogenen Folgerungen" *Jahrb. Radioakt. Elekt.*, **4**, 411.

Einstein, A. (1911) "Über den Einfluss der Schwerkraft auf die Ausbreitung des Lichtes" *Ann. Physik*, **35**, 898.

Einstein, A. (1915) "Erklärung der Perihelbewegung des Merkur aus der allgemeinen Relativitätstheorie" *Sitz. Preuss. Akad. Wiss.*, **11**, 831.

Einstein, A. (1916) "Die Grundlage der allgemeinen Relativitätstheorie", *Ann. Physik*, **49**, 769.

Einstein, A. (1917a) "Über die spezielle and die allgemeine Relativitätstheorie", Friedrich Vieweg & Sohn, Braunschweig.

Einstein, A. (1917b) "Kosmologische Betrachtungen zur allgemeinen Relativitätstheorie", *Akad. Wiss.* **1**, 142.

Einstein, A. (1936) "Lens-like action of a star by the deviation of light in the gravitational field", *Science*, **84**, 506.

Einstein, A., and Straus, E. G. (1945) "The Influence of the Expansion of Space on the Gravitation Fields Surrounding the Individual Stars" *Reviews of Modern Physics*, **17**, Issue 2–3, 120–124.

Eisenhauer, F., Schödel, R., Genzel, R. et al. (2003) "A geometric determination of the distance to the galactic center". *Astrophys. J.*, **597**, L121.

Eisenstein, D. J., Zehavi, I., Hogg, D. W. et al. (2005) "Detection of the baryon acoustic peak in the large-scale correlation function of SDSS luminous red galaxies", *Astrophys. J.*, **633**, 560, astro-ph/0501171

Eke, V. R., Cole, S., and Frenk, C. S. (1996) "Cluster evolution as a diagnostic for Ω", *Mon. Not. Roy. Astron. Soc.*, **282**, 263.

Ekholm, T., Baryshev, Yu., Teerikorpi, P., Hanski, M., and Paturel, G. (2001) "On the quiescence of the Hubble flow in the vicinity of the Local group: a study using galaxies with distances from the Cepheid PL-relation" *Astron. Astrophys.*, **368**, L17.

Ellis, G. F. R. (2003) "Cosmology: The shape of the Universe", *Nature*, **425**, 566.

Fall, S. M. and Jones, B. J. T. (1976) "Isotropic cosmic expansion and the Rubin-Ford effect". *Nature*, **262**, 457.

Fang, L. L., Mo, H. J., and Ruffini, R. (1991) "The cellular structure of the universe and cosmological tests" *Astron. Astrophys.*, **243**, 283.

Feng, J. L. (2010), "Dark matter candidates from particle physics and methods of detection". *Ann. Rev. Astron. Astrophys.*, **48**, 495.

Ferreras, I., Melchiorri, A. and Silk, J. (2001) "Setting new constraints on the age of the Universe", *Mon. Not. Roy. Astron. Soc.*, **327**, L47.

Forman, W. R. (1970) "A Reduction of the Mass Deficit in Clusters of Galaxies by Means of a Negative Cosmological Constant" *Astrophys. J.*, **159**, 719.

Freedman, W. L., Madore, B. F., Gibson, B. K. et al. (2001) "Final results from the Hubble Space Telescope Key Project to measure the Hubble constant", *Astrophys. J.*, **553**, 47.

Freedman, W. L. and Madore, B. F. (2010) "The Hubble constant", Annu. Rev. Astron. Astrophys. **48**, 673.

Freire, P. C. C. and Wex, N. (2010), "The orthometric parametrization of the Shapiro delay and an improved test of general relativity with binary pulsars," *Mon. Not. Roy. Astron. Soc.*, **409**, 199.

Friedmann, A. A. (1923) *Die Welt als Raum und Zeit* Ostwalds Klassiker der Exakten Wissenschaften, Band 287 *The World as Space and Time*, Petrograd Academia.

Friedmann, A. A. (1922) "Über die Krümmung des Raumes" *Z. Physik*, **10**, 377 (English transl. in *Sov. Phys.-Uspekhi*, **20**, 1964)

Friedmann, A. A. (1924) "Über die Möglichkeit einer Welt mit konstanter negativer Krümmung des Raumes" *Z. Physik*, **21**, 326.

Gabrielli A., Sylos-Labini F., Joyce M., and Pietronero L. (2005) *Statistical physics for cosmic structures*, Springer Verlag.

Gamow, G. and Teller, E. (1939) "On the Origin of Great Nebulae", *Phys. Rev.*, **55**, 654.

Gamow, G. (1946) "Expanding Universe and the Origin of Elements" *Phys. Rev.*, **70**, 572.

Gaztañaga E. (2003) "Correlation between Galaxies and Quasi-stellar Objects in the Sloan Digital Sky Survey: A Signal from Gravitational Lensing Magnification?" *Astrophys. J.*, **589**, 82.

Geller, M. J., Diaferio, A., and Kurtz, M. J. (1999) "The mass profile of the Coma Galaxy Cluster", *Astrophys. J.*, **517**, L23.

Geller, M. J. and Huchra, J. P. (1989) "Mapping the Universe", *Science*, **246**, 897.

Ghigna, S., Moore, B., Governato, F., Lake, G., Quinn, T., and Stadel, J. (2000) "Density profiles and substructure of dark matter halos: converging results at ultra-high numerical resolution", *Astrophys. J.*, **544**, 616.

Girardi, M., Borgani, S., Guiricin, G., Mardirossian, F., and Mezzetti, M. (2000) "Optical luminosities and mass-to-light ratios of nearby galaxy clusters", *Astrophys. J.*, **530**, 62.

Girardi, M., Giuricin, G., Mardirossian, F., Mezzetti, M., and Boschin, W. (1998) "Optical mass estimates of galaxy clusters", *Astrophys. J.*, **505**, 74.

Gliner, E. B. (1965) *JETP* **49**, 542.

Gliner, E. B. and Dymnikova, I. G. (1975) "A nonsingular Friedmann cosmology" *Soviet Astron. Lett.*, **1**, 93.

Golfand, Y. A. and Likhtman, E. P. (1971) "On the extension of the algebra of Poincare group" *JETP Lett*, **13**, 323.

Goobar, A., Hannestad, S., Mörtsell, E., and Tu, H. (2006) "The neutrino mass bound from WMAP 3 year data, the baryon acoustic peak, the SNLS supernovae and the Lyman-α forest". *Journal of Cosmology and Astroparticle Physics* 606 (6): 19. arXiv:astro-ph/0602155

Gott, J. R. III, Turner, E. L., and Aarseth, S. J. (1979) "N-body simulations of galaxy clustering. III. The covariance function", *Astrophys. J.*, **234**, 13.

Governato, F., Moore, B., Cen, R., Stadel, J., Lake, G., and Quinn, T. (1997) "The Local Group as a test of cosmological models" *New Astronomy*, **2**, no. 2, p. 91–106.

Gromov, A., Baryshev, Yu., Suson, D., and Teerikorpi, P. (2001) "Lemaître-Tolman-Bondi model: fractality, non-simultaneus bang time and the Hubble law" *Grav. Cosmol.* **7**, 140.

Gunn, J. E. and Peterson, B. A. (1965) "On the density of neutral hydrogen in intergalactic space", *Astrophys. J.*, **142**, 1633.

Gurevich, L. E. (1975) "On the origin of the metagalaxy" *Astrophys. Space Sci.*, **38**, 67.

Gurvits, L. I., Kellermann, K. I., and Frey, S. (1999) "The "angular size—redshift" relation for compact radio structures in quasars and radio galaxies" *Astron. Astrophys.*, **342**, 378.

Guth, A. H. (1981) "Inflationary universe: a possible solution to the horizon and flatness problems", *Phys. Rev.*, **D23**, 347.

Guth, A. H. (1997) *The inflationary universe. The quest for a new theory of cosmic origins*, Perseus Books, Addison-Wesley, Reading.

Haggerty, M. J. and Wertz, J. R. (1972) "On the redshift–magnitude relation in hierarchical cosmologies" *Mon. Not. Roy. Astron. Soc.*, **155**, 495.

Hall, A. (1894) "A suggestion in the theory of Mercury", *Astron. J.*, **14**, 49.

Hamilton, W. R. (1834) "On a general method in dynamics" and "Second essay on a general method in dynamics" Collected Papers, Vol. II. Cambridge, 1940, 103–211.

Hamuy, M., Phillips, M. M., Maza, J., Suntzeff, N. B., Schommer, R. A., and Aviles, R. (1995) "Hubble diagram of distant type IA supernovae". *Astron. J.*, **109**, 1.

Hansen, B. M. S., Brewer, J., Fahlman, G. G. et al. (2002) "The white dwarf cooling sequence of the globular cluster Messier 4", *Astrophys. J.*, **574**, L155.

Harrison, E. R. (1970) "Fluctuations at the threshold of classical cosmology", *Phys. Rev.*, *D*, **1**, 2726.

Hartwick, F. D. A. (2011) "The Velocity Field around Groups of Galaxies" *Astron. J.*, **141**, 198.

Hesser, J. E., Harris, W. E., Vandenberg, D. A., Allwright, J. W. B., Shott, P., and Stetson, P. B. (1987) "A CCD color-magnitude study of 47 Tucanae", *Publ. Astron. Soc. Pac.*, **99**, 739.

Hewitt, J. N., Turner, E. L., Schneider, D. P., Burke, B. F., Langston, G. I., and Lawrence, C. R. (1988) "Unusual Radio Source MG1131 + 0456: a Possible Einstein Ring." *Nature*, **333**, 537.

Hicken, M., Wood-Vasey, W. M., Blondin, S. et al. (2009) "Improved dark energy constraints from 100 new CfA superÙnova Type Ia light curves." *Astrophys. J.*, **700**, 1097–1140.

Hilbert, David (1915) „Grundlagen der Physik, Erste Mitteilung, vorgelegt in der Sitzung" vom 20. November 1915 p. 395. English translation http://www.springerlink.com/content/t2681418480nq841

Hill, V., Plez, B., Cayrel, R. et al. (2002) "First stars. I. The extreme *r*-element rich, iron-poor halo giant CS31082-001: Implications for the *r*-process site(s) and radioactive cosmochronology", *Astron. Astrophys.* **387**, 560.

Hinshaw, G., Weiland, J. L., Hill, R. S. et al. (WMAP Collaboration), (2009) "Five-Year Wilkinson Microwave Anisotropy Probe Observations: Data Processing, Sky Maps, and Basic Results". *Astrophys. J. Suppl.*, **180**, 225.

Hobson, M. P., Efstathiou, G., Lasenby, A. N., and Bacon, D. (2006) "General Relativity", by M. P. Hobson, G. P. Efstathiou and A. N. Lasenby, pp. 590. Cambridge University Press, February 2006. ISBN-10: 0521829518. ISBN-13: 9780521829519.

Hoekstra, H., Franx, M., Kuijken, K. et al. (2001) "Weak-lensing study of low-mass galaxy groups: implications for Ω_m", *Astrophys. J.*, **548**, L5.

Hoekstra, H., Franx, M., Kuijken, K., Carlberg, R. G., and Yee, H. K. C. (2003) "Lensing by galaxies in CNOC2 fields", *Mon. Not. Roy. Astron. Soc.*, **340**, 609.

Hoekstra, H., Franx, M., Kuijken, K., and van Dokkum, P. G. (2002) "HST large-field weak lensing analysis of MS2053-04: study of the mass distribution and mass-to-light ratio of x-ray luminous clusters at $0.22 < z < 0.83$", *Mon. Not. Roy. Astron. Soc.*, **333**, 911.

Hoekstra, H., Yee, H. K. C., and Gladders, M. D. (2004) "Properties of Galaxy Dark Matter halos from Weak lensing" *Astrophys. J.*, **606**, 67.

Hoffman Y., Martinez-Vaquero L. A., Yepes G., and Gottlöber S. (2008) "The local Hubble flow: is it a manifestation of dark energy?" *MNRAS*, **386**, 390.

Höflich, P., Wheeler, J. C., and Thielemann, F. K. (1998) "Type Ia Supernovae: Influence of the Initial Composition on the Nucleosynthesis, Light Curves, and Spectra and Consequences for the Determination of Ω_M and Λ" *Astrophys. J.*, **495**, 617.

Hogg, David W., Eisenstein, Daniel J., Blanton, Michael R. et al. (2005) "Cosmic homogeneity demonstrated with luminous red galaxies". *Astrophys. J.*, **624**, 54.

Holmberg, J. and Flynn, C. (2004) "The local surface density of disc matter mapped by Hipparcos", *Mon. Not. Roy. Astron. Soc.*, **352**, 440.

Hooley, A., Longair, M. S., and Riley, J. M. (1978) "The angular diameter-redshift test for quasi-stellar radio sources with large redshifts", *Mon. Not. Roy. Astron. Soc.*, **182**, 127.

Howell, A., Sullivan, M., Conley, A., and Carlberg, R. (2007) "Predicted and Observed Evolution in the Mean Properties of Type Ia Supernovae with Redshift". *Astrophys. J.*, **667**, L37.

Howell, D. A. (2011) "Type Ia supernovae as stellar endpoints and cosmological tools." *Nat. Commun.* 2:350 doi: 10.1038/ncomms1344.

Hoyle, F. (1948) "A New Model for the Expanding Universe", *Mon. Not. Roy. Astron. Soc.*, **108**, 37.

Hoyle, F. (1953) "On the fragmentation of gas clouds into galaxies and stars" *Astrophys. J.*, **118**, 513.

Hoyle, F. (1959) The relation of radio astronomy to cosmology. in *Radio Astronomy, IAU Symp. 9*, ed. R. N. Bracewell, p. 529.

Hoyle, F., Burbidge, G., and Narlikar, J. (2000) *A Different Approach to Cosmology* (Cambridge University Press, Cambridge). http://arxiv.org/pdf/astro-ph/9606055v1.pdf

Hoyle, F. and Sandage, A. (1956) "The Second-Order Term in the Redshift-Magnitude Relation". *Pub. Astron. Soc. Pac.*, **68**, 301.

Hu, W. and White, M. (2001) "Power Spectra Estimation for Weak Lensing", *Astrophys. J.*, **554**, 67.

Hu, W. and White, M. (2004) "The Cosmic Symphony", *Sci. Am.* February issue, 44.

Hubble, E. P. (1925) "Cepheids in spiral nebulae" *Observatory*, **48**, 139.

Hubble, E. P. (1929) "A relation between distance and velocity among extra-galactic nebulae," *Proceedings of the National Academy of Sciences*, **15**, 168–173.

Huchra, J. P. and Geller, M. J. (1982) "Groups of galaxies. I – Nearby groups" *Astrophys. J.*, **257**, 423.

Huchra, J. P., Vogeley, M. S., and Geller, M. J. (1999) "The CfA Redshift Survey. Data for the South Galactic cap", *Astrophys. J. Suppl.*, **121**, 287.

Hulse, R. A. (1994) "The discovery of the binary pulsar", *Rev. Mod. Phys.*, **66**, 699.

Hulse, R. A., and Taylor, J. H. (1975) "Discovery of a pulsar in a binary system", *Astrophys. J.*, **195**, L51.

Jaakkola, T. (1971) "On the Redshifts of Galaxies" *Nature*, **234**, 534.

Jaakkola, T. and Moles, M. (1976) "The type-redshift relation in the Vaucouleurs' groups of galaxies" *Astron. Astrophys.*, **53**, 389.

Jackson, J. C. (1970) "The dynamics of clusters of galaxies in universes with non-zero cosmological constant, and the virial theorem mass discrepancy" *Mon. Not. Roy. Astron. Soc.*, **148**, 249.

Jackson, J. C. and Dogdson, M. (1997) "Deceleration without dark matter." *Mon. Not. Roy. Astron. Soc.*, **285**, 806.

Jackson, J. C. and Jannetta, A. L. (2006) "Legacy data and cosmological constraints from the angular-size/redshift relation for ultra-compact radio sources." *J. Cosmol. Astroparticle Phys.*, **0611**, 002.

Jägers, W. J. (1987) "0.6 GHz mapping of extended radio galaxies. III. 3C66B, NGC1265, 3C129, DA240, 3C236, 4C48.29, IC708 & IC711, 4CT51.29.1, 3C310, Abell2256, 3C402, and 3C465", *Astron. Astrophys Supp.* **71**, 603.

Janes, K. A. and Phelps, R. L. (1994) "The galactic system of old star clusters. The development of the galactic disk", *Astron. J.*, **108**, 1773.

Jarosik, N., Bennett, C. L., Kogut, A., Komatsu, E. et al. (2011) "Seven-Year Wilkinson Microwave Anisotropy Probe (WMAP1) Observations: Sky Maps, Systematic Errors, and Basic Results", *The Astrophysical Journal Supplement Series*, 192, Issue: 2, page 14. For other "seven year" papers see http://lambda.gsfc.nasa.gov/product/map/dr4/map_bibliography.cfm

Jeans, J. H. (1902) "The stability of a spherical nebula" *Phil Trans*, **199A**, 1 and 49.

Jerjen, H. and Tammann, G. A. (1993) "The Local Group motion towards Virgo and the microwave background", *Astron. Astrophys.*, **276**, 1.

Jimenez, R. and Padoan, P. (1998) "The ages and distances of globular clusters with the luminosity function method: the case of M5 and M55" *Astrophys. J.*, **498**, 704.

Jimenez, R., Thejll, P., Jorgensen, U. G., MacDonald, J., and Pagel, B. (1996) "Ages of globular clusters: a new approach", *Mon. Not. Roy. Astron. Soc.*, **282**, 926.

Jõeveer, M. and Einasto, J. (1978) "Has the Universe a cell structure", *The Large scale Structure of the Universe*", IAU Symp. 79, Eds. M. S. Longair and J. Einasto, Reidel, Dordrecht, p. 241.

Jõeveer, M., Einasto, J., and Tago, E. (1978) "Spatial distribution of galaxies and clusters of galaxies in the Southern Galactic hemisphere", *Mon. Not. Roy. Astron. Soc.*, **185**, 357.

Kahn, F. D. and Woltjer, L. (1959) "Intergalactic matter and the Galaxy" *Astrophys. J.*, **130**, 705.

Karachentsev, I. D., Sharina, M. E., Dophin, A. E., and Grebel, E. K. (2003) "Distances to nearby galaxies around IC342", *Astron. Astrophys.*, **408**, 111.

Karachentsev, I. D., Kashibadze, O. G., Makarov, D. I., and Tully, R. B. (2009) "The Hubble flow around the Local Group" *Mon. Not. Roy. Astron. Soc.*, **393**, 1265.

Kellermann, K. I. (1993) "The cosmological deceleration parameter estimated from the angular-size/redshift relation for compact radio sources". *Nature*, **361**, 134

Kessler, R., Becker, A. C., Cinabro, D. et al. (2009) "First-year sloan digital sky survery-II supernova results: hubble diagram and cosmological parameters." *Astrophys. J. Suppl. Ser.*, **185**, 32.

King, I. (1966) "The structure of star clusters. III. Some simple dynamical models", *Astron. J.*, **71**, 64.

Kirkman, D., Tytler, D., Suzuki, N., O'Meara, J. M., and Lubin, D. (2003) "The cosmological baryon density from the deuterium-to-hydrogen ratio in QSO absorption systems: D/H toward Q1243 + 3047", *Astrophys. J. Supp.*, **149**, 1.

Kirshner, R. P., Oemler, A. Jr., Schechter, P. L., and Shectman, S. A. (1981) "A million cubic megaparsec void in Bootes", *Astrophys. J.*, **248**, L57.

Klinkhamer F. R. and Volovik G. E. (2009) "Vacuum energy density kicked by the electroweak crossover," *Phys. Rev. D* **80**, 083001.

Kneib, J.-P., Hudelot, P., Ellis, R. S. et al. (2003) "A wide-field Hubble Space Telescope study of the Cluster CL0024 + 1654 at $z = 0.4$. II. The cluster mass distribution", *Astrophys. J.* **598**, 804.

Kocevski, D. D. and Ebeling H. (2006) "On the origin of the Local Group's peculiar velocity". *Astrophys. J.*, **645**, 1043.

Kolb, E. W. and Turner, M. S. (1990) *The Early Universe* Addison-Wesley, Reading.

Komatsu, E., Smith, K. M., Dunkley, J. et al. (2011) "Seven-year Wilkinson Microwave Anisotropy Probe (WMAP) Observations: Cosmological Interpretation", *Astrophys. J. Supp.*, **192**, 18.

Korn, A. J., Grundahl, F., Richard, O. et al. (2006) "A probable stellar solution to the cosmological lithium discrepancy," *Nature* **442**, 657–659.

Kovalevsky, J. (1998) "First Results from HIPPARCOS" *Ann. Rev. of Astron. and Astrophys.*, **36**, 99.

Kowalski, M., Rubin, D., Aldering, G. et al. (2008) "Improved cosmological constraints from new, old, and combined supernova data sets". *Astrophys. J.*, **686**, 749.

Krauss, L. M. and Chaboyer, B. (2003) "Age estimates of globular clusters in the Milky Way: constraints on cosmology", *Science*, **299**, 65.

Krauss, L. M. and Schramm, D. N. (1993) "Angular diameters as a probe of a cosmological constant and Ω", *Astrophys. J.* **405**, L43.

Kuzmin, G. G. (1952) "Proper movements of the galactic-equatorial A and K stars of the perpendicularly galactic plane and dynamic density of the Galaxy". *Tartu Astron. Obs. Publ.*, **32**, 5.

Kuzmin, G. G. (1955) "On the question of the size of the dynamic parameters of C and density of matter in the vicinity of the Sun" *Tartu Astron. Obs. Publ.*, **33**, 3.

Kuzmin V. A. (1970) "CP-noninvariance and baryon asymmetry of the universe" *JETP Lett.* **12**, 228.

Kuzmin V. A., Rubakov V. A., and Shaposhnikov M. E. (1985) "On anomalous electroweak baryon-number non-conservation in the early universe" *Phys. Lett. B.*, **155**, 36.

Lagrange, J.-L. (1788) *Mecanique Analytique*. Courcier (reissued by Cambridge University Press, 2009; ISBN 9781108001748)

Lahav, O., Lilje, P. B., Primack, J. R., and Rees, M. J. (1991) "Dynamical effects of the cosmological constant" *Mon. Not. Roy. Astron. Soc.*, **251**, 128.

Laine, S., Zheng, J.-Q., and Valtonen, M. J. (2004) "Improved models for the evolution of the Coma cluster of galaxies", *Astron. J.*, **127**, 765.

Lake, K. (2005) "The flatness problem and Λ" *Phys. Rev. L.*, **94**, t1102L

Lanczos, K. (1922) "Bemerkung zur de Sitterschen Welt", *Phys. Z.*, **23**, 539.

Lanczos, K. (1923) "Über die Rotverschiebung in der de Sitterschen Welt", *Z. Physik*, **17**, 168.

Le Verrier, U. J. J. (1859) "Theorie du movement de Mercure", *Ann. Obs. Imp. Paris.*, **5**, 104.

Leff, H. (2002) "Teaching the photon gas in introductory physics" *American Journal of Physics*, **70**, 792.

Lehto, H. J. and Valtonen, M. J. (1996) "OJ 287 Outburst Structure and a Binary Black Hole Model", *Astrophys. J.*, **460**, 207.

Lemaître, G. (1927) "Un univers homogene de masse constante et de rayon croissant, redant compte de la vitesse radiale des nebuleuses extra-galactiques", *Ann. Soc. Sci. Bruxelles*, **A47**, 49.

Lewis, A. D., Ellingson, E., Morris, S. L., and Carlberg, R. G. (1999) "X-ray mass estimates at $z \approx 0.3$ for the Canadian network for observational cosmology cluster sample", *Astrophys. J.*, **517**, 587.

Liddle, A. R. (1999) "An Introduction to Cosmological Inflation in High Energy Physics and Cosmology," 1998 Summer School, ICTP, Trieste, Italy, 29 June–17 July 1998. Edited by A. Masiero, G. Senjanovic, and A. Smirnov. World Scientific Publishers, p. 260.

Liebert, J. (1980) "White dwarf stars". *Ann. Rev. Astron. Astrophys.*, **18**, 363.

Lifshitz, E. M. (1946) "Cosmological Perturbations" *J. Physics* (Moscow) **10**, 116.

Lin, H., Kirshner, R. P., Shectman, S. A. et al. (1995) "The Power Spectrum of Galaxy Clustering in the Las Campanas Redshift Survey", Clustering in the Universe, Proceedings of the 30th Rencontres de Moriond, Moriond Astrophysics Meeting, held in Les Arcs, Savoie, France, March 11–18, 1995. Edited by S. Maurogordato, C. Balkowski, C. Tao, and J. Tran Thanh Van. Paris: Editions Frontiers, 1995., p. 35.

Lindblad, B. (1925) "Star-streaming and the structure of the stellar system" *Arkiv f. Matematik, Astronomi och Fysik* **19A**, no. 21, 1.

Linde, A. (1990) *Particle Physics and Inflationary Cosmology*, Harwood Acad. Press, NY.

Liske, J., Grazian, A., Vanzella, E. et al. (2008) "Cosmic dynamics in the era of Extremely Large Telescopes" *Mon. Not. Roy. Astron. Soc.*, **386**, 1192.

Loeb A. (1998) "Direct Measurement of Cosmological Parameters from the Cosmic Deceleration of Extragalactic Objects" *Astrophys. J.*, **499**, L111.

Luminet, J.-P., Weeks, J. R., Riazuelo, A., Lehoucq, R., and Uzan, J.-P. (2003) "Dodecahedral space topology as an explanation for weak wide-angle temperature correlations in the cosmic microwave background" *Nature*, **425**, 593. Figure is from http://arxiv.org/pdf/astro-ph/0310253v1.pdf

Luzzi, G., Shimon, M., Lamagna, L. et al. (2009) "Redshift dependence of the cosmic microwave background temperature from Sunyaev-Zeldovich measurements". *Astrophys. J.*, **705**, 1122.

Lynden-Bell, D. (1967) "Statistical mechanics of violent relaxation in stellar systems", *Mon. Not. Roy. Astron. Soc.*, **136**, 101.

Lynden-Bell, D., Faber, S. M., Burstein, D. et al. (1988) "Spectroscopy and photometry of elliptical galaxies. V—Galaxy streaming toward the new supergalactic center". *Astrophys. J.*, **326**, 19.

Lynds, R. (1971) "The absorption-line spectrum of 4C05.34", *Astrophys. J.*, **164**, L73.

Lynds, R. and Petrosian, V. (1989) "Luminous arcs in clusters of galaxies", *Astrophys. J.*, **336**, 1.

Macciò A. V., Governato, F., and Horellou. C. (2005) "The signature of dark energy on the local Hubble flow" *Mon. Not. Roy. Astron. Soc* **359**, 941.

Maeda, K., Benetti, S., Stritzinger, M. et al. (2010) "An asymmetric explosion as the origin of spectral evolution diversity in type Ia supernovae". *Nature*, **466**, 82.

Makarov, D. I. and Karachentsev, I. D. (2011), Galaxy groups and clouds in the local ($z \sim 0.01$) Universe. *Mon. Not. Roy. Ast. Soc.*, **412**, 2498.

Malmquist, K. G. (1920) "A study of the stars of spectral type A", *Medd. Lund. Obs. Ser.* II, Nr. 22.

Marochnik L., Usikov D., and Vereshkov G. (2010) Dark energy as a macroscopic effect of quantum gravity. *Bull. Amer. Astron. Soc.*, Vol. 42, 441.

Martinez, V. J. and Saar, E. (2002) *Statistics of the galaxy distribution*, Chapman & Hall/CRC, Boca Raton, Florida.

Martinez-Vaquero L. A., Yepes G., Hoffman Y., Gottlöber S., and Sivan M. (2009) "Constrained simulations of the local universe—II. The nature of the local Hubble flow." *MNRAS* **397**, 2070.

Masso E. and Rota F. (2002) "Primordial Helium Production in Charged Universe." *Phys. Lett.* **B545**, 221.

Matese, J. J., Whitman, P. G., Innanen, K. A., and Valtonen, M. J. (1995) "Periodic modulation of the Oort cloud comet flux by the adiabatically changing galactic tide", *Icarus*, **116**, 255.

Mattig, W. (1958) "Über den Zusammenhang zwischen Rotverschiebung und scheinbarer Helligkeit" *Astron. Nachr.*, **284**, 109.

Mattsson, T. (2010) "Dark energy as a mirage". *Gen. Rel. Grav.*, **42**, 567.

McGaugh, S. S., Schombert, J. M., de Blok, W. J. G., and Zagursky, M. J. (2010) "The Baryon Content of Cosmic Structures," *Astrophys. J. Lett.* **708**, L14.

Mellier, Y. (1999) "Probing the universe with weak lensing", *Ann. Rev. Astron. Astrophys.*, **37**, 127.

Melott, A. L., Einasto, J., Saar, E., Suisalu, I., Klypin, A. A., and Shandarin, S. F. (1983) "Cluster analysis of the non-linear evolution of large scale structure in an axion/gravitino/photino-dominated universe" *Phys. Rev. Lett.*, **51**, 935.

Merrifield, M. R. (1992) "The rotation curve of the Milky Way to 2.5 R_0 from the thickness of the H I layer". *Astron. J.*, **103**, 1552.

Michelson, A. A. and Morley, E. W. (1887) "On the Relative Motion of the Earth and the Luminiferous Ether". *American Journal of Science*, **34**, 333.

Mignard, F. (1998) "Main Properties of the HIPPARCOS Catalogue" *New Horizons from Multi-Wavelength Sky Surveys, Proceedings of the 179th Symposium of the International Astronomical Union*, held in Baltimore, USA August 26–30, 1996, Kluwer Academic Publishers, edited by Brian J. McLean, Daniel A. Golombek, Jeffrey J. E. Hayes, and Harry E. Payne, p. 399.

Miley, G. K. (1971) "The radio structure of quasars—a statistical investigation", *Mon. Not. Roy. Astron. Soc.*, **152**, 477.

Miley, G. K., Perola, G. C., van der Kruit, P. C., and van der Laan, H. (1972) "Active galaxies with radio tails in clusters" *Nature*, **237**, 269.

Mohayaee, R. and Tully, R. B. (2005), "The Cosmological Mean Density and Its Local Variations Probed by Peculiar Velocities". *Astrophys. J.*, **635**, L113.

Mohr, J. J., Mathiesen, B., and Evrard, A. E. (1999) "Properties of the intracluster medium in an ensemble of nearby galaxy clusters", *Astrophys. J.*, **517**, 627.

Mollerach, S. and Roulet, E. (2002) *Gravitational Lensing and Microlensing*, World Scientific, New Jersey.

Moore, B., Governato, F., Quinn, T. Stadel, J., and Lake, G. (1998) "Resolving the structure of Cold Dark Matter halos", *Astrophys. J.*, **499**, L5.

Mullis, C. R., Rosati, P., Lamer, G. et al. (2005) "Discovery of an x-ray luminous galaxy cluster at $z = 1.4$", *Astrophys. J.*, in press.

Navarro, J. F., Frenk, C. S., and White, S. D. M. (1997) "A Universal Density Profile from Hierarchical Clustering" *Astrophys. J.*, **490**, 493.

Newcomb, S. (1895) "The elements of the four inner planets and the fundamental constants", Washington, D. C., U. S. Govt. Office.

Niemi, S.-M. and Valtonen, M. (2009) "The origin of redshift asymmetries: how ΛCDM explains anomalous redshift" *Astron. Astrophys.*, **494**, 857.

Nilson, P. (1973) *Uppsala general catalogue of galaxies*, Uppsala Astr. Obs., Uppsala.

Nilsson, K., Valtonen, M. J., Kotilainen, J., and Jaakkola. T. (1993) "On the redshift-apparent size diagram of double radio sources", *Astrophys. J.*, **413**, 453.

Nilsson, K., Valtonen, M. J., Zheng, J.-Q., Byrd, G. G., Korhonen, H., and Andersen, M. I. (2000) "Deducing the orbit of the radio galaxy 3C 129" in *Small Galaxy Groups*, ASP Conference Series, **209**, 408, eds. M. Valtonen and C. Flynn.

Nurmi, P., Heinmmki, P., Teerikorpi, P. and Chernin, A. (2010) "Normalized Hubble Diagrams for Simulated Galaxy Groups," Hunting for the Dark: The Hidden Side of Galaxy Formation, Edited by V. P. Debattista and C. C. Popescu, AIP Conferences Proceeding, Volume 1240, p. 419–420.

Ohm, E. A. (1961) "Project Echo: Receiving System" *Bell System Technical Journal.* **40**, Issue 4.

Okun, L. B. (1985) *Particle Physics: The Quest for the Substance of Substance* Harwood Academic Pubs., London-Paris-New York.

Oort, J. (1932) "The force exerted by the stellar system in the direction perpendicular to the galactic plane and some related problems" *Bull. Astron. Inst. Netherlands*, **6**, 249.

Oort, J. (1965) "Stellar Dynamics", *Galactic Structure*, edited by A. Blaauw and M. Schmidt.

Öpik, E. J. (1915) "The densities of visual binary stars" *Bull. Soc. Astron. Russie*, **21**, 150.

Öpik, E. J. (1922) "An estimate of the distance of the Andromeda Nebula", *Astrophys. J*, **55**, 406.

Ostriker, J. P., Peebles, P. J. E., and Yahil, A. (1974) "The size and mass of galaxies, and the mass of the universe" *Astrophys J.*, **193**, L1.

Ostriker, J. P. and Steinhardt, P. J. (1995) "The observational case for a low density Universe with non-zero cosmological constant" *Nature*, **377**, 600.

Paturel, G. and Teerikorpi, P. (2005) "The extragalactic Cepheid bias: significant influence on the cosmic distance scale", *Astron. Astrophy.*, **443**, 883.

Peacock, J. A. (1999) *Cosmological Physics*, Cambridge University Press. Cambridge, UK.

Peebles P. J. E. (1984) "Tests of cosmological models constrained by inflation" *Astrophys. J.*, **284**, 439.

Peebles, P. J. E. (1980) *The Large Scale Structure of the Universe*, Princeton Univ. Press, Princeton.

Peebles, P. J. E. (1982) "Large-scale background temperature and mass fluctuations due to scale-invariant primeval perturbations", *Astrophys. J.*, **263**, L1.

Peebles, P. J. E. (1993) *Principles of Physical Cosmology*, Princeton Univ. Press, Princeton.

Peirani, S. and de Freitas Pacheco, J.A. (2008) "Dynamics of nearby groups of galaxies: the role of the cosmological constant." *Astron. Astrophys*, **488**, 845.

Penzias, A. A. and Wilson, R. W. (1965) "A Measurement of Excess Antenna Temperature at 4080 Mc/s", *Astrophys. J.*, **142**, 419.

Percival, W. J., Cole, S., Eisenstein, D. J. et al. (2007a) "Measuring the Baryon Acoustic Oscillation scale using the Sloan Digital Sky Survey and 2dF Galaxy Redshift Survey", *Mon. Not. Roy. Astron. Soc.*, **381**, Issue: 3, Page 1053.

Percival, Will J., Nichol, R. C., Eisenstein, D. J. et al. (2007b) "Measuring the Matter Density Using Baryon Oscillations in the SDSS" *Astrophys. J.*, **657**, 51.

Perlmutter, S., Aldering, G., Goldhaber, G. et al. (1999) "Measurements of Omega and Lambda from 42 High-Redshift Supernovae". *Astrophys. J.*, **517**, 565.

Peterson, S. D. (1979) "Double galaxies. II—Data analysis and a galaxian mass (M/L) determination", *Astrophys. J.*, **232**, 20.

Petrovskaya, I. and Teerikorpi, P. (1986) "Rotation curve of the outer parts of our Galaxy from neutral hydrogen 21 cm line profiles". *Astron. Astrophys.*, **163**, 39.

Pietronero, L. (1987) "The fractal structure of the Universe: correlations of galaxies and clusters and the average mass density". *Physica A.*, **144**, 257.

Pound, R. V. and Rebka, G. A. (1959) "Gravitational redshift in nuclear resonance", *Phys. Rev. Lett.*. **3**, 439.

Pound, R. V. and Rebka, G. A. (1960) "Apparent weight of photons", *Phys. Rev. Lett.*, **4**, 337.

Prause, N., Reimers, D., Fechner, C., and Janknecht, E. (2007) "The baryon density at $z = 0.9$–1.9. Tracing the warm-hot intergalactic medium with broad Lyman α absorption." *Astron. Astrophys.*, **470**, 67–72.

Pskovskii, Yu. P. (1977) "Light curves, color curves, and expansion velocity of type I supernovae as functions of the rate of brightness decline" *Astronomicheskii Zhurnal*, **54**, 1188–1201. *Soviet Astronomy*, **21**, 675.

Qian, Y.-Z., Sargent W. L. W., and Wasserburg, G. J. (2002) "The Prompt Inventory from Very Massive Stars and Elemental Abundances in Ly α Systems", *Astrophys. J.*, **596**, L61.

Qian, Y.-Z. and Wasserburg, G. J. (2002) "Determination of Nucleosynthetic Yields of Supernovae and Very massive Stars from Abundances in Metal-Poor Stars", *Astrophys. J.*, **567**, 515.

Qian, Y-Z. and Wasserburg, G. J. (2003) "Hierarchical Structure Formation and Chemical Evolution of Damped Ly α Systems", *Astrophys. J.*, **596**, L9.

Rauch, M. and Haehnelt, M. G. (1995) "Omega_baryon and the geometry of intermediate-redshift Lyman alpha absorption systems", *Mon. Not. Roy. Astron. Soc.*, **275**, 76.

Rees, M. J. (1986) "Lyman absorption lines in quasar spectra: evidence for gravitationally-confined gas in dark minihaloes", *Mon. Not. Roy. Astron. Soc.*, **218**, 25P.

Refsdal, S. (1964) "The gravitational lens effect", *Mon. Not. Roy. Astron. Soc.*, **128**, 295.

Riess, A. G., Filippenko, A. V., Challis, P. et al. (1998) "Observational evidence from supernovae for an accelerating universe and a cosmological constant", *Astron. J.*, **116**, 1009.

Riess, A. G., Nugent, P. E., Gilliland, R. L. et al. (2001) "The Farthest Known Supernova: Support for an Accelerating Universe and a Glimpse of the Epoch of Deceleration" *Astrophys. J.*, **560**, 49.

Riess, A. G., Press, W. H., and Kirshner, R. P. (1995) "Using Type Ia supernova light curve shapes to measure the Hubble constant" *Astrophys. J.*, **438**, L17.

Riess, A. G., Strolger, L.-G., Casertano, S. et al. (2007) *Astrophys. J.* **656**, 98. Also see http://arxiv.org/abs/astro-ph/0611572. http://www.astro.ucla.edu/~wright/sne_cosmology.html

Riess, A. G., Strolger, L.-G., Tonry, J. et al. (2004) "Type Ia supernova discoveries at $z > 1$ from the Hubble Space Telescope: Evidence for past deceleration and constraints on dark energy evolution", *Astrophys. J.*, **607**, 665.

Rindler, W. (1977) *Essential Relativity: Special, General, and Cosmological*, 2nd ed., Springer-New York.

Rines, K., Geller, M. J., Kurtz, M. J., Diaferio, A., Jarrett, T. H., and Huchra, J. P. (2001) "Infrared mass-to-light profile throughout the infall region of the Coma cluster", *Astrophys. J.*, **561**, L41.

Riotto A. and Trodden M. (1999) "Recent Progress in Baryogenesis" *Ann. Rev. Nuc. Part. Sci.*, **49**, 35.

Robaina, A. R. and Cepa, J. (2007) "Redshift-distance relations from type Ia supernova observations. New constraints on grey dust models". *Astron. Astrophys.*, **464**, 465.

Roberts, M. S. and Whitehurst, R. N. (1975) "The rotation curve and geometry of M31 at large galactocentric distances", *Astrophys. J.*, **201**, 327.

Robertson, H. P. (1935) "Kinematics and world-structure" *Astrophys. J.*, **82**, 284.

Robertson, H. P. (1936) "Kinematics and world-structure II & III", *Astrophys. J.*, **83**, 187 and 257.

Rood, H. J. (1974) "Empirical properties of the mass discrepancy in groups and clusters of galaxies", *Astrophys. J.*, **188**, 451.

Rood, H. J. and Sastry, G. N. (1971) "'Tuning Fork' classification of rich clusters of galaxies", *Publ. Astron. Soc. Pacific*, **83**, 313.

Roukema, B. F., Lew, B., Cechowska, M., Marecki, A., and Bajtlik, S. (2004) "A hint of Poincaré dodecahedral topology in the WMAP first year sky map" *Astron. Astrophys.* **423**, 821.

Rubakov V. A. and Shaposhnikov M. E. (1996) "Electroweak baryon number non-conservation in the early Universe and in high-energy collisions" *Physics Uspekhi*, **39**, 461.

Rubin, V. C. and Ford W. K. Jr. (1970) "Rotation of the Andromeda Nebula from a spectroscopic survey of emission regions" *Astrophys. J.*, **159**, 379.

Rubin, V. C., Ford, W. K. Jr. and Thonnard, N. (1978) "Extended rotation curves of high-luminosity spiral galaxies IV. Systematic dynamical properties, Sa through Sc" *Astrophys. J.*, **225**, L107.

Rubin, V. C., Roberts, M. S., Thonnard, N., and Ford, W. K., Jr. (1976) "Motion of the Galaxy and the Local Group determined from the velocity anisotropy of distant ScI galaxies. II. The analysis of the motion" *Astron. J.*, **81**, 719.

Sachs, R. K. and Wolfe, A. M. (1967) "Perturbations of a cosmological model and angular variations of the microwave background", *Astrophys. J.*, **147**, 73.

Sakharov, A. D. (1966) "The Initial Stage of an Expanding Universe and the Appearance of a Nonuniform Distribution of Matter" *JETP Lett.* **22**, 241.

Salpeter, E. E. (1964) "Accretion of interstellar matter by massive objects", *Astrophys. J.*, **140**, 796.

Sandage, A. (1953) "The color-magnitude diagram for the globular cluster M3", *Astron. J.*, **58**, 61.

Sandage, A. (1961) The ability of the 200-inch telescope to discriminate between selected world models. *Astrophys. J.*, **133**, 355.

Sandage, A. (1962) "The change of redshift and apparent luminosity of galaxies due to the deceleration of the expanding universes" *Astrophys. J.*, **136**, 319.

Sandage, A. (1970) "Main-sequence photometry, color-magnitude diagrams, and ages for the globular clusters M3, M13, M15, and M92", *Astrophys. J.*, **162**, 841.

Sandage, A. (1999) "Bias properties of extragalactic distance indicators. VIII. H_0 from distance-limited luminosity class and morphological type-specific luminosity functions for Sb, Sbc, and Sc galaxies calibrated using Cepheids" *Astrophys. J.*, **527**, 479.

Sandage, A., Tammann, G. A., and Saha, A. (1998) "The time scale test for Ω: the inverse Hubble constant compared with the age of the universe", *Phys. Rep.*, **307**, 1.

Sandage, A. (1995) "Practical Cosmology: Inventing the Past", in *The deep universe*, eds. Binggeli, R. Buser (Berlin: Springer-Verlag), pp. 1–232.

Sandage, A., Tammann, G. A., and Hardy E. (1972) "Limits on the local deviation of the Universe for a homogeneous model" *Astrophys. J.*, **172**, 253.

Sandage A., Tammann G. A., Saha A., Reindl B., Macchetto, F. D., and Panagia N. (2006) "The Hubble constant: A summary of the HST program for the luminosity calibration of Type Ia Supernovae by means of Cepheids", *Astrophys. J.* **253**, 843.

Sargent, W. L. W., Young, P. J., Boksenberg, A., Shortridge, K., Lynds, C. R., and Hartwick, F. D. A. (1978) "Dynamical evidence for a central mass concentration in the galaxy M87", *Astrophys. J.*, **221**, 731.

Sarkar, D., Amblard, A., Holz, D. E., and Cooray, A. (2008) "Lensing and Supernovae: Quantifying the Bias on Dark Energy Equation of State". *Astrophys. J.*, **678**, 1.

Schaefer, B. E. (2006), "The Hubble Diagram to Redshift >6 from 69 Gamma-Ray Bursts," *Astrophys. J.*, **660**:16 arXiv:astro-ph/0612285.

Schmidt, R. W., Allen, S. W., and Fabian, A. C. (2001) "CHANDRA observations of the galaxy cluster Abell 1835", *Mon. Not. Roy. Astron. Soc.*, **327**, 1057.

Schwarzschild, K. (1916) "Über das Gravitationsfeld eines Massenpunktes nach der Einsteinschen Theorie", *Sitz. Acad. Wiss. Physik-Math Kl.*, **1**, 189.

Scranton, R., Mènard, B., Richards, G. T., and Nichol, R. C. (2005) "Detection of Cosmic Magnification with the Sloan Digital Sky Survey" *Astrophys. J.*, **633**, 589.

Seljak, U., Makarov, A., McDonald, P. et al. (2005) "Cosmological parameter analysis including SDSS Lyα forest and galaxy bias: constraints on the primordial spectrum of fluctuations, neutrino mass, and dark energy", *Phys. Rev. D*, **71**, Issue 10, id. 103515 (PhRvD Homepage), astro-ph/0407372

Shapiro, I. I., Counselman, C. C. III, and King. R. W. (1976) "Verification of the principle of equivalence for massive bodies", *Phys. Rev. Lett.*, **36**, 555.

Shectman, S. A., Landy, S. D., Oemler, A. Jr., Tucker, D. L., Lin, H., Kirshner, R. P., and Schechter, P. L. (1996) "The Las Campanas Redshift Survey", *Astrophys. J.*, **470**, 172.

Sherwin, B. D., Dunkley, J., Das, S. et al. (2011) "Evidence for Dark Energy from the Cosmic Microwave Background Alone Using the Atacama Cosmology Telescope Lensing Measurements", *Phys Rev. Lett.*, **107**, 021302.

Shklovskii, I. S. (1964) "Physical conditions in the gaseous envelope of 3C273", *Astron. Zh.*, **41**, 801.

Shmaonov, T. A. (1957) *Pribory i Teknika Eksperienta* (*Experimental Devices and Methods*) Moscow.

Shu, F. H. (1978) "On the statistical mechanics of violent relaxation", *Astrophys. J.*, **225**, 83.

Shvartsman, V. F. (1969) "Density of relict particles with zero rest mass in the universe" *JETP Lett.*, **9**, 315.

Silk, J. (1967) "Fluctuations in the primordial fireball", *Nature*, **215**, 1155.

Slipher, V. M. (1917) "A spectroscopic investigation of spiral nebulae" *Proc. Amer. Phil. Soc.*, **56**, 403.

Slipher, V. M. (1917) "Radial velocity observations of spiral nebulae" *Observatory*, **40**, 304.

Smoot, G. F., Bennett, C. L., Kogut, A. et al. (1991) "Preliminary results from the COBE differential microwave radiometers: Large angular scale isotropy of the cosmic microwave background", *Astrophys. J.*, **371**, L1.

Smoot, G. F., Bennett, C. L., Kogut, A. et al. (1992) "Structure in the COBE differential microwave radiometer first-year maps", *Astrophys. J.*, **396**, L1.

Sokoloff, D. D. (1971) "Topology of Models of the Universe", *Sov. Phys. Dokl.* **15**, 1112.

Sokoloff, D. D. and Shvartsman, V. F. (1974) "An estimate of the size of the universe from a topological point of view" *Sov. Phys. JETP*, **39**, 196.

Sokoloff, D. D. and Starobinsky, A. A. (1975) "Globally inhomogeneous 'spliced' universes" *Sov. Astron.*, **19**, 629.

Songaila, A. (1998) "The Redshift evolution of the metagalactic ionizing flux inferred from metal line ratios in the Lyman Forest" *Astron. J.*, **115**, 2184.

Sparke, L. S. and Gallagher, J. S. (2000) *Galaxies in the Universe: and Introduction*, Cambridge University Press, Cambridge.

Spergel, D. N., Bean, R., Doré, O. et al. (2007) "Three-Year Wilkinson Microwave Anisotropy Probe (WMAP) Observations: Implications for Cosmology" *Astrophys, J. Supp.*, **170**, 377.

Spergel, D. N., Verde, L., Peiris, H. V. et al. (2003), "First year *Wilkinson Microwave Anisotropy Probe (WMAP)* observations: determination of cosmological parameters",

Astrophys. J. Suppl., **148**, 175. For latest values and references go to http://lambda.gsfc.nasa.gov/product/map/dr4/pub_papers/sevenyear/basic_results/wmap_7yr_basic_results.pdf

Squires, G., Kaiser, N., Babul, A. et al. (1996) "The dark matter, gas, and galaxy distributions in Abell 2218: a weak gravitational lensing and x-ray analysis", *Astrophys. J.*, **461**, 572.

Stachel, J. (2002) *Einstein from 'B' to 'Z'*, Birkhäuser, Boston.

Strauss, M. A., Yahil, A., Davis, M., Huchra, J. P., and Fisher, K. (1992) "A redshift survey of IRAS galaxies. V. The acceleration of the Local group", *Astrophys. J.*, **397**, 395.

Sulentic J. W. (1984) "Redshift differentials in a complete sample of galaxy groups". *Astrophys. J.*, **286**, 442.

Sullivan, M., Guy, J., Conley, A. et al. (2011) "SNLS3: constraints on dark energy combining the supernova legacy survey three year data with other probes." *Astrophys. J.* (in press).

Surdej, J., Claeskens, J. F., Crampton, D. et al. (1993) "Gravitational lensing statistics based on a large sample of highly luminous quasars". *Astron. J.*, **105**, 2064.

Sylos Labini F., Vasilyev N. L., Baryshev Yu. V., and Lopez-Corredoira M. (2009a) "Absence of anti-correlations and of baryon acoustic oscillations in the galaxy correlation function from the Sloan Digital Sky Survey DR7". *Astron. Astrophys.*, **505**, 981.

Sylos Labini F., Vasilyev N. L., and Baryshev Yu. V. (2009b) "Breaking the self-averaging properties of spatial galaxy fluctuations in the Sloan Digital Sky Survey—Data release six". *Astron. Astrophys.*, **508**, 17.

Tammann G. A. (1979) "Precise determination of the distances of galaxies" In *Scientific Research with the Space Telescope* (IAU Colloq. No. 54), p. 263.

Tammann, G. A., Sandage, A., and Yahil, A. (1980) "The determination of cosmological parameters" *Physical Cosmology*, eds. R. Belian, J. Adouze, and D. N. Schramm, North-Holland, Amsterdam, p. 53.

Tarenghi, M., Tifft, W. G., Chincarini, G., Rood, H. J., and Thompson, L. A. (1978) "The structure of the Hercules Supercluster", *The Large scale Structure of the Universe*, IAU Symp. 79, Eds. M. S. Longair and J. Einasto, Reidel, Dordrecht, p. 263.

Taylor, J. H. and Weisberg, J. M. (1989) "Further experimental tests of relativistic gravity using the binary pulsar PSR 1913 + 16", *Astrophys. J.*, **345**, 434.

Teerikorpi, P. (1975) "On the Effect of the Luminosity Selection on the Redshift-distance Relationship", *Astron. Astrophys.*, **45**, 117.

Teerikorpi, P. (1978) "A Study of Galactic Absorption as Revealed by the Rubin et al. Sample of Sc Galaxies". *Astron. Astrophys.*, **64**, 379.

Teerikorpi, P. (1997) "Observational Selection Bias Affecting the Determination of the Extragalactic Distance Scale", ARA&A, **35**, 101.

Teerikorpi, P. (2000) "Evidence for the class of the most luminous quasars II. Variability, polarization, and the gap in the M_V distribution". *Astron. Astrophys.*, **353**, 77.

Teerikorpi, P. (2001) "Dark haloes in Karachentsev's sample of binary galaxies". *Astron. Astrophys.*, **371**, 470.

Teerikorpi, P. (2002) "Dusty haloes of galaxies at intermediate redshifts". *Astron. Astrophy.*, **386**, 865.

Teerikorpi, P. (2003) "Evidence for the class of the most luminous quasars IV. Cosmological Malmquist bias and the Λ term". *Astron. Astrophys.*, **399**, 829.

Teerikorpi, P. (2011) "On Öpik's distance evaluation method in a cosmological context", *Astron. Astrop.*, **531**, A10.

Teerikorpi, P., Bottinelli, L., Gouguenheim, L., and Paturel, G. (1992), "Investigations of the local supercluster velocity field: I. Observations close to Virgo, using Tully-Fisher distances, and the Tolman-Bondi expanding sphere". *Astron. Astrophys.* **260**, 17.

Teerikorpi P. and Chernin A. D. (2010) The Hubble diagram for a system within dark energy: the location of the zero-gravity radius and the global Hubble rate *Astron. Astrophys.*, **516**, A93.

Teerikorpi, P., Chernin, A. D., and Baryshev, Yu. V. (2005) "The quiescent Hubble flow, local dark energy tests, and pairwise velocity dispersion in a $\Omega = 1$ universe" *Astron. Astrophys.*, **440**, 791.

Teerikorpi, P., Chernin, A. D., Karachentsev, I. D., and Valtonen M. (2008) "Dark energy in the environments of the Local Group, the M81 group, and the Cen A group: the normalized Hubble diagram" *Astron. Astrophys.*, **483**, 383.

Teerikorpi, P., Gromov, A., and Baryshev, Yu. (2003) "Limits on dark energy—matter interaction from the Hubble relation for two-fluid FLRW models" *Astron. Astrophys.*, **407**, L9.

Teerikorpi, P., Valtonen, M., Lehto, K., Lehto, H., Byrd G., and Chernin, A. (2009) *The Evolving Universe and the Origin of Life—The Search for Our Cosmic Roots.* (Springer, New York 2009).

Terrell, J. (1977) "The luminosity distance equation in Friedmann cosmology" *Am. J. Phys.*, **45**, 869.

Theureau, G., Hanski, M., Ekholm, T. et al. (1997) "Kinematics of the Local universe. V. The value of H_0 from the Tully–Fisher B and D_{25} relations for field galaxies". *Astron. Astrophys.*, **322**, 730.

Tifft, W. G. and Gregory, S. A. (1976) "Direct observations of the large-scale distribution of galaxies" *Astrophys. J.*, **205**, 696.

Tifft, W. G. and Gregory, S. A. (1978) "Observations of the large scale distribution of galaxies", *The Large scale Structure of the Universe*, IAU Symp. 79, Eds. M. S. Longair and J. Einasto, Reidel, Dordrecht, p. 267.

Tolman, R. C. (1930) On the estimation of distances in a curved universe with a non-static line element. *Proceedings of the National Academy of Sciences of the United States of America*, **16**, 511.

Totsuji, H. and Kihara T. (1969) "The Correlation Function for the Distribution of Galaxies". *Publ. Astr. Soc. Japan*, **21**, 221.

Tran, K.-V. H., Kelson, D. D., van Dokkum, P., Franx, M., Illingworth, G., and Magee, D. (1999) "The velocity dispersion of MS1054-03: A massive galaxy cluster at high redshift", *Astrophys. J.*, **522**, 39.

Treu, T. and Koopmans, L. V. E. (2002) "The internal structure and formation of early-type galaxies: the gravitational lens system MG2016 + 112 at $z = 1.004$", *Astrophys. J.*, **575**, 87.

Tropp, E. A., Frenkel, V. Ya., and Chernin, A. D. (1988) *Alexander Alexandrovich Friedmann: His work and life*. Moscow: Nauka (in Russian).

Tropp, E. A., Frenkel, V. Ya., and Chernin, A. D. (1993) *Alexander Alexandrovich Friedmann: His work and life*. Cambridge University Press, Cambridge, England.

Tucker, D. L., Oemler, A., Kirshner, R. P. et al. (1997) "The Las Campanas Redshift Survey galaxy-galaxy autocorrelation function" *Mon. Not. Roy. Astron. Soc.*, **285**, L5. Figure 14.12a modified from separate source after this article's figure.

Tully R. B. and Fisher, J. R. (1977) "A new method of determining distances to galaxies" *Astron. & Astrophys.*, **54**, 66.

Tully, R. B. and Fisher, J. R. (1978) "A tour of the Local Supercluster", *The Large Scale Structure of the Universe*, IAU Symp. 79, Eds. M. S. Longair and J. Einasto, Reidel, Dordrecht, p. 214.

Tully, R. B. and Shaya, E. J. (1984), "Infall of galaxies into the Virgo cluster and some cosmological constraints". *Astrophys. J.*, **281**, 31.

Tyson, J. A., Kochanski, G. P., and Dell'Antonio, I. P. (1998) "Detailed mass map of CL 0024 + 1654 from strong lensing", *Astrophys. J.*, **498**, L107.

Tyson, J. A., Wenk, R. A., and Valdes, F. (1990) "Detection of systematic gravitational lens galaxy image alignments: mapping dark matter in galaxy clusters", *Astrophys. J.*, **349**, L1.

Valageas, P., Schaeffer, R., and Silk, J. (2002) "The phase-diagram of cosmological baryons", *Astron. Astrophys.*, **388**, 741.

Valtonen, M. J. and Byrd, G. G. (1979), "A binary model for the Coma cluster of galaxies", *Astrophys J.*, **230**, 655.

Valtonen, M. J. and Byrd, G. G. (1986), "Redshift asymmetries in systems of galaxies and the missing mass", *Astrophys. J.*, **303**, 523.

Valtonen, M. J., Byrd, G. G., McCall, M. L., and Innanen, K A. (1993) "A revised history of the Local Group and a generalized method of timing", *Astron. J.*, **105**, 886.

Valtonen, M. J. and Karttunen, H. (2005) *The Three-body Problem*, Cambridge Univ. Press, Cambridge.

Valtonen, M. J., Niemi, S., Teerikorpi, P. et al. (2008) "Redshift asymmetry as an indication of dark energy around groups of galaxies" *American Astronomical Society, DDA meeting*, 39.0103V

Valtonen, M. J., Lehto, H. J., Nilsson, K. et al. (2006) "A massive binary black-hole system in OJ287 and a test of general relativity", *Nature*, **452**, 851.

Valtonen, M. J., Mikkola, S., Lehto, H. J., Gopakumar, A., Hudec, R., and Polednikova, J. (2011) "Testing the black holeno-hair theorem with OJ287", *Astrophys. J.*, **742**, 22.

Van Albada, T. S. and Sancisi, R. (1986) "Dark matter in spiral galaxies", *Roy. Soc. Phil. Trans. Ser. A.*, **320**, 447.

Vikhlinin, A., Markevitch, M., Forman, W., and Jones, C. (2001) "Zooming in on the Coma cluster with CHANDRA: Compressed warm gas in the brightest cluster galaxies", *Astrophys. J.*, **555**, L87.

von Seeliger, H. von (1895), *Astronomische Nachrichten*, **137** (1895), 129–136: and **138** (1895), 51–54, 255–258.

Wagoner, R. V., Fowler, W. A., and Hoyle, F. (1967) "On the synthesis of elements at very high temperatures", *Astrophys. J.*, **148**, 3.

Walker, A. G. (1936) "On Milne's theory of world-structure" *Proc. Lon. Math. Soc.*, **42**, 90.

Walsh, D., Carswell, R. F., and Weymann, R. J. (1979) "0957 + 561A,B—twin quasistellar objects or gravitational lens" *Nature*, **279**, 381.

Weinberg, S. (1977) *The First Three Minutes* (basic Books, New York), p. 26.

Weinberg, D. H., Davè, R., Katz, N., and Kollmeier, J. A. (2003) "The Lyman-α forest as a cosmological tool", *The emergence of cosmic structure*, AIP Conf. Proc., **666**, 157.

Wess, J. and Zumino, B. (1977) "Superspace formulation of supergravity" *Physics Letters* B, **66**, Issue 4, p. 361–364.

Weyl, H. (1951) *Symmetry*, Princeton Univ. Press, Princeton.

White, S. D. M., Frenk, C. S., and Davis, M. (1983) "Clustering in a neutrino-dominated universe", *Astrophys. J.*, **274**, L1.

White, S. D. M., Navarro, J. F., Evrard, A. E., and Frenk, C. S. (1993) "The baryon content of galaxy clusters—a challenge to cosmological orthodoxy", *Nature*, **366**, 429.

Whiting, A. B. (2003) "The kinematic state of the Local Volume" *Astrophys. J.*, **587**, 186.

Will, C. M. (1993) *Theory and Experiment in Gravitational Physics* Cambridge UP, Cambridge.

Wilson, G., Kaiser, N., and Luppino, G. A. (2001b) "Mass and light in the Universe", *Astrophys. J.*, **556**, 601.

Wilson, G., Kaiser, N., Luppino, G. A., and Cowie, L. L. (2001a) "Galaxy halo masses from galaxy-galaxy lensing", *Astrophys. J.*, **555**, 572.

Wirén, S., Zheng, J.-Q., Valtonen, M. J., and Chernin, A. D. (1996) "Computer simulations of interacting galaxies in compact groups and the observed properties of close binary galaxies", *Astron. J.*, **111**, 160.

Wood-Vasey, W. M., Miknaitis, G., Stubbs, C. W. et al. (2007) "Observational constraints on the nature of dark energy: first cosmological results from the ESSENCE supernova survey," *Astrophys. J.*, **666**, 694.

Wu, K., Lahav, O., and Rees M. (1999) "The large-scale smoothness of the Universe". *Nature*, **397**, 225.

Yonehara, A., Umemura, M., and Susa, H. (2003) "Quasar Mesolensing—Direct Probe to Substructures around Galaxies" *Pub. Astron. Soc. Japan*, **55**, 1059.

Yoshii, Yu. and Takahara F. (1988) "Galactic evolution and cosmology: probing the cosmological deceleration parameter", *Astrophys. J.*, **326**, 1.

Zeldovich, Ya. B. (1965) "Survey of Modern Cosmology" *Adv. Astron. Astrophys.* **3**, 241.

Zeldovich, Ya. B. (1968) "Physical Limitations on the Topology of the Universe" *Physics Uspekhi* **11**, 381.

Zeldovich, Ya. B. (1972) "A hypothesis, unifying the structure and the entropy of the universe", *Mon. Not. Roy. Astron. Soc.*, **160**, 1P.

Zeldovich, Ya. B. and Novikov, I. D. (1967) "Physical Limitations on the Topology of the Universe" *JETP Lett.*, **6**, 236.

Zeldovich, Ya. B. and Novikov, I. D. (1983) *The Structure and Evolution of the Universe* The Univ. Chicago Press, Chicago and London.

Zheng, J.-Q., Valtonen, M. J., and Byrd, G. G. (1991) "Maffei 1 as an interloper of the Local Group of galaxies and the mass of the Local Group" *Astron. Astrophys.*, **247**, 20.

Zheng, J.-Q., Valtonen, M. J., and Chernin, A. D. (1993) "Computer simulations of interacting galaxies in compact groups and the observed properties of triple galaxies", *Astron. J.*, **105**, 2047.

Zwicky, F. (1933) "Die Rotverschiebung von extragalaktischen Nebeln", *Helv. Phys. Acta.*, **6**, 110.

Zwicky, F. (1937) "Nebulae as gravitational lenses", *Phys. Rev.*, **51**, 290.

Zwicky, F., Herzog, E., Karpowicz, M., Kowal, C. T., and Wild, P. (1961–1968) *Catalogue of galaxies and of clusters of galaxies*, CalTech, Pasadena.

Index